Surgical Ethics

Alberto R. Ferreres
Editor

Surgical Ethics

Principles and Practice

 Springer

Editor
Alberto R. Ferreres
Department of Surgery
University of Buenos Aires
Buenos Aires
Argentina

Department of Surgery
University of Washington
Seattle, WA
USA

ISBN 978-3-030-05963-7   ISBN 978-3-030-05964-4   (eBook)
https://doi.org/10.1007/978-3-030-05964-4

This Springer imprint is published by the registered company Springer Nature Switzerland AG
The registered company address is: Gewerbestrasse 11, 6330 Cham, Switzerland

Preface

This textbook *Surgical Ethics: Principles and Practice* is devoted to disseminate the foundations and implications of ethics in the surgical arena, the so-called surgical ethics. Theoretical ethics attempts to understand the underlying grounds, assumptions, and concepts of ethical systems; meanwhile, practical ethics is related to the application of ethical standards in everyday surgical practice and care. This book attempts to provide acting surgeons in the different fields with a thorough and deep practical insight of the field of ethics as well as tools for solving ethical conflicts in everyday care.

Surgery is characterized by some unique features: it harms before healing, penetrates the patient's body, and thus is highly invasive; decision-making is performed many times under conditions of uncertainty and is prone to risks, errors, accidents, complications, and sequelae. Thus, surgery is a moral practice, and as such, the surgeon becomes a moral fiduciary agent for his or her patient. Trust is paramount to build an effective and beneficial patient-surgeon relationship, placing the patient at the center of our fiduciary care.

Ethics lies at the core of surgical professionalism: surgeons should not only achieve surgical competence and diligence but also need to be ethically and morally reliable. In this way, an outstanding quality of care will be offered to all members of our society.

I want to specially thank all the coauthors, leaders in their fields, who have done an amazing job. Their collaboration has been outstanding, and without their participation, this book would not be in your hands today.

My eternal gratitude to Dr. Carlos Pellegrini, a true beacon in my academic life and instrumental in allowing me to contribute to this field at the University of Washington, Department of Surgery. He and his wife Kelly have been long-standing friends, mentors, and supporters.

A special mention to our Editor Mr. Prakash Jagannathan, whose role was fundamental and was highly effective in guiding our efforts.

Buenos Aires, Argentina   Alberto R. Ferreres

Contents

Contributors

Maria S. Altieri, MD, MS Division of Bariatric, Foregut and Advanced GI Surgery, Department of Surgery, Stony Brook Medicine, Stony Brook, NY, USA

John Alverdy, MD University of Chicago, Department of Surgery, Chicago, IL, USA

Peter Angelos, MD, PhD Department of Surgery, The University of Chicago, Chicago, IL, USA

The MacLean Center for Medical Ethics, The University of Chicago, Chicago, IL, USA

Alexis G. Antunez University of Michigan Medical School, Ann Arbor, MI, USA

Karen Brasel, MD, MPH Department of Surgery, Oregon Health and Science University, Portland, OR, USA

Douglas Brown, PhD Department of Surgery, Washington University in St. Louis School of Medicine, St. Louis, MO, USA

Darren S. Bryan, MD Department of Surgery, The University of Chicago, Chicago, IL, USA

The MacLean Center for Medical Ethics, The University of Chicago, Chicago, IL, USA

Miguel A. Caínzos Department of Surgery, Hospital Clínico Universitario, Santiago de Compostela, Spain

Georgina D. Campelia, PhD Department of Bioethics & Humanities, University of Washington School of Medicine, & UW Medicine Ethics Consultation Service, Seattle, WA, USA

Oscar Cano-Valderrama, MD, PhD Department of Surgery, Hospital Clínico San Carlos, Madrid, Spain

Amtul R. Carmichael Queens Hospital, Burton-on-Trent, UK

University of Aston, Birmingham, UK

Imventarza Oscar Cesar Hospital Argerich- Hospital Garrahan, Buenos Aires, Argentina

Jonathan K. Chica, MD St. Louis University, St. Louis, MO, USA

Edward E. Cho, MD Department of Surgery, Methodist Richardson Medical Center, Richardson, TX, USA

Susana Ciruzzi, JD, PhD Ethics Committee "Prof Juan P Garrahan" Hospital and "Dr Alfredo Lanari" Institute, University of Buenos Aires, Buenos Aires, Argentina

Mary Condron, MD Department of Surgery, Oregon Health and Science University, Portland, OR, USA

Rojas Luis Daniel EAIT (Transplant Institute of Buenos Aires City), Buenos Aires, Argentina

Sagar S. Deshpande University of Michigan Medical School, Ann Arbor, MI, USA

Inmaculada Domínguez-Serrano, MD, PhD Department of Surgery, Hospital Clínico San Carlos, Madrid, Spain

Gerald Dubowitz University of California, San Francisco, San Francisco, CA, USA

Denise M. Dudzinski, PhD, MTS Department of Bioethics & Humanities, University of Washington School of Medicine, & UW Medicine Ethics Consultation Service, Seattle, WA, USA

E. Christopher Ellison, MD, FACS Department of Surgery, The Ohio State University, Columbus, OH, USA

Alberto R. Ferreres, MD, PhD, JD, MPH Department of Surgery, University of Buenos Aires, Buenos Aires, Argentina
Department of Surgery, University of Washington, Seattle, WA, USA

Judith C. French, PhD Cleveland Clinic Lerner College of Medicine of the Case Western Reserve University, Cleveland, OH, USA
Department of General Surgery, Cleveland Clinic, Cleveland, OH, USA

Sabha Ganai, MD, PhD, MPH, FACS Southern Illinois University School of Medicine, Department of Surgery, Springfield, IL, USA

Salustiano Gonzalez-Vinagre Department of Surgery, Hospital Clínico Universitario, Santiago de Compostela, Spain

Richard Jacobson, MD University of Chicago, Department of Surgery, Chicago, IL, USA
Rush University Medical Center, Department of Surgery, Chicago, IL, USA

D. Rohan Jeyarajah Department of Surgery, Methodist Richardson Medical Center, Richardson, TX, USA

Jason D. Keune, MD, MBA, FACS St. Louis University, St. Louis, MO, USA

Charles W. Kimbrough, MD MPH Department of Surgery, The Ohio State University, Wexner Medical Center, Columbus, OH, USA

Piroska K. Kopar, MD Surgical Critical Care, Acute Care and Trauma Surgery, Yale School of Medicine, New Haven, CT, USA

Cassandra C. Krause, MD, MA Department of Surgery, Loma Linda University Health, Loma Linda, CA, USA

Laurel Mulder, MD Rush University Medical Center, Department of Surgery, Chicago, IL, USA

Jukes P. Namm, MD, FACS Department of Surgery, Center for Christian Bioethics, Loma Linda University Health, Loma Linda, CA, USA

Rosa Angelina Pace, MD Ethics Committee Italian Hospital, Buenos Aires, Argentina

Timothy M. Pawlik, MD, MPH, PhD Department of Surgery, The Urban Meyer III and Shelley Meyer Chair for Cancer Research, The Ohio State University, Wexner Medical Center, Columbus, OH, USA

Carlos A. Pellegrini, MD UW Medicine, University of Washington, Seattle, WA, USA

Aurora D. Pryor, MD, FACS Division of Bariatric, Foregut and Advanced GI Surgery, Department of Surgery, Stony Brook Medicine, Stony Brook, NY, USA

Anthony M. Roche University of Washington, Seattle, WA, USA

H. Alejandro Rodriguez, MD Department of Surgery, University of Washington, Seattle, WA, USA

Kerstin Sandelin Department of Molecular Medicine and Surgery, Karolinska Insitutet, Stockholm, Sweden

Andrew G. Shuman, MD, FACS, FRCSEd (Hon) Department of Otolaryngology-Head and Neck Surgery, University of Michigan Medical School, Ann Arbor, MI, USA

Center for Bioethics and Social Sciences in Medicine, University of Michigan Medical School, Ann Arbor, MI, USA

Steven M. Steinberg, MD, FACS Division of Trauma, Critical Care and Burn, The Ohio State University, Columbus, OH, USA

Antonio J. Torres, MD, PhD, FACS, FASMBS Department of Surgery, Hospital Clínico San Carlos, Madrid, Spain

Christian J. Vercler, MD, MA, FACS, FAAP Department of Surgery, University of Michigan, Ann Arbor, MI, USA

Center for Bioethics & Social Sciences in Medicine, University of Michigan, Ann Arbor, MI, USA

Anji E. Wall, MD, PhD Stanford University, Division of Abdominal Transplantation, Palo Alto, CA, USA

R. Matthew Walsh, MD, FACS Cleveland Clinic Lerner College of Medicine of the Case Western Reserve University, Cleveland, OH, USA

Department of General Surgery, Cleveland Clinic, Cleveland, OH, USA

Andrew L. Warshaw Massachusetts General Hospital and Partners HealthCare, Massachusetts General Hospital, Harvard Medical School, Boston, MA, USA

Richard I. Whyte, MD, MBA Harvard Medical School, Boston, MA, USA

Douglas E. Wood University of Washington, Seattle, WA, USA

Robert M. Zollinger Department of Surgery, The Ohio State University, Columbus, OH, USA

Part I

Principles and Foundations of Surgical Ethics

History and Development of Medical Ethics In the West

Georgina D. Campelia and Denise M. Dudzinski

Key Points

- In Western medicine, medical ethics evolved from an amorphous idea in ancient medicine to a distinct field of study in the twentieth century.
- The different eras of medical ethics are marked by struggles and responses, as medical ethics moves from (1) oaths of faith and fidelity grounded in the authority of higher powers (state, church, crown) to (2) oaths of decorum grounded in professional consensus and contractual agreements within the medical profession and finally to (3) ethical codes formulated collaboratively and grounded in moral reasoning.
- Many of the same virtues and obligations have defined medical ethics throughout the centuries (e.g., beneficence, compassion, confidentiality, fidelity, trustworthiness, respect, integrity, and justice), but their meaning and application have changed with evolving medical knowledge and technology, societal perceptions and understanding, and historical events.

Introduction: From Professional Ethics to Bioethics

"We have to be there at the birth of ideas, the bursting outward of their force: not in books expressing them, but in events manifesting this force, in struggles carried on around ideas, for or against them." — Michel Foucault[1]

Medical Ethics

The application of moral reasoning in the setting of clinical practice and medical research.

Today it is also distinguished by subfields:

1. Theoretical bioethics
2. Clinical ethics
3. Surgical ethics

Philosopher, social theorist, and historian Michel Foucault reminds us that terms like "medical ethics" do not simply name an idea. Rather the meaning and import of a term can be found in the interactions and even conflict that produces

---

[1] Foucault, M. "Les Reportages d'Idees", in *Corriere Della Sera* (Milan, 12 Nov. 1978; repr. In Dilder Eribon, *Michel Foucault*), 1989; Tr.1991.

G. D. Campelia (✉) · D. M. Dudzinski
Department of Bioethics & Humanities, University of Washington School of Medicine, & UW Medicine Ethics Consultation Service, Seattle, WA, USA
e-mail: gdcamp@uw.edu

© Springer Nature Switzerland AG 2019
A. R. Ferreres (ed.), *Surgical Ethics*, https://doi.org/10.1007/978-3-030-05964-4_1

3

and results from the terminology. Today, "medical ethics" refers to the *application of moral reasoning in the setting of clinical practice and medical research.* This involves the use of moral theories (e.g., utilitarianism), moral principles (e.g., respect for autonomy), and virtues (e.g., trustworthiness) to help guide the medical field.

With this in mind, our history begins before Thomas Percival coined the term "medical ethics" in his 1803 book *Medical Ethics.* We consider how ethics' influence on medicine has shifted over several centuries in the West. Medical ethics is an ancient professional ethos, but its path to a distinctive field of study is marked by collaboration and heroism, as well as episodes of deep conflict and violence.

What conflicts or historical shifts informed medical ethics as we understand it today and how has such naming distinguished the field itself? We chart the transformation of the amorphous, unnamed *idea* of medical ethics, beginning with some of the earliest evidence of the character and oaths of the profession to its establishment as a distinctive field of study informing patient care.

Emergence of Medical Ethics, But Not Yet Bioethics

It is thus an open question whether a subject like "medical ethics" existed before it had a designation. Could medical ethics really have existed before 1803, if no one had used an expression designating this concept? – Robert B. Baker & Laurence B. McCullough.[2]

Baker and McCullough are right to question whether medical ethics existed before it was named. Words matter, but as Foucault observes above, ideas begin to form long before they are named. Surgery, for instance, was not defined by written accounts of the first surgical interventions but materialized through the development of tools and techniques, through conversations and arguments about the trade, in the course of accidents and mistakes that cost lives, and, perhaps

most importantly, in response to patients in need of the cures and remedies specific to the art (fractures, head wounds, wounds requiring sutures).[3]

Likewise, medical ethics took shape in the day-to-day practice of medicine and the nature of the physician's relationship with a patient. It highlighted the importance of physicians' moral character. Physicians responded to injury and trauma because society called on them to heal. As philosopher and bioethicist Albert Jonsen articulates, the ethical norms go hand in hand with the healer's role in restoring order:

> Illness is seen as the consequence of a knowing or an unknowing infraction of the order and law of nature or society; the healer must apply the remedies that restore order and reintegrate the sick person into conformity with that order... The work of the healer must not only be correct, that is, the proper remedy for the illness is used, but it must also be right and good, done in conformity with rules, customs, and beliefs that constituted the meaning of life for the society. [29, p. 6]

The healer must never intend harm, so the art and practice are grounded in beneficence. But, for much of medicine's history, scientific knowledge and effective remedies were inadequate, so the power to heal eluded many. Doctors who failed to live up to the identity of healer were perceived as quacks because their treatments rarely restored health and were more likely to harm. They also faced retributive justice as defined by legal codes. The Code of Hammurabi is one of the oldest known set of laws guiding social justice and punishment, including medical malpractice [7]. The law specifies, for instance, that "If a physician make a large incision with the operating knife, and kill him, or open a tumor with the operating knife, and cut out the eye, his hands shall be cut off" [10, p. Code 218].

Prior to the nineteenth century, duties of physicians were defined in oaths and codes, and they were rarely grounded in moral reasoning (e.g., one ought not do *X because* it takes advantage of a patient's vulnerability and so harms the patient). Rather, these oaths were grounded in social expectation, moral beliefs, and valued character traits. This is not to say that they were not serious

---

[2] Baker and McCullough [4].

[3] See, e.g., Ellis [13], Gawande [16].

commitments to beliefs about what was morally right. As Baker articulates,

> Oaths were taken so seriously that signing a loyalty oath was taken as sufficient evidence of loyalty and could even secure pardons. Conversely, refusal to sign a loyalty oath was tantamount to treason... [3, p. 40]

In fact,

> Oaths, vows, and promises are quintessential deontological acts: they bind the person to his word, as testified before a higher being... Oaths were taken solemnly and observed stringently. [30, p. 4]

Deontology
An ethical framework based on duty and obligation (as opposed to consequences).
Examples:
Divine command theory – e.g., "One ought not lie because the gods command truthfulness."
Kantian ethics – e.g., "One ought not lie because rationality demands truthfulness."

However, these codes of conduct are a better reflection of laws and customs that already existed in society than of considered judgments using moral frameworks. As such, medical ethics did not yet have the weight of moral reasoning and justification.

Moral Frameworks
Structured approaches to what one should do or who one should be that are grounded in reasoned argumentation.
*Deontology *Utilitarianism *Virtue Theory
*Communitarianism *Feminist Ethics *Care Ethics *Casuistry *Contract Theory *Principlism

---

Ethics as validated by *custom* rather than *moral argument* is exemplified in one of our earliest examples of ethics in Western medicine: the Hippocratic Oath (written in approximately the fifth century BCE). First printed as part of the *Corpus Hippocraticum* in 1526 in Venice, it is typically attributed to later generations of physicians rather than Hippocrates himself [25]. The oath proclaims the same foundational principles that continue to characterize the "right" and the "good" of the profession of medicine. Famously, the oath obligates "First do no harm" (Latin, *Primum non nocere*) [22], known today as the principle of non-maleficence [5].

The power and motivation of the Hippocratic Oath is the physician's personal or societal duty rather than more principled obligations to patients (such as a duty to respect patient values *because* human beings, especially vulnerable patients, are inherently morally worthy). Early physicians were primarily bound by fidelity to the gods and goddesses: "I swear by Apollo the physician, and Asclepius, and Hygieia and Panacea and all the gods and goddesses as my witnesses..." [22]; and to their teachers: "To hold him who taught me this art equally dear to me as my parents... to look upon his offspring as equals to my own siblings, and to teach them this art... by the set rules, lectures, and every other mode of instruction..." [22]. The health and well-being of one's patients come only after these other obligations and are in honor of the gods: "In purity and according to divine law will I carry out my life and my art" [22]. The practice of medicine was essentially an act of faith. The ethical obligations of the physician (e.g., do no harm) fall out of and are beholden to the healer's broader societal duties and relationships (e.g., to the gods/goddesses or to one's teachers) as opposed to being grounded in moral reasoning (e.g., utilitarianism – creating the greatest good for the greatest number). This is not to say that moral values and reasoning are no longer interwoven with social custom but rather that early medical ethics was *grounded in* social custom.

Today, medical ethics is grounded in forms of moral reasoning that are particular to the medical profession. While this does not characterize early medical ethics, a similar structure of moral justification is found in ancient Greek ethics. Philosophers like Plato (427–347 BCE) and Aristotle (384–322 BCE) present an early demand for a more objective approach to morality, one in which the duties that spring from one's relationship to society are scrutinized and subordinated to obligations to the "right" and the "good." Both philosophers make significant connections between medicine and ethics, often invoking medicine as a parallel methodological model.

In Plato's *Republic*, for instance, Socrates questions his companions about whether it is just to act according to what is advantageous for oneself and employs medicine as an example:

> Now tell me, is the doctor in the precise sense, of whom you recently spoke, a money-maker or one who cares for the sick? Speak about the man who is really a doctor. (Plato, *Republic*, 341c)[4]

Similarly, Aristotle uses medicine (here as a kind of contrast) to gain a better understanding of the *moral* good:

> Let us go back to the good we are looking for—what might it be? For it appears to be one thing in one activity or sphere of expertise, another in another: it is different in medicine and in generalship, and likewise in the rest. What then is the good that belongs to each? … In medicine this is health… for it is for the sake of this that they all do the rest. The consequence is that if there is some one end of all practical undertakings, this will be the practicable good… (Aristotle, *NE*, 1097a15-24)[5]

While Plato and Aristotle differ in their philosophical approaches, both sought an understanding of moral goodness as separate from social custom. As is apparent in Aristotle's passage above, there is a clear distinction between the practice of medicine as a form of τέχνη/technê (i.e., art or craft) and the practice of virtue as the manifestation of εὐδαιμονία/eudaimonia (i.e., flourishing, happiness, or living well). εὐδαιμονία/eudaimonia is the end or moral goal every human being properly strives for. Health may be required to achieve it, because illness often confounds human flourishing. But it is eudaimonia that is the final or complete end of human life. In other words, a proper identification of a healer might require the successful

healing of a wound (as we see in the above codes), and we might say health and healing are important because they lead to happiness (εὐδαιμονία/eudaimonia). But being or acting virtuously (i.e., manifesting eudaimonia) cannot be feigned (*NE* II.4) and cannot be justified according to some further end. Virtue demands moral reasoning to articulate it as such and defend it against rational counter-arguments.

Through the early eighteenth century, decorum, faith, and obedience dictated physician behavior [3, 29, 30, 34]. The virtues and duties of the medical profession continued to be defined largely by the trade's customs and broader social norms. Beginning in the sixteenth century, medical ethics was increasingly formalized by medical licensure and its legal-ethical codes. But the ethical demands of these codes remained grounded in social norms (e.g., decorum), the institutions themselves (e.g., the church), and consensus more than analytic arguments. For instance, some of the earliest oaths of healers regarded labor and delivery. In 1555 in England, Bishop Bonner constructed an oath required for midwifery licensure ([3, pp. 19–21]. This medical-ethical oath included duties to protect mother and child, e.g., "ye shall not suffer any chylde to be murdered, maymed, or otherwise hurtydem as nygh," as well as duties in interest of the church, e.g., "when of necessity ye shall chrystyn any chylde, ye shall use pure and cleane water, nother mixt with rose water, damaske water, or otherwise altered or confected" [3, p. 21; 8, pp. 165–166].

These oaths were designed in the interest of the church and the crown, which often trumped the interests of the women in labor [3, p. 23]. For instance:

> Item, ye shall never consent nor agree that any woman be delivered secretly, but in the presence of two or three lyghtes, if she travell nyght. [3, p. 21; 15]

Throughout history, societal order (e.g., the church, state, and/or crown) has demanded punishment for pregnancies that fell outside of social norms (e.g., because the woman was unmarried and bearing children out of wedlock). As such,

---

[4]Here, and throughout, we use Allan Bloom's translation of Plato's *Republic* [42].

[5]Here, and throughout, we use the Broadie and Rowe translation of Aristotle's *Nicomachean Ethics* [2].

the well-being of the pregnant woman might have required delivering in secret. But in healers' oaths at this time, this was prohibited because it violated a social order that required reporting unwed mothers, 'bastard children', and 'true paternity'. As such, the woman's interests in confidentiality were subordinate to the interests of the church and crown, which included the dual efforts to protect children through proper christening and protect social order through proper identification of paternity.

Over the course of the sixteenth, seventeenth, and eighteenth centuries, as medicine became more institutionalized and commercialized [30, pp. 43–44], the power and meaning of the healer's duties began to shift and conflict ensued. As commerce developed in Western Europe, physician-patient relationships were changed by the formation of payment contracts and fixed sums for services calculated by professionals rather than based on the patient's ability to pay or satisfaction with the services provided [30, p. 44]. As such, physicians, communities, and the state all struggled to reconcile conflict between the goals of healing and procurement of income [30, pp. 44–45]. Socrates' question is echoed, is the true physician a healer or a money-maker?

At the same time, the oaths of physicians, grounded as they were in the power of the church and crown, transformed with the Catholic-Protestant struggle. In Scotland, for instance, the Presbyterian (Scotland's national religion) resistance to Anglicization led to the revision of university oaths and inquisitions to insure the oaths' obligations were fulfilled [3, p. 41]. In 1688, a Presbyterian inquisition sought to expunge faculty and administration at Edinburgh University of non-Presbyterians [3, p. 42; 39]. It was a contest of belonging, and it mattered less that you belonged to the medical profession than that you belonged to the right church or state.

We see the influences of these struggles and the Enlightenment in the oaths of the eighteenth century. Medical practitioners, scholars, and the broader community questioned the authority of religious institutions in the Age of Reason. And this conflict brought oaths built upon contracts within the profession itself rather than commanded by the church and crown. This means that the force of the obligation no longer came from duty to a higher power or simple fidelity to mentors but from agreement among members of the medical profession, attesting to the greater social standing and influence of physicians.

One of the first instances of this shift can be found in the Edinburgh University Medical Oath, Circa 1732–1735:

> [I A.B. do solemnly declare that I will] practice physic cautiously, chastely, and honourably; and faithfully to procure all things conducive to the health of the bodies of the sick; and lastly, never, without great cause, to divulge anything that ought to be concealed, which may be heard or seen during the professional attendance. To this oath let the Deity be my witness. [47, pp. 50-51]

The place of the deity, here, comes only after the duties of the art or trade, and the deity is a witness to rather than the source of the obligation. The influence of this oath on Western medicine, particularly in the Americas, was significant. Many physicians in the colonies spent part of their training at the University of Edinburgh [3, p. 38]. And medical education in America was modeled off of the Scottish example [3, p. 39].

Moving through the eighteenth and nineteenth centuries, a more contractarian model of medical ethics develops. While many of the virtues are similar (honor, beneficence, fidelity), medicine shifts to a "learned and gentlemanly profession" [30, p. 57] as defined within the profession and agreed to by medical professionals. In 1766, America's first medical society was founded: New Jersey Medical Society. Its foundational document, Instruments of Association, pledged "never [to] enter any house in quality of our profession, nor undertake any case, either in physic or in surgery, but with the purest intention of giving the utmost relief and assistance that our art shall enable us, which we will diligently and faithfully exert for that purpose" [26, pp. 309–311]. The moral demand is grounded in the art itself (as defined by the medical institution) and agreement among its practitioners, rather than in deference to a societal leader or broader societal customs.

This same shift is visible in the oaths of other foundational medical societies. For instance, the Oath of the Medical Society of the State of New York, 1807, states:

> I, A. B., do solemnly declare, that I will honestly, virtuously, and chastely, conduct myself in the practice of physic and surgery, with the privilege of exercising which profession I am now to be invested; and that I will, with fidelity and honour, do every thing in my power for the benefit of the sick committed to my charge. [23]

In these oaths, it is not the character of the physician that shifts significantly, as there is still a focus on honesty, honor, fidelity, and beneficence. Likewise, the oath remains a "[distillation] of the ethical ideals of a community" [3, p. 37]. The transformation is in the source and power of the demand. The demand continues to be grounded in social agreements and norms, but the source of the demand comes from the institution of medicine itself. The oaths were thus defined through social agreement among medical professionals and made powerful by the nature of this social contract.

As the ethical obligations of the profession became increasingly specified and formalized in the oaths/contracts of medical institutions, several scholars began to theorize the character and ethos of medicine that took shape in those ethical codes. As the oaths become subject to moral reflection and reason, very quickly, the *idea* of medical ethics emerges in the discourse. John Gregory (1724–1773), a practicing physician who also taught philosophy at King's College, invoked moral reasoning to justify the virtues of the profession. Influenced by philosophers such as David Hume and Adam Smith, Gregory accepted the historical virtues of the profession (e.g., honor, patience, humanity) but sought to ground them in the moral sentiment of sympathy with the patient and a desire to relieve suffering [18, 30, p. 60]. Sympathy motivates the physician obligation to respond so as to relieve or heal. The other virtues, such as honor and patience, are required in service of this goal.

Then in 1803, with the publication of Thomas Percival's book *Medical Ethics: Or a Code of Institutes and Precepts Adapted to the Professional Conduct of Physicians and Surgeons*, the term "medical ethics" is coined. While not yet the field of medical ethics as we know it today, Percival's work, like Gregory's, is grounded in philosophical thought, investigating and justifying a physician's moral role rather than merely articulating consensus. While his rules are sometimes consistent with historically integrated principles of the profession, he draws on philosophical theory and renders those rules accountable to reason or, at least, reflection.

Unlike other products of the Enlightenment, Percival sought a balance between philosophy, religion, and medicine [41, p. 2266]. His thinking was influenced by philosophers and theologians alike, and this is reflected in his focus on virtue. A virtue theoretical approach, which begins from a conceptualization of the sort of person one should be rather than what actions one is obliged to take or avoid, had been common to both philosophy (Plato, Aristotle, Hume) and theology (St. Ambrose, Aquinas, Maimonides) over the preceding centuries. Percival's writings reveal a shift toward the same kind of justification sought by Plato and Aristotle. He uses the virtue of beneficence, for instance, to justify cost containment and fair distribution of resources (e.g., more rural dispensaries for the poor) [41, p. 2267]. As Pellegrino explains, "his worry is less with maintaining professional secrecy than with potential harm to the patient" [41, p. 2267]. In this regard, Percival appealed to reason rather than custom and shifted the central concern of the profession to the patient.

But alongside rational justification for these ethical duties, the public questioned the motives of the institution of medicine. For instance, New Jersey colonists denounced the formation and authority of the New Jersey Medical Society with accusations of scheming and swindling the public [3, pp. 10–11]. And this was neither the beginning nor the end. Inside the profession arose critiques of self-interest and quackery, such as Yale professor Worthington Hooker's 1850 publication *Lessons from the History of Medical Delusions*. More forceful were the violent public protests in Edinburgh (1725) and New York (1788), which erupted against the practice of "body snatching." Body snatching involved robbing graves to supply cadavers for study [3, pp. 57–58, 60]. In New York, bodies were typically taken from the graves of slaves and the poor [36], in part because they were not able to afford the iron cages and security personnel that safeguarded the bodies of the wealthy [21, 35]. Knowledge of the practice was met with building disgust and anger and eventually led to a group of citizens storming New York Hospital. These conflicts between physicians and the community were reflective of a kind of moral dissonance between medical practice and social expectations, and the struggle demanded a new understanding of the ethical obligations of physicians.

As we turn toward the mid-nineteenth century, we continue to see these clashes between the profession and the public play out in the practice and research of individual physicians. Dr. Marion Sims (1813–1883), for instance, remains immortalized in a statue in Central Park, NYC, as the founder of modern surgical gynecology. Yet, the techniques that he invented and are still in use today came from the objectification and torture of African American female slaves. One of his "patients," Anarcha Wescott, underwent 30 operations to repair vesicovaginal fistulas [40]. Some defend Sims, arguing that he was well-intentioned insofar as he sought cures when there were none for the gruesome effects of the repeated rapes, pregnancies, and births that African American female slaves were forced to endure [54]. And yet, however willingly these women cooperated with Dr. Sims, they were enslaved so their participation could never have been truly voluntary, especially since the injuries they suffered were due to their being raped by white male slave owners. Notably Dr. Sims did not conduct these surgical experiments on middle-class and wealthy white women, perpetuating a pattern of exploitation of enslaved, poor, and vulnerable people. Medicine, along with every other human enterprise, has a tendency to focus on the *beneficial outcomes* of Sims' discoveries (techniques that have helped many women) without acknowledging the *means* of discovery (human exploitation and violence).

So, it is not surprising that out of both (1) theoretical work like that of Gregory's and Percival's

and (2) clashes between physicians and their communities arose the implementation of medical ethics in professional codes. In the American Medical Association's first Code of Ethics [11], the contractarian model of medical ethics continues but with a shift toward Percival's focus on the patient and stronger moral language. The code prominently displays the importance of duties and their corresponding rights, the reciprocity of vulnerability and protection, and the universality of moral obligation. This shift is visible in the Table of Contents:

- CHAPTER I. – Of the duties of physicians to their patients, and of the obligations of patients to their physicians.
- ART. I. – Of the duties of physicians to their patients. ART. II. – Of the obligations of patients to their physicians.
- CHAPTER II. – Of the duties of physicians to each other, and to the profession at large.
- ART. I. – Of the duties of physicians for the support of professional character.
- ART. II. – Of the duties of physicians in reward to professional services to each other.
- ART. III. – Of duties of physicians in regard to vicarious offices.
- ART. IV. – Of the duties or physicians in consultations.
- ART. V. – Of the duties of physicians in cases of interference with one another.
- ART. VI. – Of the duties of physicians when differences occur between them.
- ART. VII. – Of the duties of physicians in regard to pecuniary acknowledgements.
- CHAPTER III. – Of the duties of the profession to the public, and of the obligations of the public to the profession.
- ART. I. – Of the duties of the profession to the public. ART. II. – Of the obligations of the public to physicians. [11, p. 91]

And similarly in each declaration of ethical obligation, for instance,

§ 5. A physician ought not to abandon a patient because the case is deemed incurable; for his attendance may continue to be highly useful to the patient, and comforting to the relatives around him, even in the last period of a fatal malady, by alleviating pain and other symptoms, and by soothing mental anguish. To decline attendance, under such circumstances, would be sacrificing to fanciful delicacy and mistaken liberality, that moral duty, which is independent of, and far superior to all pecuniary consideration. [11, p. 94]

As Jonsen notes:

Here, then, medical ethics becomes in substance that very American political fiction, a contract with mutual rights and duties among the contracting parties: doctors, patients, and society. An intriguing idea but, in this context, an odd one: it is only the physicians who have written the contract. [30, p. 70]

So, even as these codes remained grounded in the agreement among medical professionals, the language slowly became infused with scholarly work on medical ethics and its demand for accountability to reason. This, in addition to medical and scientific progress, heightened the power and trustworthiness of the physician. Anesthesia, for instance, developed in early mid-nineteenth century marked a dramatic shift in the ability of the surgeon to heal without causing further suffering. Likewise, the development of statistical methods in epidemiology led to far greater accuracy in identifying pathogens responsible for epidemics (e.g., cholera) [34].

These evolutions in medicine and medical ethics, marked by both progress and pitfalls, lead us to modern-day medical ethics. But as we approach the twentieth century, the conflict and violence that infused medical progress, along with its attendant resolutions, are what truly characterize the emergence of medical ethics as a distinctive field. It was struggle (as Foucault recognizes), more than the softer persuasions of scholars like Percival, that ultimately forced the materialization of the idea.

The Beginning of Bioethics

"Respect every living being, in principle, as an end in itself and treat it accordingly wherever it is possible" — Fritz Jahr.[6]

---

[6] See Jahr [28].

The twentieth century is riddled with both significant medical advancements and moral-medical failures in which medicine was enlisted in exploitative social programs and law enforcement. Often the two went hand in hand.

During this time, scientific advances dramatically changed the trustworthiness, allure, and power of the medical profession. Vaccinations became more commonplace, and many were developed for the first time (including influenza, typhoid, polio, measles, chickenpox, and tuberculosis). The development of antibiotics and antivirals made significant strides against communicable disease. This included the discovery of penicillin in 1928 by Alexander Fleming. Antibiotics also affected the safety of the already rising numbers of surgical interventions. Breakthroughs were made in surgical interventions, including the first laparoscopic surgery (Hans Christian Jacobaeus 1910), the first splenectomy (Hermann Schloffer 1916), the first open heart surgery (Henry Souttar 1925), and the first gender affirming surgery (1931). The practice of blood transfusion was developed in 1914. In 1929 at Boston Children's Hospital, Philip Drinker and Charles McKhann published on their successful use of an artificial respirator (known as the "iron lung") for patients with paralytic polio. And we see the first successful organ transplantations in the 1940s.

But these incredible advances in medicine often came at costs that unjustly affected some social groups more than others. For example, in 1907 Indiana passed the first law authorizing forced sterilization "of confirmed criminals, idiots, imbeciles and rapists" (1907 Indiana Eugenics Law, Chap. 215, H. 364). This practice became more prevalent, and, ultimately, over 60,000 individuals underwent compulsory sterilization in the USA [9, 12, 43].

This practice was common elsewhere in the world as well. "Between 1934 and 1944 (when the population was 73 million) German doctors sterilized at least 400,000 persons, including the mentally ill, the mentally disabled, the deaf, persons with tuberculosis, homosexuals, gypsies, and, of course, Jews" [43, p. 358]. The infamous medical experimentations of the Holocaust culminated in the Nuremberg Trials (1945–46) and the Nuremberg Code. But, the USA too was guilty of similar crimes. And African Americans were often the target. Some of the more well-known incidents include the Tuskegee Study (1932–1974) in which black men in the USA were deceived into thinking they were being treated for syphilis and other blood conditions, when in fact they were not [52]. Later, in 1951, Henrietta Lacks unknowingly became a vehicle for medical advancement through the creation of the first immortalized cell line, when cells from her tumor biopsy went on to be used, and continue to be used, in medical research without her consent [49]. While few subjects at this time would have been informed, at least not in the sense of "informed consent" today, Ms. Lacks' case, like the Tuskegee experiments, demonstrates how medical advancements were built on the backs of black and/or poor people without fair opportunities for these communities to benefit from the treatments they were involved in developing.

Another fairly well-known example is the Manhattan Project (1942–57), in which participants were injected with plutonium, uranium, and possibly other radioactive elements. The goal of the project was a better understanding of the already known risks of ongoing exposure to workers in the Manhattan Project (without their knowledge of said risks) (Georgetown Bioethics Archive). As such, this project failed to aim at the good of the relevant patient population (or at least minimize harm) in two senses: (1) failure of clinical equipoise (i.e., "genuine uncertainty on the part of the… investigator regarding the comparative therapeutic merits of each arm of a trial." [14, p. 141]) insofar as there was no immediate benefit to participants (e.g., a trial of a new curative treatment that is expected to help at least some), but rather known harm would be caused to the subjects, and (2) any potential long-term benefit of the knowledge was likely outweighed by the significant risk of the exposure itself (already known from observational data on workers).

Some lesser known examples targeted the disabled and prisoners. In 1943, researchers at University of Cincinnati Hospital kept 16 mentally disabled patients in refrigerated cabinets for 120 hours at 30 degrees Fahrenheit in order to "study the effect of frigid temperature on mental

disorders" [17]. In 1950 Dr. Joseph Strokes of the University of Pennsylvania infected 200 female prisoners with viral hepatitis to study the disease [24, p. 91]. And in 1963–1973, the University of Washington performed high-dose radiation tests on prisoners' testicles to find a sterility dose [44].

Such brutally immoral experimentations were not contained by US borders. In 1906, Dr. Richard Strong of Harvard conducted cholera experiments in the Philippines, killing 13 prisoners [1]. In 1940, US doctors infected thousands of Guatemalans with venereal disease [55]. And across the world we see similar experimentation with the effects of nuclear and biological weaponry. For instance, Japanese armed forces (Unit 731) field tested weapons with the plague, anthrax, and a number of other pathogens on Chinese prisoners [33].

Mistreatment and violence against society's most vulnerable populations are not new and were sometimes normalized in medicine as it was in the broader society. While scholars like Percival initiated a distinct field of medical ethics, it was in the struggles of the twentieth century that bioethics emerged as it is known today. It began with several declarations, reports, and articles that attempted to respond to the atrocities above. Theologians were enlisted first to help doctors, quickly followed by philosophers. In papers in 1926 and 1927, Fritz Jahr, a German protestant theologian, first coined the term "bioethics" (German "Bio-Ethik," directly translated as "Bio-Ethics") [27, 28]. Jahr's use of the term was broad with the goal of "safeguard[ing] all living nature from pointless destruction" [32]. His moral imperative, though, resounded in modern bioethics: "Respect every living being, in principle, as an end in itself and treat it accordingly wherever it is possible" [28].

Ethical Imperative
A demand, command, or rule that is grounded in moral reason and obligated of moral agents.

For example, physicians should respect their patients' autonomous decisions.

So, it was not just the struggle and violence but the reactions and pushback that came from other physicians, the public, theologians, and philosophers. Bioethics took shape both in recognition of crimes society did not want to commit again and also in a deepening commitment to physicians' moral roles and social responsibilities.

The 1948 Declaration of Geneva was formulated by the World Medical Association and was built on the explicit ethical obligations to respect the lives and liberties of patients, e.g., "I WILL MAINTAIN the utmost respect for human life from its beginning even under threat and I will not use my medical knowledge contrary to the laws of humanity" [56].

Henry K. Beecher, an American anesthesiologist, is famous for his 1966 article "Ethical and Clinical Research," which criticized the human experimentations taking place since WWII [6]. In the opening paragraph he states:

> Evidence is at hand that many of the patients in the examples to follow never had the risk satisfactorily explained to them, and it seems obvious that further hundreds have not known that they were the subjects of an experiment although grave consequences have been suffered as a direct result of experiments described here. [6, p. 1354]

Here, Beecher articulates two of the central principles later developed in Beauchamp and Childress' *Principles of Biomedical Ethics*: (1) beneficence and (2) respect for autonomy. The idea that subjects of experimentation should be informed about the risks (including known harms) of their participation in the research is built on a principle of respect for autonomy as an essential component of humanity.

The resistance and response to the suffering of so many human subjects throughout the twentieth century quickly became reflected in practice and scholarship as clinicians, theologians, and philosophers joined forces to analyze, critique, and promote ethics in medicine. Suddenly, medical ethics went from a somewhat vague term studied by a small subset of physicians to an interdisciplinary specialization that traversed the bounds of the professions. It became formalized in articles, reports, and even institutions. In 1971, André Hellegers established the first Center for

Bioethics as part of the Kennedy Institute at Georgetown University [19].

Then, in 1979, the National Commission for the Protection of Human Services of Biomedical and Behavioral Research published the Belmont Report, named after the house in which the authors convened to write it [31]. This report, collaboratively written by MDs, PhDs, and JDs, defines three ethical principles which have become central to the practice of medical research: (1) respect for persons, (2) beneficence, and (3) justice [46]. The principles were *justified* through philosophical reasoning and then *applied* to the field of medicine. As the report declares:

> The expression "basic ethical principles" refers to those general judgments that serve as a basic justification for the many particular ethical prescriptions and evaluations of human actions. Three basic principles, among those generally accepted in our cultural tradition, are particularly relevant to the ethics of research involving human subjects: the principles of respect of persons, beneficence and justice. [46]

Bioethics, accordingly, moved away from (1) the standards of practice grounded in social order and higher authorities and (2) the decorum and duties contractually agreed to within the profession, which we saw in earlier forms of medical ethics. While consensus is still valued and societal norms still influences, bioethics seeks clinical and ethical justification from multiple viewpoints (internal and external to the profession of medicine). The applications of ethical principles and virtues must be grounded in reasoned arguments that can stand up to critique.

Now, moral theories and approaches long utilized in philosophy and theology are brought to bear in medicine. These include principlism, virtue ethics, deontology, consequentialism, communitarianism, phenomenology, etc. In the past 50 years, we've seen a shift in medicine from a focus on *protecting* one's patient to *respecting* patient autonomy. The focus on respect for persons changes the way medicine interfaces with the population it serves, creating more of a partnership than had existed in the past centuries. The conflict of the nineteenth and twentieth centuries thus demanded multidisciplinary engagement, and physicians often welcomed it. Bioethics became an academic discipline with its own educational programs and scholarship, but it remains fundamentally interdisciplinary.

Conclusion: The Role of the Bioethicist

In the end, while advances in medicine in the nineteenth and twentieth centuries began to instill greater trust of and dependence on medical providers by large swaths of Western populations, this trust was continuously abused in underrepresented populations and challenged by the public. Medical ethics, including today's bioethics, emerged in the course of this struggle through collaboration between clinicians, theologians, patients, attorneys, social workers, philosophers, and other theorists. Now the field is distinct from medicine itself and yet defined by ongoing conflict and uncertainty over the meaning and application of the very same terms that have defined the character of medicine since the ancient Greeks: beneficence, compassion, confidentiality, fidelity, trustworthiness, respect, integrity, and justice.

The change has been both big and small, fast and slow. It is not so much the recognized virtues and obligations of medical professionals that have shifted but how and to whom they are applied. The principles of beneficence and nonmaleficence, for example, are evidenced in the Hippocratic Oath, but there they were unquestioned norms and subordinate to the dictates of the gods. Today, these same principles and virtues are grounded in moral theory (utilitarianism, communitarianism, etc.), and their application must stand up to critique both internal and external to the profession.

This shift is not surprising given the increasing influence of empiricism and positivism in the eighteenth to twentieth centuries. But, as contemporary bioethicists argue, we must cautiously recognize and remember the ongoing influence of detrimental social norms. Though bioethics exists as its own discipline, it is not immune to stereotypes, shortsightedness, and implicit bias. Implicit bias is something that will not go away [48], and it can reside in ideas that we mistakenly believe to be scientifically or theoretically distinct from such social norms. Feminist ethicists, for instance, have criticized rights-based approaches and the principle of respect for autonomy for their inherent understanding of persons as individualistic, atomistic, and independently/freely able to choose their paths in life [20, 37, 45, 50, 51, 53]. If we are not careful to acknowledge how practical-theoretical frameworks affect different individuals and communities in the practice of medicine, we will continue to exclude, disrespect, and cause harm to the same underrepresented populations who survived the twentieth century's abuses.

Concluding Remarks

- The transformation of medical ethics in the West is marked by collaboration and heroism, as well as episodes of deep conflict and violence.
- Medical ethics takes shape in different eras, beginning with (1) oaths of faith and fidelity grounded in the authority of higher powers (state, church, crown), then (2) oaths of the medical institution grounded in consensus of the profession, and finally (3) ethical codes formulated collaboratively and grounded moral reasoning.
- Many of the same virtues and obligations have defined medical ethics throughout the centuries (e.g., beneficence, compassion, confidentiality, fidelity, trustworthiness, respect, integrity, and justice), but their meaning and application in the medical setting are subject to disagreement and will require moral critique and reflection. Bioethicists are built to fulfill this role but only in ongoing interprofessional and interdisciplinary collaboration.

The twentieth and twenty-first centuries have brought with them a host of life-sustaining (e.g., implantable cardioverter defibrillators, ventricular assist devices, extracorporeal life support), curative (e.g., deep brain stimulation, targeted gene therapy), and diagnostic (e.g., preimplantation genetic diagnosis, direct to consumer genetic testing) technologies that are likely to proliferate into the future. The interdisciplinary and deliberative nature of bioethics will help medicine and society to critically reflect on ever-evolving technologies and reinforce robust critique and justification without assuming that one can have a completely objective or "god's eye" view of the moral obligations and virtues of medical practice. In this way, bioethicists will continue to play a social role, offering ethical-theoretical expertise, different perspectives, and much needed reflective pauses in a fast-paced research and practice climate.

Bibliography

1. 1906: Richard Strong, MD, cholera. 2017. from Alliance for Human Research Protection. http://ahrp.org/1906-richard-strong-md-cholera/.
2. Aristotle. Nicomachean Ethics (trans: Broadie S and Rowe C, editors: Broadie S and Rowe C). Oxford: Oxford University Press; 2002.
3. Baker R. Before bioethics: a history of American medical ethics from the colonial period to the bioethics revolution. New York, NY: Oxford University Press; 2013.
4. Baker RB, McCullough LB. What is the history of medical ethics? In: Baker RB, McCullough LB, editors. The Cambridge world history of medical ethics. Cambridge, UK: Cambridge University Press; 2009.
5. Beauchamp TL, Childress JF. Principles of biomedical ethics. New York, NY: Oxford University Press; 2012.
6. Beecher H. Ethics and clinical research. N Engl J Med. 1966;274:1354–60.
7. Belli SR, Melvin M. The evolution of MEdical malpractice law. In: Vevaina JR, Bone RC, Kassoff E, editors. Legal aspects of medicine. New York, NY: Springer-Verlag; 1989.
8. Bonner BE. Mid wife's Oath. In: Helm W, editors. Curious miscellaneous fragments of various subjects more particularly relative to english history from the year 1050, to the year 1701, compiled from British Writers during that period. London: Baldwin, Cradock and Joy; [1555] 1815.
9. Bruinius H. Better for all the world: the secret history of forced sterilization and America's quest for racial purity. New York, NY: Alfred A. Knopf; 2006.
10. The Code of Hammurabi. (c. 1754 BC). Retrieved from http://avalon.law.yale.edu/ancient/hamframe.asp.
11. Code of Medical Ethics of the American Medical Association. Philadelphia: American Medical Association; 1847.
12. Daniel K. In the name of eugenics: genetics and the uses of human heredity. New York, NY: Alfred A. Knopf; 1985.
13. Ellis H. A history of surgery. London: Greenwich Medical Media; 2000.
14. Freedman B. Equipoise and the ethics of clinical research. N Engl J Med. 1987;317(3):141–5.
15. Garnet R. The book of oaths. In: Aughterson K, editor. Renaissance woman: a sourcebook: constructions of femininity in England. New York, NY: Routledge; [1649]1995. p. 212–214.
16. Gawande A. Two hundred years of surgery. N Engl J Med. 2012;366:1716–23.
17. Goldman D, Murray M. Studies on the use of refrigeration therapy in mental disease with report of sixteen cases. J Nerv Ment Dis. 1943;97(2):152–65.
18. Gregory J. Observations on the duties and offices of a physician and on the method of promoting enquiry in philosophy. London: Strahan and Cadell; 1770.
19. Harvey JC. André Hellegers, the Kennedy institute, and the development of bioethics: the American–European connection. In: Garrett JR, Jotterand F, Ralston DC, editors. The development of bioethics in the United States. Dordrecht: Springer Science & Business Media; 2013. p. 37–54.
20. Held V. Care and justice in the global context. Ratio Juris. 2004;17:141–55.
21. Highet MJ. Body Snatching & Grave Robbing: bodies for science. Hist Anthropol. 2006;16(4):415–40.
22. Hippocrates. The Hippocratic Oath (trans: North M): National Library of Medicine; 2002. Retrieved from https://www.nlm.nih.gov/hmd/greek/greek_oath.html.
23. History of the Medical Society of the State of New York. In: Walsh JJ, editor. New York, NY: Medical Society of the State of New York; 1907.
24. Hornblum AM. Acres of skin: human experiments at Holmesburg prison. New York, NY: Routledge; 1998.
25. Iniesta I. Hippocratic Corpus. Br Med J. 2011;342(7803):929.
26. Instruments of Association and Constitutions of the New Jersey Medical Society. In: Rogers FB, Sayre AR, editors. The healing art: a history of the medical Society of new Jersey. Trenton, NJ: The Medical Society of New Jersey; 1966.
27. Jahr F. Life sciences and the teaching of ethics. In: Miller IM, Sass H-M, editors. Essays in bioethics. Berlin: LIT Verlag Münster; 1926, 2013.
28. Jahr F. Bio-Ethik. Eine Umschau über die ethischen Beziehungen des Menschen zu Tier und Pflanze. Kosmos Handweiser für Naturfreunde. 1927;24(1):2–4.
29. Jonsen AR. The birth of bioethics. New York, NY: Oxford University Press; 1998.
30. Jonsen AR. A short history of medical ethics. New York, NY: Oxford University Press; 2000.

31. Jonsen AR. Celebrating Professor Al Jonsen: Hastings Center Beecher Award Ceremony. Paper presented at the The 19th Annual Meeting of the American Society for Bioethics and Humanities. Kansas City, MO; 2017.

32. Kalokairinou EM. Fritz Jahr's bioethics imperative: its origin, point, and influence. JAHR. 2016;7/2(14):149–56.

33. Kristof N. Unmasking horror -- a special report.; Japan confronting gruesome war atrocity. The New York Times; 1995. Retrieved from http://www.nytimes.com/1995/03/17/world/unmasking-horror-a-special-report-japan-confronting-gruesome-war-atrocity.html?pagewanted=all.

34. Kuhse H, Singer P. What is bioethics? A historical introduction. In: Kuhse H, Singer P, editors. A Copanion to bioethics: second edition. Oxford: Blackwell Publishing Ltd; 2009.

35. Lovejoy B. Body Snatchers of Old New York. Lapham's Quarterly; 2013.

36. Lovejoy B. The Gory New York City Riot that Shaped American Medicine. Smithsonian; 2014. Retrieved from http://www.smithsonianmag.com/history/gory-new-york-city-riot-shaped-american-medicine-180951766/.

37. Mackenzie C, Stoljar N. Relational autonomy: feminist perspectives on autonomy, agency, and the social self. New York, NY: Oxford University Press; 2000.

38. The Manhattan Project: A New and Secret World of Human Experimentation. Retrieved 18 Sept 2017, from Georgetown University. https://bioethicsarchive.georgetown.edu/achre/final/intro_3.html.

39. Monro A. Presbyterian inquisition as it was lately Practised Against the Professors of the College of Edinburgh, Agust and September, 1690, in which the Spirit of Presbytery and their Present Method of Procedure is Plainly Discovered, Matter of Fact by Undeniable Instances Cleared, and Libels Against Particular Persons Discussed. London: J. Hindmarsh; 1691.

40. Ojanuaga D. The medical ethics of the "father of gynaecology", Dr J Marion Sims. J Med Ethics. 1993;19(1):28–31.

41. Pellegrino ED. Percival's *Medical Ethics*: The moral philosophy of an 18th-century English gentleman. Arch Intern Med. 1986;146(11):2265–9.

42. Plato. The Republic of Plato. (trans: Bloom A, editor: Bloom A). United States of America: Basic Books; 1968.

43. Reilly P. Eugenics and involuntary sterilization. Annual Review Genomics and Human Genetics. 2015;16:351–68.

44. Richards EP. Advisory Committee on Human Radiation Experiments Report, Chapter 9; 1995.

Retrieved from https://biotech.law.lsu.edu/research/reports/ACHRE/chap9_2.html.

45. Robinson F. Human rights and the global politics of resistance: feminist perspectives. Rev Int Stud. 2003;29(S1):161–80.

46. National Commission for the Protection of Human Subjects of Biomedical and Behavioral Research. The Belmont report: ethical principles and guidelines for the protection of human subjects of research. United States of America: The Commission; 1979.

47. Ryan M. A manual of medical jurisprudence, compiled from the best medical and legal works; being an analysis of a course of lectures on forensic medicine, Annually delivered in London. 2nd ed. Philadelphia: Carrey and Lea; 1832.

48. Sabin JA, Rivara FP, Greenwald AG. Physician implicit attitudes and stereotypes about race and quality of medical care. Med Care. 2008;46(7):678–85.

49. Skloot R. The immortal life of Henrietta lacks. New York, NY: Crown Publishers; 2010.

50. Tietjens Meyers D, editor. Feminists rethink the self. Boulder: Westview Press; 1997.

51. Tronto J. Moral boundaries: a political argument for an ethic of care. New York, NY: Routledge; 1993.

52. The Tuskegee Timeline. Retrieved 18 Sept 2017, from Centers for Disease Control and Prevention; 2015. https://www.cdc.gov/tuskegee/timeline.htm

53. Urban Walker M. Moral understandings: feminist study in ethics. New York, NY: Routledge; 1998.

54. Wall LL. The medical ethics of Dr. J Marion Sims: a fresh look at the historical record. Journal of Medical Ethics. 2005;32(6):346–50.

55. Walter M. First, do harm. Nature. 2012;482:148–52.

56. World Medical Association. Declaration of Geneva; 1948. Retrieved September 2017, from World Medical Association: https://www.wma.net/policies-post/wma-declaration-of-geneva/.

Suggested Literature

Baker R. Before bioethics: a history of American medical ethics from the colonial period to the bioethics revolution. New York, NY: Oxford University Press; 2013.

Eckenwiler LA, Cohn FG, editors. The ethics of Bioethis: mapping the moral landscape. Baltimore: Johns Hopkins University Press; 2007.

Jonsen AR. The birth of bioethics. New York, NY: Oxford University Press; 1998.

Jonsen AR. A short history of medical ethics. New York, NY: Oxford University Press; 2000.

The History of Surgical Ethics

Jukes P. Namm and Cassandra C. Krause

Key Points

- The unique issues that are encountered in surgery have made surgical ethics distinct from medical ethics.
- Surgical ethics has defined the surgical profession throughout the history and evolution of modern surgery.
- As emphasized by Gregory and Percival, surgical ethics revolves around the surgeon-patient relationship.

Introduction

Medical ethics dates back to ancient Greece, beginning with the writings of Hippocrates around the fourth century BCE who is given credit for writing one of the earliest works on the principles of nonmaleficence, physician decorum, and, of course, the Hippocratic Oath [1]. Although there is uncertainty regarding the origin of surgical ethics, some would suggest that it

J. P. Namm (✉)
Department of Surgery, Center for Christian Bioethics, Loma Linda University Health, Loma Linda, CA, USA
e-mail: jpnamm@llu.edu

C. C. Krause
Department of Surgery, Loma Linda University Health, Loma Linda, CA, USA
e-mail: ckrause@llu.edu

is as old as the art of surgery itself, which can be traced back to ancient Mesopotamia and Egypt. In addition to embalming, the ancient Egyptians performed eye surgery, reduced fractures, performed wound care, and even placed prostheses. On the tomb of Nenkh-Sekhmet, chief of the physicians during the 5th dynasty, is written: "Never did I do evil towards any person" [2]. Much of Greek medicine was influenced by the ancient Egyptians. However, surgical ethics as a field did not formally emerge until the advent of modern surgery in the nineteenth century and the establishment of surgery as a profession.

It is through the lens of surgical history, especially within the context of modern surgery, that one can truly appreciate the role of ethics in the unique issues that surgeons have faced in the care of their patients. From the eighteenth century onward, ethical issues distinct to surgery such as informed consent, fee splitting, itinerant surgery, transplantation, and surgical innovation have necessitated the emergence of a distinct offshoot of medical ethics termed surgical ethics through the leadership of individuals and the surgical profession.

Birth of Modern Ethics

During the time of Hippocrates (460–370 BCE), medicine became a respected profession as a result of its influence in establishing a standard requiring physicians to be accountable for their

© Springer Nature Switzerland AG 2019
A. R. Ferreres (ed.), *Surgical Ethics*, https://doi.org/10.1007/978-3-030-05964-4_2

actions [1]. During the dark ages, the art of medicine suffered as there was little accountability and physicians had little structure to guide them. Medieval physicians were motivated more by expedience than by beneficence. However, society in the eighteenth century witnessed resurgence in the profession of medicine through the emergence of medical ethics by the work of two physicians, John Gregory and Thomas Percival. These two individuals prepared the way for the birth of modern medical ethics and laid the groundwork for many of the ethical principles that physicians uphold today.

John Gregory (1724–1773)

John Gregory was a physician and moralist from eighteenth century Scotland. During this time, modern surgery was in its infancy with the establishment of the Company of Surgeons in 1745 and the Royal College of Surgeons of London in 1800 that served to distinguish surgeons from barbers [3]. The practice of medicine was extremely competitive and lacked a professional code of medical ethics. As a result, expedience and self-interest prevailed with a concomitant decline in medical competency.

Gregory helped redefine medicine as a profession by calling for physicians to set aside self-interest and to shift their focus back onto the patient [4]. He emphasized the physician's fiduciary relationship to the patient and urged physicians to draw from a sense of sympathy: "The chief of these moral qualities required of a physician, is humanity; that sensibility that makes us feel for the distresses of our fellow-creatures, and which, of consequence, incites in us the most powerful manner to relive them" [1]. He not only emphasized the virtues of truth telling and patient confidentiality, which were progressive at the time, but he expected physicians to be knowledgeable in their treatment of patients and to continuously seek improvement in their area of practice. Medicine at that time was largely based on authority and custom rather than scientific knowledge. Therefore, Gregory challenged physicians to base medical decisions on evidence

rather than dogma [5]. Furthermore, he believed that if patients were sufficiently educated regarding the physician's recommendations, they would become more motivated to comply through a therapeutic relationship [6].

Although there is little mention of surgery in his writings, Gregory does comment on the decorum of a surgeon stating that a good operator needs a resolute, collected mind, a good eye, and a steady hand [3, 7]. His work helped redirect the focus of medicine onto the relationship between the physician and patient which influenced many individuals, including Thomas Percival [1, 8].

Thomas Percival (1740–1804)

Through the influence of Gregory, Thomas Percival further pushed for the development of medical ethics. He was a strong proponent in the decorum of physicians and equality of treatment for all patients regardless of class. In his book *Medical Ethics,* he wrote about a physician's duties, professional conduct, relationships with apothecaries, and duties relative to the law [9]. He shifted the focus from a more physician-centric profession to a patient-centric profession. He stated: "The feelings and emotions of the patients, under critical circumstances, require to be known and to be attended to, no less than the symptoms of their disease" [9].

The American Medical Association (AMA) was founded in 1847 through the influence of Percival [1]. It was his ideas on the professional responsibility toward patients laid the foundation for the 1847 AMA Code of Ethics, which became the first national code of medical ethics [5, 10]. Like Gregory, Percival also mentions little regarding surgery, but he was openly opposed to itinerant surgery, emphasizing the importance of consultations before a surgeon operated on a patient [1, 3, 7]. He encouraged collaboration between surgeons and physicians with a consensus agreement prior to any important surgical operation [9]. And similarly to modern mortality and morbidity conferences, he also encouraged physicians to reflect on cases and to learn from them: "An account of every case of operation, which is rare, curious, or instructive should always be drawn up by the physician or surgeon, to whose charge it devolves, and entered in a register kept for the purpose, but open only to the physicians and hospital of the charity" [9].

Though both Gregory and Percival did not elaborate explicitly on surgical ethics, they were the first to recognize surgery as a separate profession and, more importantly, hold surgeons to the same ethical standards as physicians. Going forward, amidst the surgical advances of the nineteenth and early twentieth century that distinguished the field from the rest of medicine, it was surgeons who led the way to establish an ethical code specific to surgery.

Story Apart from Text: Robert Liston

In the 1800s, amputations were a commonly performed procedure. Before the advent of anesthesia, surgeons needed to perform surgeries as quickly as possible requiring multiple assistants to restrain the awake patient. Robert Liston, a Scottish surgeon, boasted that he could perform a leg amputation in under 2 minutes. It is reported that during a demonstration of a leg amputation, he operated so quickly that he accidently cut off his assistant's finger. Furthermore, in his fervor, he also cut through the coat of an elderly physician bystander who subsequently had a heart attack and died thinking he had been stabbed, as his coat was covered in blood. Both the patient and assistant eventually died of gangrene. So in one operation, Liston had a 300% mortality rate [11].

Birth of Modern Surgery

In the nineteenth century, two significant innovations—anesthesia and antisepsis—made it possible for surgery to dramatically progress. Prior to this, the mantra was to operate as quickly as possible with patients physically restrained and often passing out from the pain associated with the sur-

gery [12]. In 1846, John Warren and William Morton successfully anesthetized a surgical patient with ether in the Boston Ether Dome [13]. Anesthesia dramatically changed the way that surgeons approached disease as it allowed surgeons to focus more on precision and technique rather than speed. This changed not only how surgery was performed but the men who were drawn to the field. Speed and nerves became less important compared to careful planning, thoughtfulness, and refined skill. As surgery became more accepted as an academic field, its recognition as a profession subsequently grew.

The second major innovation was the advent of antisepsis. In 1850, over 90% of surgical wounds became infected, and the mortality rate for abdominal surgery was 75% [14]. In fact, many surgeons at that time considered abdominal surgery unethical [15]. In 1867, Joseph Lister applied the concepts of microbiology from Louis Pasteur and introduced the concept of hand washing before a surgical procedure using carbolic acid [13]. It did not catch on immediately, however, as it took a few decades before surgeons universally adopted the Lister sterile procedure. In fact, by 1880, only a few surgeons had adopted sterile technique. But, it soon became apparent that the patients of surgeons who practiced the Lister technique fared better and were more likely to recover [14]. Eventually, it became common practice for all surgeons to wash their hands before a procedure and later began wearing gloves and using autoclaved instruments. Thus, as the mortality rate for surgery decreased with these advances, the field of surgery became more respected, and patients started viewing surgery as a viable option when they were ill resulting in an increase in the rate of elective procedures.

As surgery became more common, this led to new developments in surgical technique and advances in the field led by William Halsted. Some of his advances included hernia repairs, mastectomies, vascular anastomosis, intestinal anastomosis, thyroid procedures, and parathyroid transplants [12]. During this time of rapid surgical innovation, there was little to no regulation. New procedures were being attempted, and individuals like Halsted had little idea of what the long-term effects of procedures would be. Many innovations were actually performed by itinerant surgeons who traveled from town to town operating on patients referred to them by the local general practitioner [16]. At this time in history, surgeons were able to operate on a fully anesthetized patient although a surgical informed consent process had yet to be clearly defined. The enthusiasm that surgeons had during this time of innovation likely led to more people undergoing procedures than perhaps needed to be done, which began to blur the lines of surgical ethics [17].

Surgical Ethics

Sir William Stokes of Scotland seems to be the first surgeon to mention the term surgical ethics. He stated in 1894: "A consideration of surgical ethics that frequently exercises the mind of the operating surgeon is the question of the principles that should guide him in dealing with cancerous growths" [17]. However, surgical ethics was merely in its infancy at the turn of the nineteenth century. Surgeons in the 1900s began recognizing the divergence in practice between medical physicians and surgeons due to advancements made in the 1800s. These created a unique set of surgical issues, which required an expansion of medical ethics. In the early twentieth century, the AMA, the ethical voice at the time, commented very little on surgical practice. Franklin Martin, a well-respected surgeon in Chicago, recognized the need for a separate governing body to oversee surgeons. Martin, along with a few other surgeons, saw the need to create a professional society of surgeons that specifically dealt with the emerging ethical issues encountered in surgery as well as a governing board that oversaw the establishment and maintenance of standards in surgery [14]. In 1913, the American College of Surgeons (ACS) was officially established.

Some of the major issues that Martin sought to address were fee splitting and itinerant surgery [14]. Fee splitting involved both the surgeon and referring physician and served as an incentive for the referring physician to recommend surgery to more patients. The concern was the significant financial conflict of interest with the potential of offering surgery for questionable if not inappropriate indications. Despite statements from the College against the unethical practice of fee splitting, it persisted up into the 1950s. In 1952, the ACS established a committee which conferred with the AMA trustees to develop a set of guidelines regarding fee splitting declaring it unethical [14].

Itinerant surgery was another issue that Martin felt needed to be addressed in the early stages of the ACS. This was the practice of surgeons performing surgeries, usually in rural areas, without ever meeting the patient. Further, they would depart after the surgery, leaving the management of any postoperative complications to the medical physician. Those in support of itinerant surgery stated that, though not ideal, it could fill the need for patients in rural areas that lacked access to hospitals and surgical care [14]. However, the ACS spoke against its practice because patient well-being was at the core of its principles [3, 14]. The ethical issue that surfaced was the importance of establishing a relationship between the surgeon and patient prior to surgery.

The College's stance against fee splitting and itinerant surgery established ethical standards for the emerging field of surgery reinforcing the importance of the patient, the surgeon-patient relationship, and the duty of the surgeon to see their patient through to recovery. Influenced by Gregory and Percival, the formation of surgery as a profession was founded on ethical principles centered on the surgeon-patient relationship. Martin and others realized the importance of establishing societal trust and maintaining that trust by creating ethical and professional standards in the field [14].

Informed Consent

In the decision for surgery, the surgeon must address the question of whether the risks of the operation outweigh the potential benefits to the patient [17]. Informed consent emerged as an important ethical issue, and although not unique to surgery, the stakes are much higher in surgery compared to other fields of medicine. Informed consent serves not only as a legal document but is the basis for which trust is formed within the surgeon-patient relationship [18–20]. It has roots back to Gregory and Percival who emphasized the relationship between physician and patient and the need to educate the patient about the diagnosis prior to treatment.

During the rapid changes in the practice of medicine throughout the twentieth century, the pendulum was swinging away from a paternalistic approach to medicine toward one more focused on respect for patient autonomy. This had obvious implications on the practice of surgeons and the process of discussing the nature of the procedure with patients prior to surgery. Previously, it was believed that patients should be informed little about the risks associated with the procedure [5]. However, just 1 year after the establishment of the ACS, the case of *Schloendorff* vs *The Society of New York* [21] served as the landmark case that defined informed consent requiring a surgeon to obtain consent before performing a procedure. The requirements of informed consent evolved over the following decades, and how much information a surgeon needed to disclose to the patient was debated. As a result, the reasonable person standard become the accepted model for informed consent, meaning the amount of information disclosed is now based on what a reasonable person would want to know regarding risks and benefits of the surgery [22].

Although informed consent seems to have become more of a legal hurdle, its primary role has always been to protect the patient. It remains an important aspect of the surgeon's fiduciary relationship and the establishment of trust that forms between the surgeon and patient. The challenge for surgeons has been to establish this trust in a relatively short period of time during their interaction with patients. Furthermore, it is also necessary to understand the patient's therapeutic goals prior to surgery since most patients are unable to actively participate in informed decision-making in the operating room.

Transplant Surgery

The era of transplant surgery opened up a whole new set of ethical issues to the field of surgery and brought to the forefront the principle of justice and allocation of resources. The first successful transplant case was in the 1950s, a living kidney donor transplant done by Joseph Murray in a set of identical twins [23]. In the early stages of solid organ transplantation, transplant recipients did not fare well, and mortality was high due to lack of safe and effective immunosuppressive agents. There was also a cloud of controversy surrounding the lack of defined death criteria and unified guidelines. Further, its outcomes were unknown raising important questions pertaining to medical futility and long-term efficacy. Even in 1975, Francis Moore, a renowned surgeon who led the way in the transplant revolution, viewed kidney transplant as still innovative and experimental [24].

A clear definition for death was critical during this time period because of the advances in artificial cardiorespiratory support. Previously, death occurred when the heart stopped, but with the ability to prolong cardiac function, a need to expand the criteria of death emerged [25]. The Uniform Declaration of Death Act drafted in 1980 and approved within the year by the American Medical Association and the American Bar Association aided in establishing guidelines for organ donation [26]. Due to the extreme scarcity of organ donors, an ad hoc committee was assembled to create a policy regarding organ donation making deceased donor transplants less controversial [25, 26]. Through this act, the concept of brain death was first introduced, defined as the irreversible cessation of brain function.

In 1984, the National Organ Transplant Act was passed, which initiated the creation of United Network for Organ Sharing (UNOS) to help ensure organ allocation in a just and fair way. Part of their requirement was transparency, keeping transplant centers accountable to the public regarding patient outcomes. They also sought to protect donors from coercion, stating: "It shall be unlawful for any person to knowingly acquire, receive, or otherwise transfer any human organ for valuable consideration for use in human transplantation" [27]. Organ donation—especially living donation—was to be an altruistic act.

As solid organ transplant has advanced with vastly improved outcomes, there still remains a scarcity of organs relative to potential recipients. One of the possible ways to meet this demand has been a push for more living donors. Living kidney donor transplant (LKDT) has been shown to be an effective transplant with minimal long-term effects or risks to the donor. The outcomes for the recipient are also better than deceased donor kidneys due to improved compatibility matching and graft function. It is reported that in the US, the number of LKDT has now surpassed the number of deceased donors [27].

The results of living donor liver transplants (LDLT) have not had the same success as LKDT. The main reason for this is the significantly increased surgical risk to the donor [28, 29]. The University of Chicago led the way in developing protocols for LDLT, publishing their protocol in 1989 for children suffering from liver failure [30]. They proposed that the donor should be a parent or close relative of the child so that the reward associated with saving a child's life would mitigate the risks associated with surgery [28]. They also established an informed consent and screening process for living partial liver donors that required surgeons, hepatologists, and psychologists to assure that donors were competent to make the decision [28]. There still exists nationwide controversy regarding LDLT due to the significant risks to the donor as well as the competing fiduciary relationship of the surgeon toward both the donor and the recipient. In 2013, only 4% of liver transplants nationwide were from living donors, and of the 166 liver transplant centers, only 43 perform LDLT [29]. As donor outcomes can be improved with emerging advances in minimally invasive techniques, these numbers are likely to increase in the future. As progress is made, surgeons will always have an ethical duty to ensure the safety of both the donor and the optimal outcome for the recipient.

Surgical Innovation

Innovation has a unique place in surgical ethics compared to medical ethics. Compared to medicine, where drugs are highly regulated by the Food and Drug Administration (FDA), many advances in surgical technique are made through innovation. Although surgical patients have benefited from the lack of cumbersome regulation to improve the quality of care, lack of appropriate regulation and oversight have also resulted harm to patients temporarily damaging the trust between society and the surgical profession.

Innovation is essential to surgical practice. Despite meticulous planning for a case, there are inevitably instances when a surgeon has to change the plan or modify a technique in order to achieve the desired results for the patient. However, surgical innovation is inherently wrought with ethical challenges. Historically, there has been little regulation regarding innovation and a concern for lack of transparency in the informed consent process [31–33]. It has only been in recent years that stricter regulation of innovative procedures has occurred, largely in part due to the events surrounding the laparoscopic revolution.

Story Apart from Text: Erich Mühe
Erich Mühe was a surgeon from Germany and the first individual to perform a laparoscopic cholecystectomy. He first became interested in laparoscopy after seeing a film on it in the 1972. He started using it for rectosigmoid polypectomies, getting practice with the instruments. In 1985, he performed the first laparoscopic cholecystectomy. Over the next year, he performed over 88 cases and presented his cases to the German Surgical Society. Unfortunately, it was not well received leaving him quite disappointed. It was not until 1992, 7 years after his first laparoscopic cholecystectomy, that he was recognized by the German Surgical Society for being a pioneer in endoscopic surgery [34].

Laparoscopic surgery took root in the late 1980s with wide adoption in the 1990s. The first laparoscopic cholecystectomy was performed in 1985 by Erich Mühe of Germany [34]. Three years later, Barry McKernan and William Saye performed the first case in the United States [34]. The novel laparoscopic cholecystectomy caught on due to the dramatic improvement in the patients' postoperative recovery compared to the open procedure. Early randomized trials comparing this new technique to the open technique were too small to reveal any significant difference in complications. And many surgeons proceeded to adopt this new technique despite lack of any data demonstrating its safety. In fact, by 1993, there were 21% more cholecystectomies being performed in New York than were being done prior to laparoscopy. However, registry data from the state of New York eventually published results indicating that the complication rates for serious injuries were 15 times more likely in laparoscopic versus open cholecystectomy [35]. This led to a moratorium on laparoscopic cholecystectomies by the ACS and a critical appraisal of the process to adopt novel surgical techniques with recommendations to carefully evaluate surgeons' training, credentialing, and privileging before performing a laparoscopic cholecystectomy [35]. It was recommended that inexperienced surgeons assist in 5 to 10 cases and do another 10 to 15 with supervision before performing it on their own [36].

Laparoscopic cholecystectomy is now the standard of care, but the lessons learned from laparoscopic innovation have led to better oversight and credentialing for future innovative techniques. Surgical innovation continues to be an important part of the surgical profession, and surgeons have taken the lead to make sure that both scientific integrity and patient safety are protected in innovation [37–39].

Surgical Ethics Today

Peter Angelos, an endocrine surgeon and renowned surgical ethicist, has described the unique relationship between a surgeon and patient: "There is a physical intimacy in the relationship between a surgeon and a patient that is unlike that found in other aspects of medical care…This vulnerability on the part of the patient demands a higher level of trust in the surgeon than is required of other relationships between patients and physicians" [40]. Miles Little, a surgeon and philosopher, further emphasizes this unique relationship between surgeon and patient through five components of the surgeon-patient relationship that includes the surgeon's power to rescue patients, the intimate proximity of surgeons and their patients especially in the operating room, the ordeal that patients endure, the physical and emotional aftermath of surgery, and the patient's desire for the surgeon's presence throughout the experience [41]. The relationship between surgeon and patient is fundamentally different than that of other physicians and their patients.

Surgical ethics is distinct from medical ethics because of three main facets [42]. The first is informed consent, which serves as the basis for the surgeon-patient relationship and represents the fiduciary relationship, which dates back to Gregory and Percival. Compared to other medical specialties, surgeons do not have the luxury of time to form this relationship and establish trust with their patients prior to surgery. The second component is the responsibility present in surgical care versus in other specialties. It is unique from other medical specialties because of the responsibility placed upon surgeons when complications occur. As sociologist Charles Bosk writes: "When the patient of an internist dies, the natural question his colleagues ask is, what happened? When the patient of a surgeon dies his colleagues ask, what did you do?" [43]. Apart from a surgeon, nobody can better understand what a surgeon and the patient have to endure through the ordeal of surgery. The third aspect that makes surgical ethics distinct is surgical innovation [42]. Since it does not fall in the realm of experimental research, innovation in surgery has benefitted from less FDA oversight than elsewhere in medicine and has driven important advances in surgery ultimately resulting in improvement in patient care. However, going forward, surgeons must not lose sight of the therapeutic relationship with their patients. The responsibility of formulating a safety net of ethical guidelines around the rapidly advancing field of surgeon lies on the shoulder of the surgical community. Society has placed its trust in the profession to regulate and maintain standards to protect the patient. This relationship has been and continues to be the starting point and foundation of surgical ethics.

Conclusion

Surgical ethics arose from the same ethical principles that undergird medical ethics. But throughout the last century, the unique aspects of surgery have necessitated the need for a distinct discipline of surgical ethics. Ethics has been intertwined in the history of surgery, and going forward, surgeons should continue to have a deep understanding of ethical principles. As surgeons, our fiduciary responsibility to society is to perform not what can be done but what should be done for our patients. Gregory and Percival believed that the physician-patient relationship should be the central focus of medical ethics, and surgery is no exception. However, the trust involved in the surgeon-patient relationship is something only a surgeon can truly understand. This trust is at the core of surgical ethics, and that is what prompted Martin and other surgeons to lay the groundwork for ethical standards in the past and why it must continue to be surgeons who lead the way for the future of surgical ethics to protect our patients and define our profession.

Concluding Remarks

- Surgical ethics, influenced greatly by medical ethics, was a driving force behind the rise of surgery as a respected profession in the United States.
- The unique dilemmas that surgeons have had to deal with such as fee splitting, itinerant surgery, informed consent, transplant, and

surgical innovation have made surgical ethics truly distinct from medical ethics.
- Trust lies at the heart of the surgeon-patient relationship and is the core of surgical ethics.
- Only a surgeon can truly understand the surgeon-patient relationship, and therefore, the future direction of surgery and surgical ethics must arise from within our own profession.

Glossary

Nonmaleficence A fundamental principle of bioethics meaning first do no harm. It is based from the writings of Hippocrates primum non nocere.

Fee splitting The practice of splitting the price the patient paid for surgery. It incentivized both the referring physician and the surgeon. It was addressed when ACS was first established as an unethical practice.

Itinerant surgery A common practice in the early twentieth century when a surgeon would come at the request of a primary care physician and perform a surgery without ever seeing the patient first. Postoperative care was also left in the hands of the primary care physician. It was addressed as unethical practice, and surgeons were barred entrance into ACS fellowship if they were known to do this practice.

Reasonable person standard Referring to the disclosure that accompanies the informed consent discussion. The reasonable person is the accepted form of disclosure meaning that the information disclosed should be in line with a hypothetical reasonable person.

Justice One of the fundamental principles of bioethics. It is based on the idea fairness and equal treatment for all involved parties.

References

1. Jonsen A. A short history of medical ethics. New York: Oxford University Press; 2000.
2. Mininberg D. In: Allen JP, editor. The legacy of ancient Egyptian medicine. The art of medicine in ancient Egypt. New York: Metropolitan Museum Press; 2005. p. 13–5.

3. Namm JP, Siegler M, Brander C, Kim TY, Lowe C, Angelos P. History and evolution of surgical ethics: John Gregory to the twenty-first century. World J Surg. 2014;38(7):1568–73.
4. McCullough L. The nature and limits of the Physician's professional responsibilities: surgical ethics, matters of conscience, and managed care. Medicine and Philosophy. 2004;29(1):3–9.
5. Jones JW, MacCullough LB, Richman BW. The ethics of surgical practice. New York, NY: Oxford University Press; 2008.
6. Beauchamp T, Faden RR. History of informed consent Reich WT encyclopedia of bioethics ed. New York, NY: Free Press; 1978. p. 1233.
7. McCullough L. John Gregory and the invention of professional medical ethics and the profession of medicine. Dordrect: Kluwer Academic Publishers; 1998.
8. Bastron D, McCullough L. What goes around, comes around: John Gregory, MD, and the profession of medicine. Proc (Bayl Univ Med Cent). 2007;20:18–21.
9. Percival T. Medical ethics. New York, NY: Leslie B Adams; 1985.
10. Thomas Percival (1740--1804). Codifier of Medical Ethics. JAMA. 1965;194(12):1319–20.
11. Inglis-Arkell E. The legend of the surgery with the 300% mortality rate: Science Direct; 2015 [cited 2017]. Available from: https://io9.gizmodo.com/the-legend-of-the-surgery-with-the-300-mortality-rate-1684894531.
12. Imber G. Genius on the edge. New York, NY: Kaplan Publishing; 2011.
13. Rise G. Modern surgery in hospitals: development of anesthesia and antisepsis. Mending bodies, saving souls a history of hospitals. New York, NY: Oxford University Press; 1888. p. 339–98.
14. Davis L. Fellowship of surgeons: a history of the American College of Surgeons. Chicago, Ill: American College of Surgeons; 1960.
15. Rosenberg C. The Care of Strangers: the Rise of America's hospital system. Baltimore and London: The Johns Hopkins University Press; 1987.
16. Risse G. Mending Bodies, Saving souls: a history of hospitals. New York and Oxford: Oxford University Press; 1999.
17. Angelos P. The right choice? Surgical ethics and the history of surgery. [cited 2017 11/8]. Available from: http://www.mdedge.com/acssurgerynews/article/141648/practice-management/right-choice-surgical-ethics-and-history-surgery.
18. Jones J, McCullough L, Richman B. Informed consent: it's not just signing a form. Thorac Surg Clin. 2005;15(4):451–60.. v
19. McKneally M, Martin D. An entrustment model of consent for surgical treatment of life-threatening illness: perspective of patients requiring esophagectomy. J Thorac Cardiovasc Surg. 2000;120(2):264–9.
20. Namm J, Siegler M, Angelos P. What is distinctive about surgical ethics. In: Ferreres A, Angelos P, Singer A, editors. Ethical issues in Surgical Care. Chicago: American College of Surgeons; 2017.
21. Schloendorff V. Society of New York Hospital. 105 NE 92. New York; 1914.
22. Beauchamp T, Childress J. Principles of biomedical ethics. 7th ed. New York, NY: Oxford University Press; 2013.
23. Brunicardi F, Anderson DK, Billiar TR, et al. Schwartz's principles of surgery. 10th ed. New York, NY: McGraw-Hill Education; 2015.
24. Moore F, et al. Surgical ethics and the dying patient. Bull Am Coll Surg. 1975;60(6):12–6.
25. Barclay WR. Guidelines for the Determination of Death. JAMA. 1981;246(19):2194.
26. Keely G, Gorsuch AM, McCabe JM, et al. Uniform determination of death act. Chicago, IL: National Conference of Commissioners On Uniform State Laws; 1980.
27. Davis C, Delmonico F. Living-donor kidney transplantation: a review of the current practices for the live donor. J Am Soc Nephrol. 2005;16(7):2098–110.
28. Singer P, Sigler M, Whitington P. Ethics of liver transplantation with living donors. N Engl J Med. 1989;321:620–2.
29. Kim P, Testa G. Living donor liver transplantation in the USA. Hepatobiliary Surg Nutr. 2016;5(2):133–40.
30. Nadalin S, Bockhorn M, Malago M, et al. Living donor liver transplantation. HPB (Oxford). 2006;8(1):10–21.
31. Michael L. Evidence-based medicine, cost containment, care effectiveness: Is it a new trilogy aimed at transforming the surgical mystique or the reality of double standards? Acta Chir Belg. 2001;101:95–100.
32. Reitsman A, Moreno JD. Ethical regulations for innovative surgery: the last frontier. J Am Coll Surg. 2002;194:792–801.
33. AM R, MJ D. Ethical regulations for innovative surgery: the last frontier? J Am Coll Surg. 2002;194(6):792–801.
34. Reynolds W. The first laparoscopic cholecystectomy. Profiles in Laparoscopy. 2001;5(1):89–94.
35. Bernard H, Hartman TW. Complications after laparoscopic cholecystectomy. Am J Surg. 1993;165:533–5.
36. The Southern Surgeons Club MM, Bennett C. The learning curve for laparoscopic cholecystectomy. Am J Surg. 1992;170(1):55–9.
37. Jones J, McCullough L, Richman B. The ethics of innovative surgical approaches for well-established procedures. J Vasc Surg. 2004;40(1):199–201.
38. McKneally M, Daar A. Introducing new technologies: protecting subjects of surgical innovation and research. World J Surg. 2003;27(8):930–4.. discussion 4-5
39. Biffl W, Spain D, Reitsma A, et al. Responsible development and application of surgical innovations: a position statement of the Society of University Surgeons. J Am Coll Surg. 2008;206(6):1204–9.
40. Angelos P. Orlo Clark and the rise of surgical ethics. World J Surg. 2009;33(3):372–4.
41. Little M. The fivefold root of an ethics of surgery. Bioethics. 2002;16(3):183–201.

42. Angelos P. Surgical ethics and the challenge of surgical innovation. Am J Surg. 2014;208(6):881–5.
43. Bosk C. Forgive and remember. 2nd ed. Chicago: The University of Chicago Press; 2003.

Suggested Literature

Angelos P. Orlo Clark and the risk of surgical ethics. 2009; 50(3): 99–134.

Litle M. The fivefold root of surgical ethics. 2002; 16(3): 183–201.
Nahrwold DL, Kernahan PJ. A century of surgeons and surgery: the American College of Surgeons 1913–2012. Chicago: American College of Surgeons; 2012.
Namm, et al. History and evolution of surgical ethics: from John Gregory to the 21st century. World J Surg. 2014;38(7):1568–73.

Surgical Ethics: Theory and Practice Background

Douglas Brown

What Does It Mean for a Practicing Surgeon to Be "Ethical"?

Consider the intentions and the struggles of a young surgeon in her first years of practice after residency [1]. Stephanie is the youngest and newest member of a team of surgeons whose practice is administered by a for-profit management company. She joined this practice with the assurance she would be fully supported in her deeply rooted resolve to care for her patients in the most beneficial and cost-effective way, with special attention to socioeconomically disadvantaged patients. She quickly discovers that numerous competing interests and expectations – some professional, others personal – pressure her to shift her focus away from her patients and their interests.

Stephanie begins most days poised to be empathetic. She is prepared to give disproportionate attention to her more vulnerable patients. She is ready to open herself to her patients' suffering to the point of risking burnout. She intends to be meaningfully present with her patients. She grips firmly her integrity. She gauges her capacity to tolerate the moral dissonance she experiences from value clashes with some of her patients. She seeks to grow professionally for patient benefit as much as for personal security.

D. Brown (✉)
Department of Surgery, Washington University
in St. Louis School of Medicine, St. Louis, MO, USA
e-mail: debrown@wustl.edu

She feels a nagging tension between her lifestyle interests and her accountability to her patients.

Stephanie would violate her integrity if she refused to look beyond each patient's presenting problem. She has already seen far too much. However, she accepts that she is not yet one of those rare surgeons who seem capable of saying "yes" to every deeply pained patient and enter yet another broken story. Fatigue, accountability to her other patients, administrative obligations, family responsibilities, reimbursement pressures, personal interests apart from medicine, and a host of other considerations force her to limit many patients' access to her time, her energy, and her heart. Instead, Stephanie triages her patients carefully to sift out the encounters in which she will enter more deeply into the patient's story, in which she will make and impose on others the sacrifices to be fully present with the patient.

Especially on her most exhausting days, Stephanie might glance enviously toward the many flourishing surgeons for whom the medical environment is most fertile. For these surgeons, a patient encounter is a sale; the patient, a consumer. Some are entrepreneurs. Lifestyle incentives motivate them. Others are researchers. Innovation and publication motivate them. They subtly sift out difficult patients from their panel of patients. They stay sufficiently detached from patient suffering to avoid any risk of being burned out. They have learned to make patients think they are present and care. They turn professional advancements into marketing tools. They lead unreflective lives. They have an easy conscience. But Stephanie is not seriously tempted to join their number.

However, Stephanie is troubled by how often she ends the day wearily thinking of the next patient as one more demand, thinking of herself as a mechanic. She ends many days numb toward patients and tired of confronting the healthcare delivery system. She feels acutely the loss of important family experiences as she does her job. She often sees little evidence that she is making a difference in the lives of vulnerable patients. She finds herself becoming apathetic to patient suffering as the day's paperwork drains her. She feels ambivalent toward patients for whom she has a dimming vision. She senses that her struggle to stay current with advancements in her specialty is posing subtle risks to patients. She is haunted by the look in her child's eyes, a look that asks, "Mom, do you care more for your patients than you do for me?" She can sound defensive. She can look disheartened.

Surgical ethics addresses the vulnerability of surgeons such as Stephanie and the many other surgeons who finish residency without such a deeply rooted, well-grounded resolve to care for all patients – including the most difficult patients – in a respectful, beneficial, fair, and cost-effective way. Once in practice, they too often yield – some with initial remorse – to incentives to practice surgery in a comfortable and an entrepreneurial way that actually – if subtly – discourages them from being genuinely present with patients. They too often compromise their integrity. They too often lose any initial qualms with hedging their fiduciary responsibilities to patients. They too often are easy targets. (See Appendix 1 for a language matrix that differentiates four common professional identities found on a spectrum with "I could not care less" at one end and "I could not care more" at the other end.)

Where Is "Ethics" in the Complexities of Patient Care?

"Encounter" is one of those everyday words in medicine. To encounter is to come upon another person face to face, often unexpectedly. To encounter is to meet another person suddenly, often violently. Each day is a series of encounters – turning hallway corners, crossing lanes, reaching for an object, getting in line, looking up from a table, chasing a prize, competing for a position, etc. Encounters make concrete and visible the set of values and the sense of purpose, out of which we decide what ought to be done. Medical school is no exception. Residency is no exception. Academic medicine is no exception. Private practice is no exception.

"Ethics" examines how well we respect those we encounter [2]. To respect is to see again or afresh, to look back wanting to see more clearly. The same root verb (Latin: *re* + *specere*) has given us such related words as speculate, inspect, spectacles, and speculum. To respect someone is to be artistic, subjective, freeing, reciprocal, gentle, engaged, holistic, attentive, patient, modest, trusting, graceful, reconciling, and humanizing. But surgeons must be scientific, objective, and detached. Therein lies the ethical complexity of patient encounters. A surgeon's clinical mindset can deteriorate into being rough, indifferent, curt, suspicious, selfish, alienating, and dehumanizing – in short, into being disrespectful.

To be seen/treated by a surgeon as "the chest wound in Room One" or "the liver cancer in Room Two" or "the acute abdomen in Room Three" is not necessarily damaging. Excellent surgical care is evidence-based. The surgeon objectifies the patient with statistical associations or by concentrating on damaged or diseased body parts. Differential diagnoses reflect plausible cause and effect explanations. The surgeon necessarily focuses on the patient's immediate problem more than on the patient's larger story. The surgeon must be sufficiently detached to achieve *aequanimitas* or balance.

However, at some point, clinically competent patient encounters cease to be respectful patient encounters. At that threshold, only by a surgeon's being sufficiently disciplined to keep the "aim eye" fixed on patients as individuals worthy of respect, compassion, and fairness can a surgeon avoid the indifference that degrades patient encounters into self-serving alienation…the indifference that leaves patients bruised, manipulated, exploited, and dehumanized.

The environments for surgical education and surgical practice tend to depersonalize patient encounters. Listen to the chatter alongside rounds, note the tone in medical record entries, analyze call room conversations and physician lounge conversations, recall morbidity-mortality conferences, remember discussions about depositions or about productivity numbers….

For patient encounters to be truly respectful, a fourth professional language is required – the language of respect, compassion, and fairness – that is fundamentally distinguishable from clinical/scientific language, from risk management/legal language, and from billing/economic language. Fluency in the professional language of respect, compassion, and fairness is *not* required to successfully complete medical school, to pass postgraduate boards, to be rewarded by practice management, to secure hospital privileges, to pass recertification examinations, to be promoted, to be elected to national positions of leadership, and even to be on a hospital ethics committee. Fluency in the professional language of respect, compassion, and fairness is, however, essential for sustaining the resolve to be a humane surgeon who cares deeply about patients – especially the most difficult patients – and who brings a resolute social conscience to the practice of surgery.

Why Do Well-Intentioned Individuals Come to Conflicting Judgments About What Should Be Done?

Each individual forms a personal sense as to what is of ultimate value and what is of lesser value. These core values serve as a filter through which information is interpreted before being applied to life's decisions. Certain relationships, experiences, circumstances, and objects are thus regarded to be of such importance to an individual that she/he is prepared to suffer great loss rather than to violate them.

Judgments about what ought or ought not to be done can usually be acted upon safely without much conflict. However, some situations require a collective judgment from a number of individuals with competing goals or divergent viewpoints. In such situations, a reflective approach to decision-making – i.e., ethics – is necessary (Fig. 1). Ethics then has to do with the determination of what ought to be done in a given situation, all things considered.

Some differences in judgment can be traced to variations in reasoning patterns. (See Appendix 2 for a diagram that helps clarify many of the reasoning patterns commonly present in patient care deliberations.) For instance, one person may be

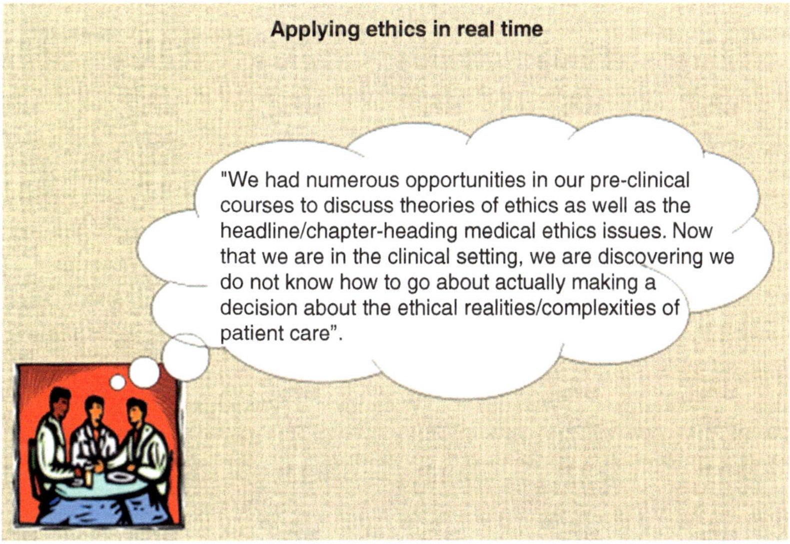

Fig. 1 The need for tools when facing ethically challenging cases

very logical, deductive, and abstract. Another person may be more intuitive, pragmatic, and relational. Other differences in judgment about what ought to be done in a given situation can be traced to variations in what is taken into consideration and the priority given to what is taken into consideration. Those conflicted about what ought to be done in a given situation may discover they are considering quite different aspects of the situation and/or they may be assigning different importance to considerations they share.

Before a thorough analysis of possible decisions can be undertaken, the participants in the decision-making process must respect each other enough to listen carefully in order to recognize and understand these differences. This approach to ethics focuses on the way we make decisions, first in reference to core values and then in reference to the interests of others affected by our decisions. It is imperative that individuals conflicted about what ought to be done cling to the "well-intentioned" assumption about each other as long as possible and only surrender this assumption after careful/thorough examination produces overwhelming evidence to the contrary.

What Are Patients and Their Families Invited to Trust?

Trust is counter-intuitive…involves risk…is needed to complete most tasks…requires courage…

"Fiduciary" in ancient Roman law denoted the transfer of a right from one person to another person with the recipient's obligation to return the right either at some future time or on the fulfillment of some condition. The fiduciary held this right as a trustee with the responsibility to exercise the right on another person's behalf. In modern surgery, "fiduciary obligation" refers to the trust patients place in their surgeons to act in their best interests. The surgeon receives the patient's trust because the surgeon possesses the special authoritative knowledge and technical skills to which the patient seeks access. Such knowledge and skills prompt the patient to seek out the surgeon in the first place. The vulnerability acknowl-

edged by the trusting patient creates a fiduciary obligation for the surgeon who accepts responsibility for the patient's care [3].

A relationship this special must be rigorously safeguarded. Accordingly, surgeons who prioritize their fiduciary obligation to patients seriously consider conflicts of interest. Surgeons are among a large and diverse work force that brings to the hospital numerous potentially conflicting priorities (Fig. 2).

Many surgeons are engaged in clinical research and in training/education healthcare learners, both being responsibilities that use patients as means to accomplish interests other than the patients' best interests. And surgeons have to navigate the availability of commercially driven surgical innovations that far too often result in eventual injury to surgical patients and even skew professional organization's technical bulletin guidelines [4].

The ethical dimensions of patient care can thus be effectively framed by asking – "What do we invite patients and families to trust?" (Fig. 3).

Each response to this centering question puts into clinically familiar language one of the four basic intentions that are foundational to surgical ethics – i.e., to avoid adding to the patient's pain/suffering (non-maleficence), to make a desired difference in the patient's well-being (beneficence), to align management plans with the patient's values and goals (self-determination), and to be fair in the use of limited resources (justice) [5]. When surgeons are able to follow through on these four intentions in an integrated way, the ethical dimension of their patients' care is sound, balanced, and in harmony, and the surgeons experience what brought them into a surgical career. For cases in which the ethical dimension of care is shaken or broken, the centering question "What do we invite patients and families to trust?" can be an effective starting point for determining which one or combination of the four intentions failed to such a degree that respect has given way to loss of confidence, suspicion, and adversarial defensiveness.

The trust upon which safe and beneficial care depends is a partnership/collaboration between surgical teams and patients (with their families

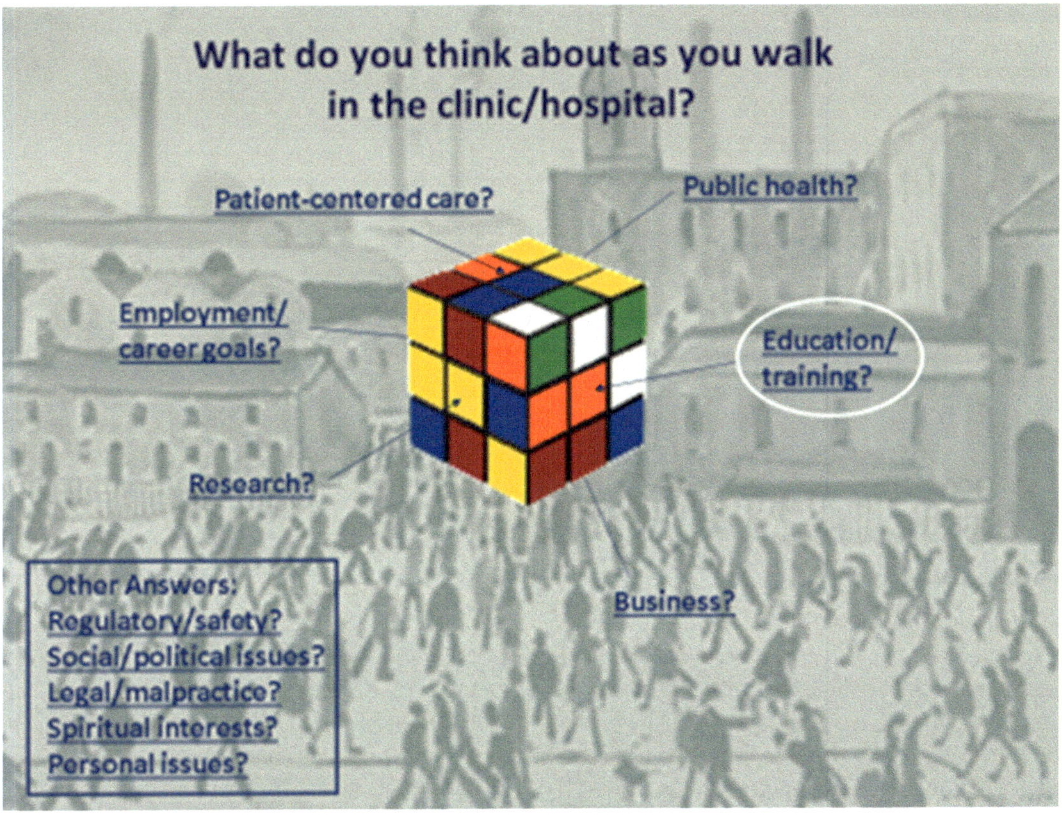

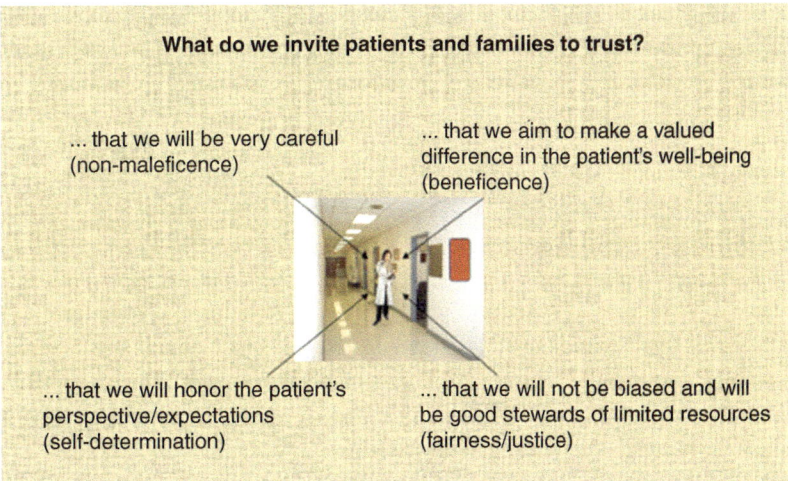

Fig. 2 The challenge to integrate multiple potentially conflicting responsibilities

Fig. 3 The clinical relevance of core ethical concepts

and friends) (Fig. 4). In order for surgeons to follow through on what they invite patients and families to trust, surgeons need their cooperation, their participation, and their assistance [6]. Thus the companion question: "What do surgeons need/expect from patients and families in order to follow through on what they invite patients and families to trust?"

As surgeons work to avoid harm, they need patients and families to provide complete and

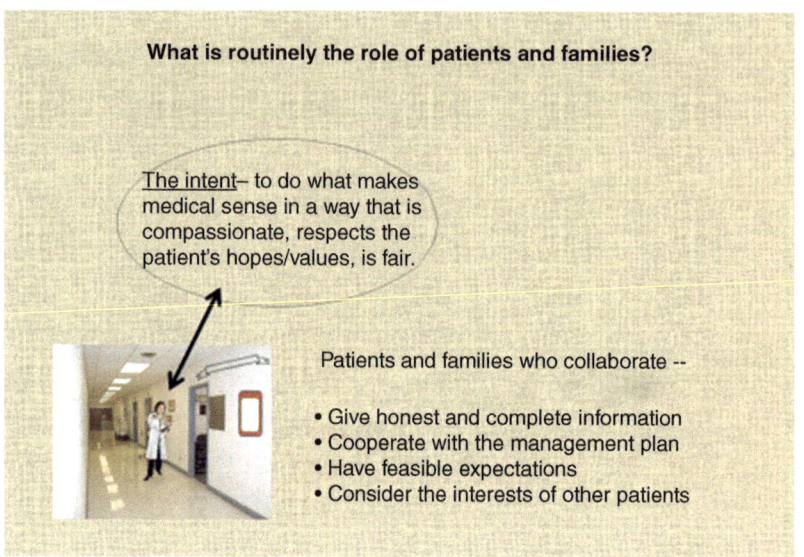

Fig. 4 The responsibilities of patients and families

reliable information. As surgeons seek to deliver beneficial outcomes, they need patients and families to make a determined effort to adhere to the management plan. As surgeons establish goals of care that align with patients' values and preferences, they need patients and families to realize there are limits to what can be achieved. As surgeons strive to be fair in the utilization of limited resources, they need patients and families to consider the interests of other patients and families. These clarifications highlight the accountability patients and families bear for following through on the four basic intentions that are foundational to surgical ethics.

When/Why Does Trust Break Down in Patient Care?

One of the cardiothoracic ICU attendings in the teaching hospital for a highly regarded medical school collaborated with the hospital's embedded ethics educator to identify vulnerabilities in patient care communication. The aim was to train the ICU staff to recognize early indications of breakdowns in patient communication before trust and respect had deteriorated, which was occurring in an alarming number of cases in the unit. They eventually focused on four recurring vulnerabilities in patient care communication – i.e., [1] the information upon which patient care

decisions are made, [2] the decision-making process, [3] the goals/expectations that influence patient care decisions, and [4] perceptions of evidence-based medical reasoning. They then developed a template for examining each vulnerability in two steps – first with a description and then with a set of assessment questions (Fig. 5).

Imagine the responsibility engineers have to ensure that bridges and buildings have structural integrity (e.g., anticipating the fatigue or fracture of materials, the initiation/growth of cracks in the materials, the limits for handling unexpected or overloading stress). Think of bridges and buildings as metaphors for the delivery of a patient's care from admission to discharge. Then reflect on the link between the integrity of the communication infrastructure upon which patient care depends and the ethical dimension of patient care.

And who is responsible for regularly assessing the communication infrastructure upon which patient care depends? The most common (and accurate) response to this question is: "We all are."

When/How Should Patients and Their Families Be Involved in Decision-Making?

Consider the following encounter. An intern writes orders for a nurse to obtain several urine samples from a patient, including one for a drug

Why the breakdowns in patient care?

Breakdown #1 – the information base

➤ Who/what are the sources of information?
➤ Is the information being exchanged honest? complete? consistent? unbiased? explained and understood within the 'big picture'?

Breakdown #2 – the decision-making process

➤ Who is involved in the decision-making?
➤ Is the decision-making process integrated? efficient? patient-centered? appropriately inclusive?

Breakdown #3 – the goals/expectations

➤ What goals/expectations are influencing the patient's care?
➤ Are the goals/expectations shared? patient-centered? fair? feasible? explicitly documented?

Breakdown #4 – perceptions of evidence-based medical reasoning

➤ What ethnic/cultural/religious/philosophical paradigms are represented?
➤ Are the paradigms compatible with an evidence-based reasoning about the patient's care?

Fig. 5 Recurring vulnerabilities in patient care communication

screen. When the nurse asks for the urine samples, he tells the patient what tests will be conducted. When a drug screen is mentioned, the patient refuses to consent. The nurse tells the intern the patient would not consent. The intern scolds the nurse for mentioning the drug screen and tells him, "I don't care that he doesn't give consent. Go back in there and get the urine sample and send it. I will deal with the patient later." The nurse instead speaks with his supervisor.

This scenario highlights the frequent disagreements in the clinical setting over when and how to involve patients and their families in decision-making. Surgeons face four questions repeatedly in every case, with each question representing a decision about whether, when, and to what extent patients, family members, and friends should be informed and share in decision-making. These four questions for presenting information to patients or their surrogates (Fig. 6) – (1) Should the information be shared? (2) Should the information be shared as an update? (3) Should the information be presented with a pause to answer questions? (4) Should the information be shared in order to reconsider the goals of care? – embody

four phases in the century-long evolution of "consent" in modern medicine [7].

Seeing these four questions being answered repeatedly in case after case calls attention to how few details in "the plan for today" the surgical team reviews on rounds each day are discussed with patients or their surrogates and also calls attention to the options other than shared decision-making when information is delivered to patients or their surrogates. The pivotal consideration for ethically sound patient care centers on the surgeon's need to keep the management plan aligned with the patient's goals, values, and preferences. Any one of these four options may be ethically justified. But each of the four options necessitates separate/distinguishable ethical reasoning – e.g., What factors influence when/how a surgical team involves patients and their families? Can a surgical team explain the ethical justification for each of the four options for involving patients, family members, and friends in decision-making? This analysis also opens discussion about the significance and the limitations of decisional capacity in determining when/how to involve patients, family members, and friends.

Fig. 6 Four recurring questions in communicating with patients and their surrogates

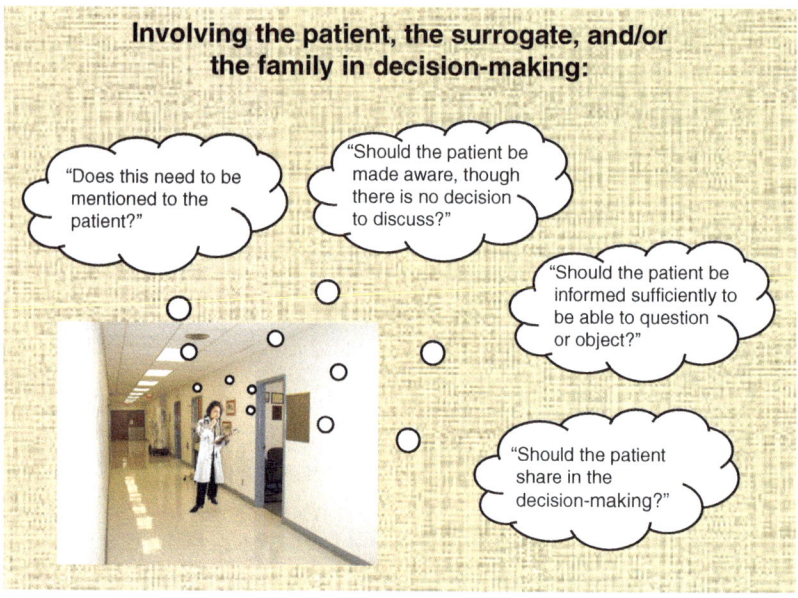

Fig. 6 Four recurring questions in communicating with patients and their surrogates

Why Is It So Hard to Keep Sense in Care at Life's End?

Two residents who were near the end of their ICU rotations were asked separately – "At any given time, how many of the management plans in the ICU make no sense to you?" The question had to do with the link between the management plans and feasible outcome/discharge expectations. Both residents responded – "50 percent."

An ethically skilled surgeon is prepared to move discussions between patients or surrogates and the surgical team toward consensus re the patient's outcome/discharge expectations. (See Appendix 3 for a comprehensive template for clarifying and documenting goals of care.) A patient's expectations may be restoration to preadmission functional status, relief from pain and suffering, survival regardless of quality of life, or survival long enough for desired closure. Quality of life outcomes that may be unacceptable to a patient include being permanently unconscious, being permanently unable to remember or make decisions or recognize loved ones, being permanently bedridden and dependent on others for activities of daily living, being permanently dependent on hemodialysis, or being permanently dependent on artificial nutrition and/or hydration (Fig. 7).

The focus of care for most surgical patients is to restore the patient to a level of function compatible with the patient's expectations, with all appropriate therapies being initiated and continued. If the surgical team concludes that such restoration cannot be achieved, further discussion with the patient and family members is needed in order to reconsider the expectations for the hospitalization. Based on this discussion, current management may not be escalated, additional interventions may not be introduced, and current life-sustaining treatments may be discontinued so as not to place undue burden on the patient. In some cases, the focus of care should shift to concentration on the patient's comfort during the dying process.

Sustaining the discussion of feasible goals of care with patients and their families is an art. Here are some effective discussion starters an ethically astute surgeon may use:

- *"What makes a day 'good' for you?"* (with attention given to how "good" is described)
- *"What are your difficult days like?"* (with attention given to how "difficult" is described)
- *"Do your good days help you make it through your difficult days?"* (with attention to indications of how firm a "yes" is and whether the good/difficult ratio is diminishing)
- *"Do you more often find yourself waking up in the morning hoping for a good day or hoping*

(Re)aligning Expectations and the Focus of Care

The patient's expectations may be: restore to preadmission living arrangement, relief from pain and suffering, "a miracle", survive regardless of quality of life, survive long enough for desired closure.

Quality of life unacceptable to the patient may include: being permanently unconscious, being permanently unable to remember, make decisions, recognize loved ones, have clear conversation, being permanently bedridden and dependent on others for activities of daily living, being permanently dependent on hemodialysis, being permanently dependent on artificial nutrition and/or hydration.

The focus of care will concentrate on the patient's comfort. Treatments that serve only to prolong the process of dying or place undue burden on the patient will not be initiated or continued.	The focus of care will be to restore the patient to a level of function compatible with the goals outlined above. All medically appropriate therapies (including transfer to an intensive care unit and other life-sustaining treatments) are to be initiated and continued. If the attending physician concludes that such restoration cannot be achieved, further discussion with the patient or surrogate will be conducted to reconsider the goals of care. Based on this discussion, current management may not be escalated, additional interventions may not be introduced, and current life-sustaining treatments may be discontinued so as not to place undue burden on the patient.

Fig. 7 A framework for aligning patient expectations with the focus of care

not to have a bad day?" (with attention to how encouraged or discouraged the patient is)

- *"What do you want me to know as I and the surgical team consider how best to take care of you?"* (with attention oriented toward acceptable or unacceptable outcomes rather than toward management plan details)
- *"What outcomes do you want to keep fighting for?"* (with attention to how feasible the outcomes are)
- *"Are you concerned that your illness will interfere with your participation in any activities or events in the near future that are especially important to you?"* (with attention to what demands these activities or events would make on the patient, to how feasible it is for the patient to participate in these activities or events, to what condition the patient hopes to have at the time of these activities or events)
- *"Do you have any questions or worries that are hard to talk about with your family or friends?"* (with reassurances that such can be discussed with you in complete confidence)
- *"Patients sometimes tell me they find themselves thinking 'that would be worse than dying.' Have*

you had this thought?" (with attention to indications regarding what such conditions would be)

Treatments that in the surgeon's best professional judgment will not have a reasonable chance of benefiting the patient (Fig. 8) and will serve only to prolong the dying process of or place undue burden on the patient should not be offered, initiated, or continued [8].

Two types of non-beneficial or futile considerations create ethical dilemmas – i.e., physiologic futility and value-based futility. A surgeon faces physiologic futility when the patient has no chance of recovery and interventions are merely prolonging the dying process. A surgeon faces value-based futility when a patient's stated goal or expectation is not achievable: a patient hopes to have meaningful conversation with loved ones, but the surgeon's medical judgment is that there is no chance the ability to have meaningful conversation will be regained.

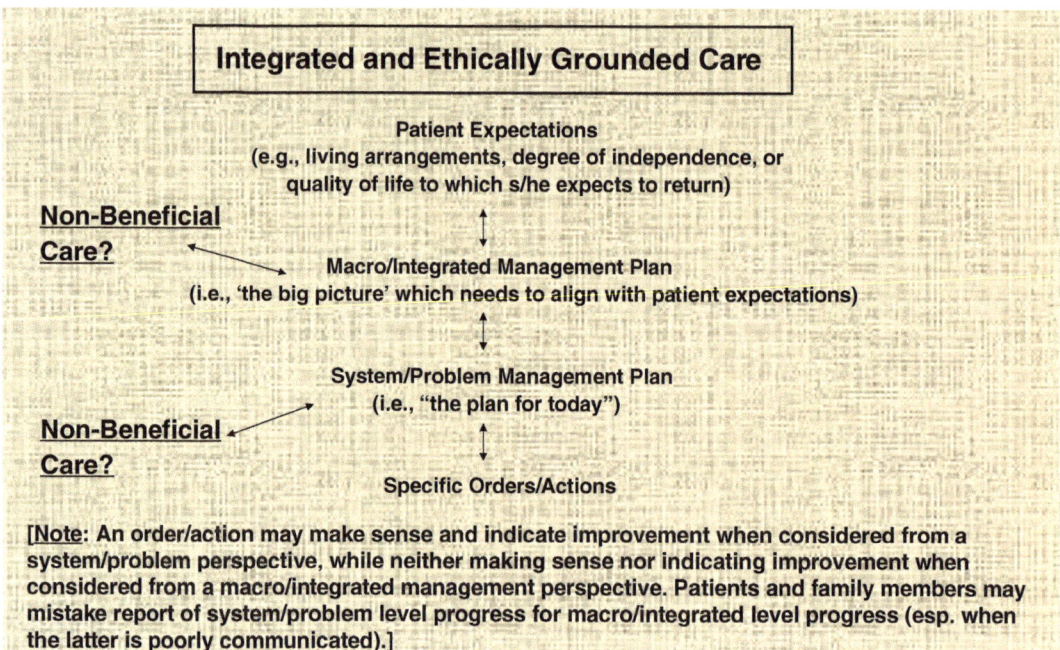

Integrated and Ethically Grounded Care

Patient Expectations
(e.g., living arrangements, degree of independence, or
quality of life to which s/he expects to return)

Non-Beneficial Care?

Macro/Integrated Management Plan
(i.e., 'the big picture' which needs to align with patient expectations)

System/Problem Management Plan
(i.e., "the plan for today")

Non-Beneficial Care?

Specific Orders/Actions

[Note: An order/action may make sense and indicate improvement when considered from a system/problem perspective, while neither making sense nor indicating improvement when considered from a macro/integrated management perspective. Patients and family members may mistake report of system/problem level progress for macro/integrated level progress (esp. when the latter is poorly communicated).]

Fig. 8 An explanation of non-beneficial ("futile") care

Many ethics consultations are triggered when surgeons face desperate patients or surrogates who demand – "Just do everything." An ethically skilled surgeon has learned not to be stymied by such appeals and instead sensitively reassures the patient or surrogate that everything will be done:

- That is medically reasonable, justifiable, and defensible
- That is standard of care
- That is consistent with the patient's values, goals, and expectations
- That is within the limits of the hospital's resources and scope of service
- That is in the patient's best interests
- That hospital policy permits
- That is legally permissible

Remember – every intervention is a "trial of treatment."

Is Concern for Justice (Ir)Relevant at the Bedside?

It is simple enough to say "I am for justice." It is much more complicated to be just. One reason is the sacrifices and the risks associated with

following through on the commitment to be just. Another reason is the reality that no single definition of what it means to be just is equally compelling and effective for all situations. (See Appendix 4 for a delineation of distinguishable descriptions of just decisions about access to and distribution of limited resources.)

Consider the experience of the leadership team for a nonprofit community health center serving a patient population burdened by generations of poverty. The leadership team takes seriously the resolve in their center's mission statement to care for patients "in a fair and gentle manner." During a strategic planning session, the leadership team realizes they had made several decisions based on different ways to determine what is a fair use of limited resources. See if you can identify the different ways to decide what is fair that are embedded in these examples of their staffing and compensation decisions.

- The community health center's seven physicians receive the same compensation. There are no productivity incentives.
  Fairness is:

- The clinical and administrative support staff members are compensated near/at their "mar-

ket" potential, with the distance from "market" increasing across the compensation spectrum to the physicians (whose compensation is at 80% of their "market" among community health center physicians).
Fairness is:

- End-of-the-year bonuses for non-physician employees are the same amount for all, whereas such bonuses for the physicians are calculated using an equation that takes tenure into consideration.
Fairness is:

- The community health center is committed to delivering the same access to and quality of care to all patients regardless of a patient's ability to pay.
Fairness is:

- The community health center gives disproportionate attention to the healthcare needs of and barriers faced by the most disadvantaged patients in the service area.
Fairness is:

- The leadership team makes decisions about the utilization of the staff and capital resources based on "public health" funding priorities.
Fairness is:

Is it legitimate to use different interpretations of what is fair? If so, what integrates the results into an experience that is considered to be fair? Should fairness be measured by the resulting harmony, balance, and reciprocity? Are the anchors for fairness [1] treating equals equally and [2] handling inequalities with disproportionate regard for the less advantaged? If so, can complacency (or resignation) about inequalities be overcome? How do inequalities at/from birth influence attempts to be fair? Should the interests, rights, and/or liberties of a few ever be sacrificed for the interests, rights, and/or liberties of the many? How far beyond those immediately affected should consequences be tracked in assessing the fairness of a decision? How should an organization's being "for profit" or "not for

profit" alter deliberations about a fair distribution of benefits and advantages?

Mass casualty events radically escalate the ethical challenges to be fair. Surgeons are forced to shift from the familiar individualized patient care paradigm to the less familiar public health paradigm. They are expected to be utilitarian (i.e., to deliver the greatest good for the greatest number). Resources have to be rationed. Patients have to be triaged. Potentially life-sustaining treatment may have to be withheld or withdrawn from one patient and given to another patient.

Surgeons' intent on being fair invites their patients and surrogates to trust that they will not discriminate against them. Following through on this intention is especially challenging when demographic variables about which they have preferences and biases are medically significant to a patient's diagnosis and management (Fig. 9).

The responsibility to recognize and discipline one's preferences and biases when caring for patients is clearly relevant at the bedside.

Surgeons also question – often expressed with frustration and/or cynicism – systems, protocols, and decisions for distributing limited resources (e.g., personnel, rooms, supplies, lifesaving interventions, VIP privileges, capital investments). They can feel complicit…but trapped/powerless. Ethically grounded surgeons look for opportunities to participate in team, department, institutional, and national efforts to assess/revise the unofficial decision-making culture and/or the official policies that raise fairness concerns about the distribution of limited resources (i.e., "organizational ethics") [9].

Consider an exercise designed to clarify one's priorities re access to and distribution of limited resources (Fig. 10) [10].

First, visualize the range of possible life circumstances represented in the ring of photographs. Second, imagine not knowing your life circumstances (e.g., your age, ethnicity, health, work, education, financial resources, nationality, etc.). Surgeons see firsthand how fragile and unpredictable one's life circumstances are. Third, without knowing which life circumstances will be your lot, explain how you would propose limited resources should be accessed and distributed.

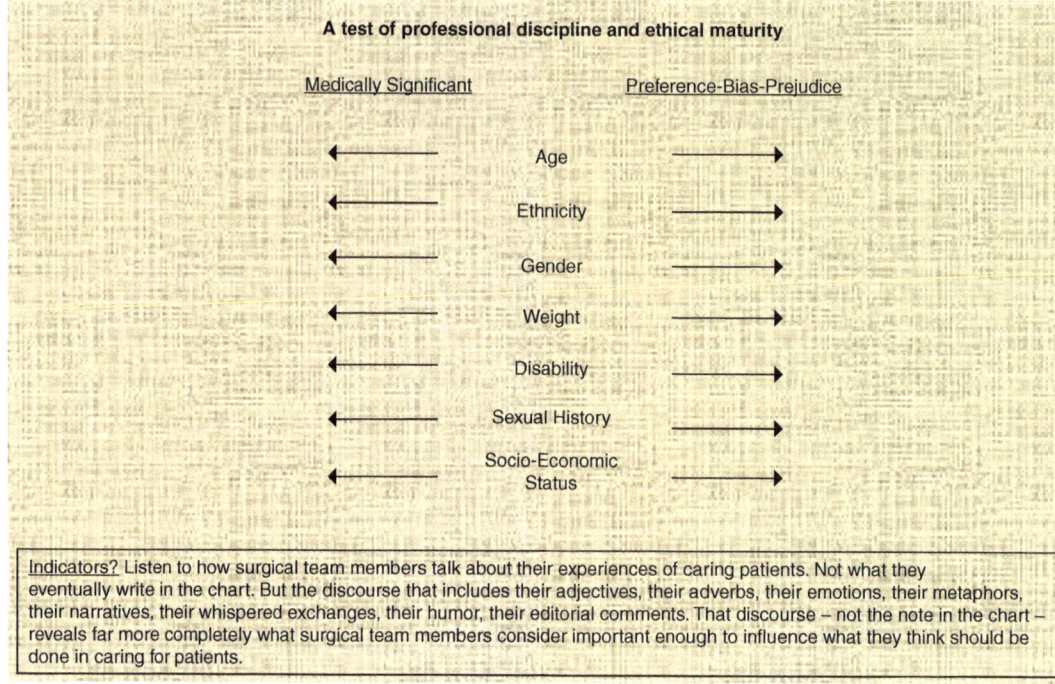

Fig. 9 Patient information that can be both medically significant and prejudicial

Fig. 10 An ethically accountable perspective on accessing and distributing limited resources

"Who Cares…Really?"

Most medical students choose to pursue a surgical career confident they will be ethically model surgeons, humane with a resilient social conscience. However, they quickly feel they are being herded through year after grinding year of preparation. They are being trained, but not necessarily educated. They are under intense supervision as they expand/strengthen their knowledge base, as they become efficient in examining patients, and as they learn to do procedures. From one stage to the next, they accommodate standards for identifying "good performance" that may have little to do with valuing patients as individuals. They finish residency still feeling the effects of chronic fatigue but anxious finally to be focusing on their own patients. Instead, for several more years – among new colleagues and under smothering fiscal scrutiny – they struggle to find their own practice style, to get out from under enormous debt, to publish, and to catch up on a long-delayed personal life. Do they receive sufficient incentives to give of themselves…to care deeply…to be truly present with their patients…to concentrate on the disadvantaged… to be reflective? [11]

"Ethics" for these surgeons is analogous to an irrigation system delivering nourishment to plants that would otherwise wither.

Appendix 1

a language matrix that differentiates four common professional identities found on a spectrum with "I could not care less" at one end and "I could not care more" at the other end.

Professional Profiles That Delineate Integrity			
"I could not care less" ◄·········	··········	··········	·········► "I could not care more"
an assault pt a victim a scam	a sale pt a consumer a business	an encounter contract with pt a profession	a meeting covenant with pt a vocation
criminal; manipulates pt; desecrates social fabric	entrepreneur acccommodates pt, capitalizes on social fabric	servant; empathetic toward pt; leavens social fabric	partner, respects pt, challenges social fabric
lifestyle is everything	lifestyle is highest priority	lifestyle is in tension with accountability to/for pts	lifestyle is integrated with accountability to/for pts
pt mix is defined by fraudulent interst	pt mix is weighted toward personal interests	pt mix is weighted toward the more vulnerable pts	pt mix is centered on the most vulnerable pts
exploits pt suffering; immune to being burned out	detached from pt suffering; avoids being burned out	burdened by pt suffering; risks being burned out	drawn into pt suffering; copes with being burned out
harmfully present pt a means only	apparently present pt primarily a means	meaningfully present pt an end and a means	fully present pt essentially an end
professional advancement sought as a cover	professional advancement sought as a marketing asset	professional advancement sought for quality of care and security	professional advancement sought as a benefit to pts
no conscience; no moral dissonance; no integrity	easy conscience; lwe moral dissonance, compromises professional integrity	pangs of conscience; underlying moral dissonance; wrestles with integrity	restless conscience; deep moral dissonance; risks self-righteous self-image and/or reputation

The Struggle			
"I could not care less?"	**"Do I care**	**. . . really?"**	**"I could not care more"**
	a task pt one more demand a job	an encounter contract with pt a profession	
	mechnaic; numb toward pt; tired of confrontions	servant; empathetic toward pt; leavens social fabric	
	experience with family sactrificed to care for pts	lifestyle is in tension with accountability to/for pts	
	patching rather than healing the more vulnerable pts	pt mix is weighted toward the more vulnerable pts	
	apathetic re pt suffering; overworked; drained by paperwork	burdened by pt suffering; risks being burned out	
	Weakly present pt a means to an end	meaningfully present pt an end a means	
	professional advancement slowed by being behind; subtle risks to pts	professional advancement sought for quality of care and security	
	guilty conscience; moral complacence; defensive sounding cynical	pangs of conscience: underlying moral dissonance; wresties with integrity	

Appendix 2

a diagram that helps clarify many of the reasoning patterns commonly present in patient care deliberations.

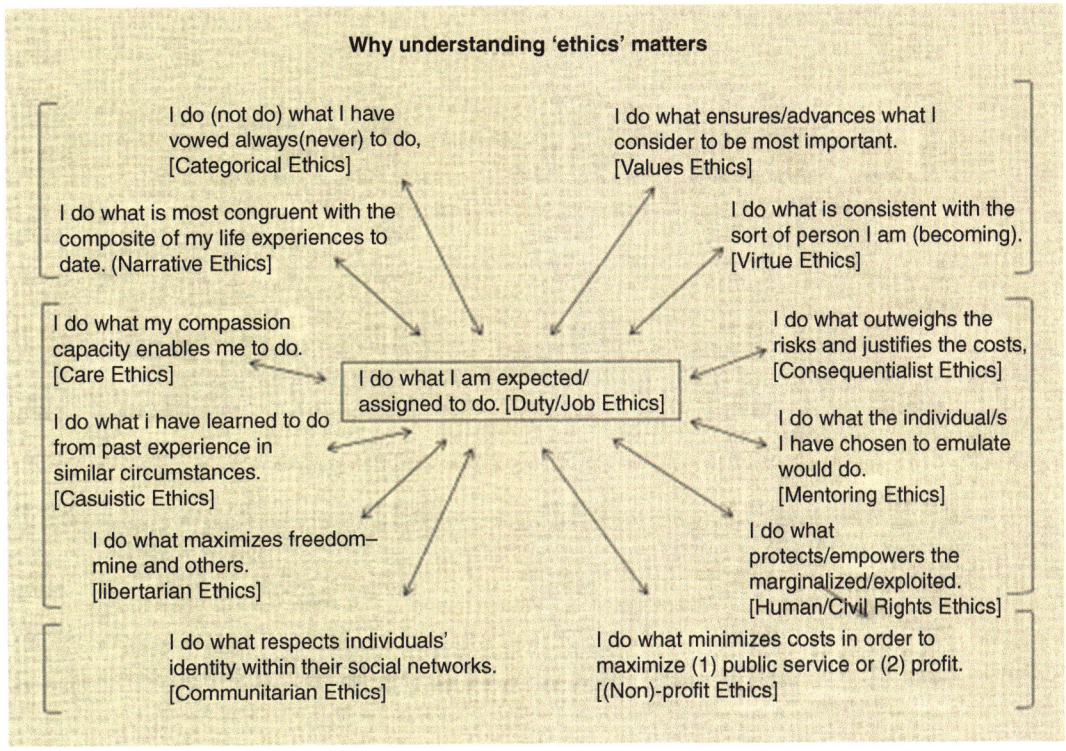

Appendix 3

a comprehensive template for clarifying and documenting goals of care.

Goals of Care -- Communication Template

PART A: Document Goals of Care

Based upon comprehensive discussion between the patient _____ (or surrogate) and the treating physician,

The following explanation best describes the patient's current goals of care:he _____

EXAMPLES include but are not limited to: "return to prior living situation at previous functional status" or "return to prior living situation after physical therapy" or "remain in my home" or "be free of pain or breathlessness" or "maintain my privacy and dignity" or "be able to interact with my loved ones" or "attend my granddaughter's graduation".

NOTE: "Do everything" is NOT a goal of care. Ask the patient (or surrogate) what 'everything' is intended to achieve.AMPLES

NOTE: To set realistic goals, the patient (or surrogate) needs a clear description of what to expect. Discuss and document if the patient wants aggressive life-support measures stopped and wants treatment instead to focus on comfort and dignity if any one or combination of the following is the most likely outcome:OTE:

____ being permanently unconscious (i.e., completely unaware of surroundings with no chance of regaining consciousness)

____ being permanently unable to remember, understand, make decisions, recognize loved ones, have conversations

____ being permanently bedridden and completely dependent on the assistance of others to accomplish daily activities__

(e.g., eating, bathing, dressing, moving)

____ being permanently dependent on mechanical ventilation

____ being permanently dependent on hemodialysis

____ being permanently dependent on artificial nutrition (tube feedings) and/or intravenous hydration for survival

____ death likely to occur within days to weeks and treatments are only prolonging the dying process

____ other (specify): _____

PART B: Document Focus of CareT

Based upon the above understanding of the patient's goals of care:

☐ The focus of care will be to restore the patient to a level of function compatible with the goals outlined above.

Specific testing and treatments will be ordered by the patient's physicians with the intent to achieve these goals.

☐ The focus of care will concentrate on the patient's comfort. Treatments that serve only to prolong the process of dying or place undue burden on the patient will not be initiated or continued.

PART C: Recommend Resuscitation StatusART

1. Based on the current condition, prognosis and comorbidities, and on weighing likely benefits, harms and goals outlined above --
    A. The treating physician **does / does not (circle one)** recommend <u>CPR</u> in the event of cardiac arrest.
    B. The treating physician **does / does not (circle one)** recommend <u>intubation</u> in the event of impending respiratory arrest.
    C. The treating physician at this time **cannot make a definitive recommendation (circle)** regarding CPR or intubation.
2. These recommendations have been discussed with the patient (or surrogate) with reassurance that if resuscitation is not performed, treatment will be provided with the goal of comfort and dignity: **Yes / No**
3. For the patient (or surrogate) who decides to be resuscitated (i.e., Code 1) despite the treating physician's recommendation against such, the treating physician has discussed the likely immediate consequences of CPR if successful: **Yes / No**
4. **Person with whom to speak if the patient lacks decisional capacity:**

Name: _____ Relation: _____ Phone Number: _____

Appendix 4

a delineation of distinguishable descriptions of just decisions about access to and distribution of limited resources.

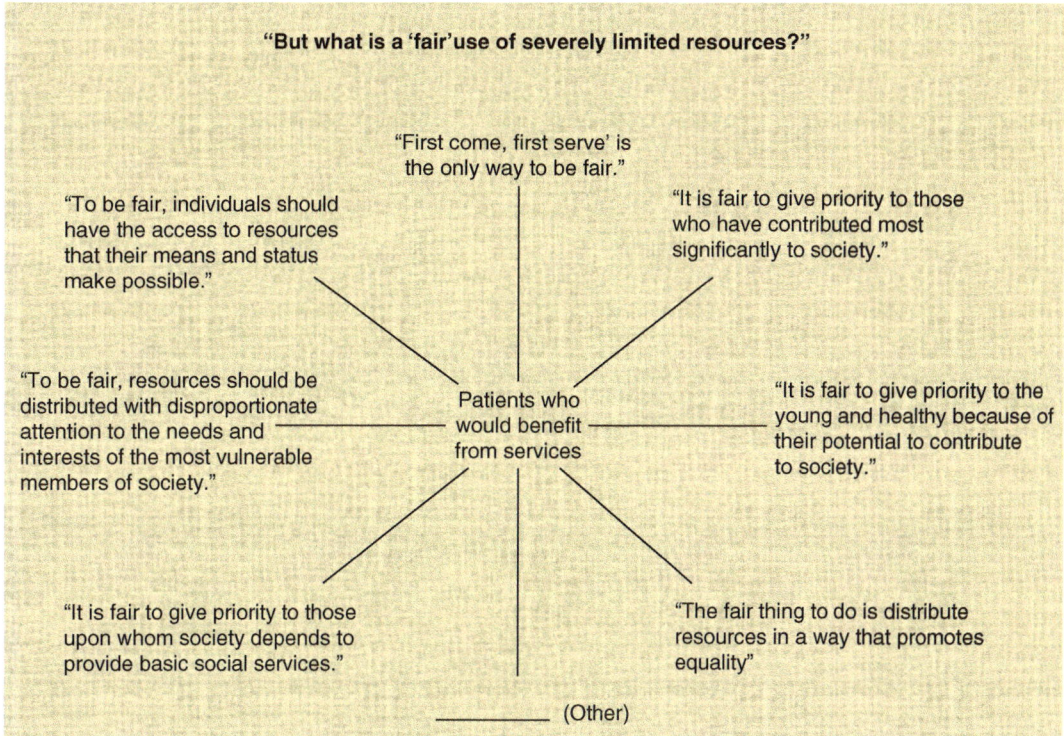

"But what is a 'fair' use of severely limited resources?"

"First come, first serve' is the only way to be fair."

"To be fair, individuals should have the access to resources that their means and status make possible."

"It is fair to give priority to those who have contributed most significantly to society."

"To be fair, resources should be distributed with disproportionate attention to the needs and interests of the most vulnerable members of society."

Patients who would benefit from services

"It is fair to give priority to the young and healthy because of their potential to contribute to society."

"It is fair to give priority to those upon whom society depends to provide basic social services."

"The fair thing to do is distribute resources in a way that promotes equality"

(Other)

References

1. Brown D. A toolkit for practical medical ethics. Virtual Mentor. 2009;11:909–14.
2. Buber M. *Ich und Du* (translation: Kaufman W). New York: Charles Scribner's Sons; 1923.
3. Wall L, Brown D. Pharmaceutical sales representatives and the patient-physician relationship. Ob Gyn. 2002;100:594–9.
4. Wall L, Brown D. Commercial pressures and professional ethics: troubling revisions to the recent ACOG practice bulletins on surgery for pelvic organ prolapsed. Inter Urogyn J. 2009;20:765–7.. plus letters to the editor
5. Beauchamp T, Childress J. Principles of biomedical ethics. 7th ed. New York: Oxford University Press; 2012.
6. Schwarze ML, Bradley CT, Brasel KJ. Surgical "buy-in": the contractual relationship between surgeons and patients that influences decisions regarding life-supporting therapy. Crit Care Med. 2010;38(3):843–8.
7. Faden R, Beauchamp T. A history and theory of informed consent. New York: Oxford University Press; 1986.
8. AMA. Code of medical ethics. Opinion. 2014;2:035.
9. Spencer EM, Mills AE, et al. Organization ethics in health care. New York: Oxford University Press; 2000.
10. Rawls J. A theory of justice. 2nd ed. Cambridge, MA: Belknap Press; 1999.
11. Wall A, Angelos P, Brown D, et al. Ethics in surgery. Curr Probl Surg. 2013;50(March):89–136.

Foundations and Principles of Surgical Ethics

Alberto R. Ferreres

Key Points
- To introduce a historical perspective of the development of surgical ethics
- To outline important personalities who influenced this development
- To specify the importance of surgical ethics
- To explain and illustrate why surgical ethics is different from Medical ethics

Surgical ethics (SE) is considered to be part of medical ethics, and both disciplines belong to the realm of bioethics. Bioethics is the branch of applied ethics that studies the philosophical, social, and legal issues arising in medicine and other life sciences and is chiefly concerned with human life and well-being [1].

The issues included in the goal of study are represented by the moral approach, the decision-making process, and the behavior of those involved in the patient surgical care, with specific emphasis on the patient-surgeon relationship.

A. R. Ferreres (✉)
Department of Surgery, University of Buenos Aires, Buenos Aires, Argentina

Department of Surgery, University of Washington, Seattle, WA, USA
e-mail: aferre17@uw.edu

The term bioethics was used for the first time by Fritz Jahr in 1926, in reference to the use of plants and animals, but remained unused until the rebirth of bioethics during the 1970s. In 1970, van Rensselaer (1911–2011), from the University of Wisconsin, Madison, called for the need of a new field to tackle the basic issues of human thriving. He started to use bioethics to describe a new field which represented the intersection of human values, biology, and medicine. Years later the medical field and profession secure the use of the term bioethics. In that sense, Andre Hellegers (1926–1979), a Dutch gynecologist and founder of the Institute of Bioethics at Georgetown University, was highly instrumental, since Georgetown's approach was the one that led the medical arena and a huge influence in the medical scenario [2].

Surgery has some unique features that make ethics in the surgical field very specific and focused on crucial topics:

- Surgery harms before healing and curing the patient. There is a therapeutic justification to prevent this procedure from being otherwise a felony or an assault.
- Every surgical procedure is invasive and represents an aggression, so the requirements for a surgical informed consent process are stringent.
- Surgery is characterized by fallibility, and 40–60% of all medical errors are performed in the operating room.

- Surgical decision-making is usually performed under conditions of uncertainty.
- Surgery is embedded by risks, accidents, complications, and sequelae [3].

Medical ethics is considered to have started in ancient Greece, being its first representative Hippocrates of Cos (460–370 BC) and finishing its influence with Galen (AD 129–199), who developed and taught his science and art in Pergamon (present Bergama, Turkey). The *Hippocratic Corpus* represents a collection of about 60 ancient Greek medical works associated with Hippocrates, but none is proven to be written by Hippocrates himself. This collection may represent the remains of a library of Cos or even a collection compiled in Alexandria, but it includes works from the Coan and Cnidian schools of ancient Greek medicine. The famous maxim "As to diseases, make a habit of two things- to help, or at least to do no harm" is mentioned in Epidemics, book I, 11. The preamble of "On the Physician" offers the physical and moral traits of the ideal physician; in the same fashion, "The Precepts" also refers to the physician's behavior; meanwhile the "Decorum" provides tips about good manners at the doctor's office or when visiting patients [4]. The Hippocratic Oath includes details regarding the physician ethics and moral behavior in its four sections: a tribute to the Gods. It is divided in four parts: a pledge to pagan deities; an itemization of positive obligations, those ones that are required; a listing of negative obligations, those which are forbidden; and a final allegiance. It also contains a laudation to the figure of the master or teacher: "To hold him who has taught me this art as equal to my parents and to live my life in partnership with him…," evidence of the master-apprentice model of training and learning, which stayed through the medieval period until the nineteenth century.

Plato (427–347 BC) developed the first germ of the informed consent process, which only took place in the case of free citizens but did not apply for slaves: "The slave doctor prescribes what mere experience suggest as if he had exact knowledge; and when he has given his orders, like a tyrant, he rushes off with equal assurance to some other servant who is ill …but the other doctor, who is a free man, attends and practices upon free men; and he carries his enquiries far back and goes into the nature of the disorder, he enters into discourse with the patient and his friends, and is at once getting information from the sick man, and also instructing him as far as he is able, and he will not prescribe for him until he has first convinced him" [5].

Galen considered mandatory the knowledge of philosophy for medical practitioners. He was not only a physician and a medical author but was also involved in philosophical issues addressing topics in the field of epistemology, causation in the world of nature, and philosophy of mind (now psychology). He was influential through the medieval times and even later, both in Europe and in the Arabic world [6].

In the Islamic world, the figure of Avicenna (980–1037) was influential in the field of medical ethics: although in his Canon there is not a specific section devoted to ethics, the whole book is filled with ethical references that focus on human beings as the target of medicine. In the Jewish tradition, Maimonides (1135–1204) was a Talmud scholar and one of the world's greatest physicians. The influence of Buddha is remarkable in India and southwest Asia, while the teachings of Confucius (551–479 BC) were influential in China. In this country the *Huangdi Neijing* (*The Inner Canon of the Yellow Emperor*) is a very old medical text in the form of dialogues between the mythical Yellow Emperor and six of his ministers, addressing traditional Chinese medicine and ethical behaviors [4].

Isaac Judaeus (circa 832–circa 942), considered the father of medieval Jewish Neoplatonism, in his "Physician guide" forwarded a word of caution against those "know it all" physicians and requested that doctors should lead an exemplary and virtuous life, free from gluttony and other vices. During the Middle Ages, medicine was influenced by the teachings and power of the Catholic Church; the care of the poor and the sick was considered a gift of charity and a service of compassionate people to God.

Arnau de Vilanova (1240–1311), originally from present-day Catalonia, was professor of medicine in Montpellier and personal physician to Kings Peter III and James II of Spain. He was a devoted admirer of Galen's teachings. He was fluent in Arabic, and his recommendations included a practice-oriented clinical care: to go beyond transmitted teachings by careful therapeutic experimentation, to draw judicious conclusions about these experiments, and to communicate these conclusions with succinct clarity [7]. Jehan Yperman (1260–1331) is considered the master of Flemish medicine, and among his recommendations, the following can be mentioned: "Physicians and surgeons must not only have a knowledge of Medicine, but must also know the books of Nature, which is called Philosophy…the doctor must also know Ethics, as this science teaches good morals" [8].

Gabriele de Zerbi (1445–1505) was professor in Padova and also practiced in Verona and was a very successful author of his time. He made special mention of the Hippocratic Oath. In his book *Advice to Physicians*, he highlighted the virtue of fidelity, since "The doctor is the faithful companion to the body of his patient, suffers with him and rejoices in his health" [9]. The German Ahasverus Fritsch (1629–1701) published in 1684 his book *The Sinning Physician*, where he described some of the types included in this definition: "those practicing medicine without sufficient learning, failing to entrust themselves and their patients to divine providence, charging the poor a fee or overcharging the rich, prolonging treatment for the sake of pain, failing to consult when appropriate or abandoning a patient or fleeing the city in contagion, advocating for a virtuous and ethical behavior" [4].

John Gregory (1724–1773) should be considered as the modern developer of medical ethics. He was born in a family of academicians and raised to become appointed First Physician to her Majesty for Scotland. He was professor of philosophy and medicine at King's College between 1746 and 1764, with a brief interval while in London, and taught practice of medicine at Edinburgh between 1766 and his death. He was influenced by Francis Bacon's philosophy of science and medicine and a contemporary of Sir David Hume and was the one responsible for the transformation of medicine from a trade into a profession. He introduced the concept of medicine as a fiduciary profession and the physician as a moral or fiduciary agent, founded in the need for the physician to know and respect the patient's best interest, putting aside any self-interest and guided by altruism. Gregory defined medicine as "the art of preserving health, prolonging life, treating diseases and making death easy" [10]. Being a fiduciary means that the physician must be in a position to know reliably the patient's interest, should be concerned primarily with protecting and promoting the patient's interests, and should be concerned only secondarily with protecting and promoting his or her own interests.

Thomas Percival (1740–1804) became a leading figure at the Manchester Infirmary and due to the conflict between Tory surgeons and Whig physicians was asked to develop a code of professional conduct to rule in these situations. His work was originally called Medical Jurisprudence and was later entitled Medical Ethics, being this the first use of this term. He introduced his opus as a system of professional ethics and represented the strongest influence for the first Code of Ethics of the American Medical Association [11].

On the other side of the Atlantic, Samuel Bard (1742–1821) was the personal physician to George Washington and his family. His commencement lecture to the first graduate class of Columbia University was about "A discourse on the duties of the physician" where he emphasized the virtues of integrity and ability and condemned ignorance and dishonesty [12]. Worthington Hooker (1806–1867) was the vice president of the American Medical Association in 1846 and a huge advocate of ethics and promoter of the first AMA Code of Ethics, based in the one developed by Percival in the United Kingdom. In his book *Physician and Patient*, published in 1848, he credited the French pathologist Auguste François Chomel, the successor of Laennec, with the first use of the dictum "Primum non nocere." He also favored keeping hope and truthfulness in the frame of the physician-patient relationship [13].

Sir William Stokes, a well-reputed Irish surgeon and past president of the Royal College of

Surgeons Ireland, delivered an introductory address to the session 1894/5 in the Meath Hospital and County Dublin Infirmary on Monday, October 8, 1894. He mentioned examples of "nimia diligentia chirurgiae," synonym of unnecessary surgery, and he should be credited for using the term surgical ethics for the first time when he mentioned: "A consideration of surgical ethics that frequently exercises the mind of the operating surgeon is the question of the principles that should guide him in dealing with cancerous growths. The question as to what constitutes justification in dealing with them in an operative way is ever present and surrounded with difficulty, as the result of such interference must end in weal or woe, satisfaction or regret to the patient as to the operator." In other words, he poses the following challenge: do the risks of the operation outweigh the potential benefits to the patient? [14].

A huge landmark in the development and increasing influence of surgical ethics was the introduction of anesthesia and the inconveniences of its early adoption, due to the unacceptable initial high rate of risks, accidents, and complications [15]. The concept of professional diligence as a core foundation for medical and surgical ethics was reintroduced by Richard Cabot (1868–1939), professor at Harvard University and a contemporary of Ernest A. Codman (1869–1940). He was concerned with the occurrence of medical errors and his research focused in the comparison of clinical diagnosis and autopsy findings, and he was a fierce advocate of the "ethics of clinical competence" [16]. The development of transplantation was very fertile in the arising of new and unknown conflicts and dilemmas brought to the table of ethical discussion. Surgical innovation has also represented a challenge in the advance of surgical ethics [17].

McCullough et al. consider that surgical ethics is founded on the recognition of the rights of surgical patients. There is no surgical procedure or intervention without the active role of a surgeon who is part of the therapeutic alliance of the patient-surgeon relationship. In this dyadic relationship which should be based on trust, the surgeon deters a role of "authority" due to his or her training, expertise, clinical judgment, and fiduciary responsibilities; meanwhile the patient deters a position of "authority" since he or she will consent and allow that upon his or her body a surgeon may act displaying the art and science of surgery [18].

Miles Little highlighted the uniqueness of surgical ethics, and his proposal of five categories characterizes this branch of ethics, which pertains and traverses the surgical field in all its disciplines. These categories are rescue, proximity, ordeal, aftermath, and presence. The first four are experienced exclusively by the patients; meanwhile the presence of the surgeon throughout the whole process of intervention and recovery should be considered as a virtue and a duty inherent to our profession [19].

(a) Rescue: there is a need in all those patients requiring a surgical consultation for rescue, submitting their desire to the power of the surgeon. This power is based on the abilities or skills needed to be displayed: the techne or technical skill, the episteme or knowledge, and phronesis or practical judgment, all *virtuous* characteristics to be displayed by an ethical surgeon.

(b) Proximity: the patient-surgeon relationship is very intense, usually short and invasive, and, of course, intimate due to the interchange of values.

(c) Ordeal: the physical and mental impact of surgery may probably remain for life and refers to the usually unpleasant and sometimes painful experience to go through a surgical intervention.

(d) Aftermath: the aftereffects are usually obvious, specially in cancer patients who after an operation are subjected to chemo- and/or radiotherapy, life controls with a tension every time a diagnostic modality is requested for assessment or restaging.

(e) Presence: patients request the presence of the surgeon and/or the surgical team in a very proactive way, with complete supervision of the recovery process and a leading role of the different team members needed during the perioperative stages.

Thus, SE acknowledges all these categories, circumstance that added to the peculiarities of the surgical therapeutic approach, are the reason why SE differentiates itself from the field of medical ethics [20].

Surgical ethics combines and provides a frame for providing solutions to all the issues, problems, decisions, conflicts, and dilemmas a surgeon confronts in the daily activity of patient care, surgical research, surgical education and training, leadership, and management. It needs be understood as a discipline of both ethics and surgery which provides an answer to the question "What ought morality to be in surgery?". Besides, all the questions about what represents good professional practice concern surgical ethics rather than surgical technique; nonetheless it must be understood that being a competent surgeon is a prerequisite to being an ethical one. So, technical mastery is necessary to fulfill the expectations of our patients and society, but this quality is not enough. The present challenge is to be a superb and a complete surgeon. The central question has changed. It is not just "What can we do for this patient?" but today's question is "What should be done for this patient?" and this is a question for surgical ethics. As stated by Pellegrini, "Surgery is a moral practice, and every surgeon is a moral agent" [21]; and this situation can only be determined by focusing on the ethical dimensions of surgical care.

In the last 30 years, the attention and focus on surgical ethics have changed dramatically; now surgical ethics lies at the core of surgical professionalism, and it must remembered that every time a surgeon is confronted to a surgical decision, he or she will need to differentiate "how to treat" questions which are a matter of medical and surgical science from "why to treat" issues, related to surgical ethics. The role of being a fiduciary for the patient's sake is a pivotal one, meaning that in the conflict between altruism and self-interest, the first one must prevail [22]. Besides, surgery is characterized by a fierce accountability: when the patient of an internist dies, the natural question his colleagues ask is "What happened?". When a patient of a surgeon dies, his colleagues ask "What did you do?". By the nature of his craft and his beliefs about it, the surgeon is more accountable than other physicians, and he or she also has much more to account for [23].

Surgery and surgical ethics rely on the principles of beneficence, nonmaleficence, respect for autonomy, and justice, as collated by Beauchamp and Childress and included in the Belmont Report [24]. They based their principle system on the one previously described by Donald Ross, in 1930 [25].

In addition to the four classic ethical principles, truthfulness, fairness, integrity, dignity and respect of people's rights, and honesty should be added, all of which have a heavy impact on every stage of surgical care.

The main characteristics of the four classic ethical principles are the following [3]:

Beneficence

Beneficence stands for acts of mercy, kindness, and charity and involves the principle of acting with the best interest of the other in mind. In contrast to utility, positive beneficence requires agents to provide benefits to others and is more an ideal than an obligation: according to the Good Samaritan's parable, we cannot demand others to act in an exceeded moral behavior.

The principle of positive beneficence supports an array of moral rules of duties: basically, to protect and defend the rights of others and to prevent harm from occurring to others. The rules of beneficence prevail positive requirements of action. David Hume argued that the obligation to benefit others arises from social interactions: "All our obligations to do good to society seem to imply something reciprocal: I receive the benefits of society and therefore ought to promote its interest." Reciprocity is the act or practice of making an appropriate and often proportional return. Examples include returning benefit with proportional benefit, harm with proportional criminal sentencing, and friendly actions with gratitude.

Nonmaleficence

Nonmaleficence is based on the dictum "Primum non nocere" ("above all, do no harm") and requires intentionally refraining from actions that would cause harm.

The principle of nonmaleficence imposes an obligation not to inflict harm on others. Though the aphorism "Primum non nocere" is often considered the fundamental principle in the Hippocratic tradition of medical ethics, it does not appear in the Hippocratic corpus exception made of the translation of a single passage: "At least do no harm." Nonetheless the Hippocratic Oath expresses obligations of nonmaleficence and of beneficence: "I will use treatment to help the sick according to my ability and judgment, but I will never use it to injure or wrong them" [27].

Since the most important rule of this principle consists in not to inflict evil or harm, it requires only intentionally refraining from actions that cause harm. Harm stands for physical harm, especially pain, suffering, disability, death, or loss of chance (survival) as well as mental harm. This obligation includes not only the duty not to inflict harm but also the duty not to impose a risk of harm. In cases of risk imposition, both law and morality recognize a standard of due care that determines whether the agent who is causally responsible for the risk is legally or morally liable as well.

On the other hand, negligence represents the lack of due care and a departure from the professional standard that determines the due care in any given situation. When addressing negligence, there is a focus on the behavior or misdemeanor that falls below a standard of due care that the law or the moral codes have established to protect others from the careless imposition of risks. The following are essential elements in a professional model of negligent care:

1. A surgeon's duty to the patient
2. Breach of duty from the surgeon
3. Harm suffered by the affected patient
4. Causality link between the breach of duty and the harm achieved

Professional malpractice is an instance of negligence that involves not adjusting to the professional standards of due care.

Respect for Autonomy

The concept of autonomy is derived from the Greek words autos (self) and nomos (rule, governance, law), referring originally to the self-determination of city-states in ancient Greece.

The concept of autonomy precludes the individual decision-making in health care and research, as patients and as subjects, in the surgical care process. There are two conditions essential for autonomy: liberty (freedom from external controlling influences) and agency (capacity and capability for intentional action). The moral requirements of respect for autonomy include autonomous action in terms of normal choosers who act intentionally, with full and complete understanding and free from external or controlling influences that determine their action.

Two philosophers have powerfully influenced the contemporary interpretations of respect for autonomy: Immanuel Kant and John Stuart Mill. Kant argued that respect for autonomy flows from the recognition that all persons have unconditional worth, each having the capacity to determine his or her own moral destiny. Mill concerned himself primarily with the individuality of autonomous agents, arguing that society should permit individuals to develop according to their own convictions so long as they do not interfere with a like expression of freedom by others or unjustifiably harm others. Kant's position entails a moral imperative of respectful treatment of persons as an end in itself [26].

The basic paradigm of autonomy in health care is represented by the surgical informed consent process and entails competent judgment. To be valid, the conditions of competent judgment must satisfy the standards of its determination. These abilities include the following spectrum: to understand one's situation and the consequences of performing or refusing therapeutic treatment, to understand all relevant information, to receive and provide rational and risk-/benefit-related reasons of doing or not, to arrive to a final and reasonable decision, and to communicate that choice to the health-care team.

Justice

Aristotle first conceptualized justice as the "rendering to each individual of what is due to him or her" and refers to the development of fairness and equality. More recent influences in biomedical ethics originate from Rawls' *A Theory of Justice*, in which he argues that a social arrangement forming a political state is a community effort to advance the good of all in society. Representing an egalitarian point of view, Rawls attempts to solve the problem of distributive justice (the socially just distribution of goods in a society) by using a variation of the familiar device of the "social contract." The resultant theory is known as "justice as fairness," from which the author derived his two principles of justice: the liberty principle and the difference principle [28].

In the health-care setting, justice primarily refers to the distribution of usually scarce resources in a fair way from a communitarian point of view. However, it includes the obligation to respect the patients' rights and morally acceptable laws from the patient's individual point of view.

The four ethical principles embrace more specific rules applicable to surgical care and provide a framework to consider systematically the ethics of surgical practice [29]. In that sense, the principle typology of surgical ethical issues includes the following:

(a) Respect for autonomy: surgical informed consent, truth telling, confidentiality, communication skills
(b) Beneficence: surgical competence, sound judgment, continuing medical education, accountability, communication skills
(c) Nonmaleficence: surgical competence, sound judgment, recognizing one's limitations, disclosure and discussion of complications and surgical errors
(d) Justice: allocation of scarce resources, legal issues, human rights

Concluding Remarks

- The foundations of surgical ethics lie in the Greek and Roman tradition, and its development has been influenced by different religions.
- Many figures have been influential, but the significant role played by John Gregory and Thomas Percival should be duly acknowledged.
- The first one to use the term surgical ethics was the Irish surgeon William Stokes in 1894.
- Miles Little, an Australian surgical oncologist, was the one who defined the five categories of surgical ethics.
- Surgery is a moral practice and every surgeon should be considered a moral fiduciary agent.
- Surgical ethics provides a frame for three aspects of the surgical practice: (a) the virtues and obligations of the surgeon in the physician-patient relationship, (b) the surgeon's professional obligations and behavior, and (c) the surgeon's responsibilities to society.

References

1. https://www.britannica.com/topic/bioethics. Accessed 18 Dec 2017.
2. Rich WT. The word "bioethics": the struggle over its earlier meanings. Kennedy Inst Ethics J. 1995;5:19–34.
3. Ferreres AR. Ethical debate: the ethics of non performing extended lymphadenectomy in patients with gastrointestinal cancer. World J Surg. 2013;37:821–8.
4. Jensen AR. A short history of Medical Ethics. New York, NY: Oxford University Press; 2000.
5. Plato. The Laws, book XI (translation by Bury RG). Cambridge, MA: Harvard University Press; 1926. Loeb Classical Library vol 187.
6. Drizis TJ. Medical Ethics in a writing of Galen. Acta Med Hist Adriat. 2008;6:333–6.
7. Vilanova A. Opera Medica Omnia, vol I (Tractatus de Intentione Medicorum). Barcelona: University of Barcelona; 2015.
8. Bullough VL. The development of Medicine as a profession: the contributions of the medieval university to modern medicine. New York, NY: Hafner Publishing Company; 1966. p. 94.
9. French A. The medical ethics of Gabriele de Zerbi. In: Wear A, Roger K, French A, et al., editors. The medical renaissance of the sixteenth century.

Cambridge, UK: Cambridge University Press; 1984. p. 84.

10. McCullough LB, editor. John Gregory's writings of medical Ethics and philosophy of medicine. Dordrecht: Kluwer Academic Publishers; 1998.

11. Percival T. Medical Ethics, or a code of institutes and precepts, adapted to the professional conduct of physicians and surgeons. Cambridge, UK: Cambridge University Press; 2014.

12. Bard S. Discourse on the duties of a physician with some sentiments on the usefulness and necessity of a public hospital (King's College Commencement, may 16, 1769) Published as Advice to those gentlemen who receive the first medical degree conferred by the university. New York, NY: Robertson, 1769. p. 2–3.

13. Hooker W. Physician and patient. London: Forgotten Books; 2015.

14. Stokes W. The ethics of operative surgery. Dublin: Printed for the author by John Falconer; 1894.

15. Jackson SH, Van Norman G. Anesthesia, anesthesiologists and modern medical ethics. In: Eger EIII, Saidman LJ, Westhorpe RN, editors. The wondrous story of anesthesia. New York, NY: Springer; 2014. p. 205–18.

16. Cabot RC. The meaning of right and wrong. New York, NY: Macmillan; 1933.

17. Tung T, Organ CH. Ethics in surgery: historical perspective. Arch Surg. 2000;135:10–3.

18. McCullough LB, Jones JW, Brody BA. Surgical Ethics. New York, NY: Oxford University Press; 1998.

19. Little M. The fivefold root of an ethics of surgery. Bioethics. 2002;16:183–201.

20. Little M. Invited commentary: is there a distinctively surgical ethics? Surgery. 2001;129:668–71.

21. Pellegrini CA. Presidential address: the surgeon of the future: anchoring innovation and science with moral values. Bull Am Coll Surg. 2013;98:8–14.

22. Jonsen AR. Watching the doctor. New Engl J Med. 1983;308:1531–5.

23. Bosk C. Forgive and remember: how to manage medical failure. Chicago: University of Chicago Press; 1979.

24. Beauchamp TL, Childress JF. Principles of biomedical Ethics. 4th ed. New York, NY: Oxford University Press; 1994.

25. Ross WD. The right and the good. Oxford: Oxford University Press; 1930.

26. Kant I. Groundwork for the metaphysic of morals. New Haven: Yale University Press; 2002. p. 52.

27. Smith CM. Origin and uses of "primum non nocere": above all, do no harm! J Clin Pharmacol. 2005;45:371–7.

28. Rawls J. (1999) A theory of justice (Revised Edition). Cambridge, MA: Harvard University Press; 1999.

29. Adedeji S, Sokol DK, Palser T, et al. Ethics of surgical complications. World J Surg. 2009;33:732–7.

The Ethical Challenges of Surgical Leadership

E. Christopher Ellison

Key Points
- Definition of leadership and followers
- Definition of ethics and why it is fundamental to leadership
- Definition of ethical leadership and organizations
- Assessment of bad leadership
- The major ethical challenges of leadership
- Ethical decision-making

The Ethical Challenges of Surgical Leadership

> What I must do, is all that concerns me; not what people think. This rule, equally arduous in actual and intellectual life, may serve for the whole distinction between greatness and meanness. It is harder because you will always find those who think they know what your duty is better than you.
>
> *Ralph Waldo Emerson* [1]

Not a week goes by without organizational crises making the news headlines. In these instances a common thread seems to be a failure of duty and adhering to ethical principles. Recent examples include the recent data breach at Equifax impacting the personal identity of nearly 140 million Americans and the apparent delayed public response by the leadership [2, 3]; the US Justice Department unsealing an indictment in May of 2015 charging 14 world soccer figures, including officials of Fédération Internationale de Football Association (FIFA) [4]; or the National Collegiate Athletic Association (NCAA) investigation in 2017 concerning an athletic shoe manufacturer and claims of money being paid directly to college basketball players or their families in a few programs by that manufacturer [5]. Health care is not immune from such crises. For example, the Tampa, Florida-based health insurer WellCare announced in August of 2010 that it would pay $137.5 million to the federal government to settle ongoing investigations and $200 million to settle a class action suit related to alleged Florida Medicaid fraud [6]. Nor is surgery immune. An example is the poor handling of the Dr. Michael Swango affair by several institutions in the USA in the early 1980s. These organizations were criticized for not bringing forward suspicions of patient harm being committed by this would-be murderer [7]. Another example is represented by issue of surgeons' potential conflict of interest with support from the pharmaceutical and the surgical technology industry. Finally, compliance investigations involving professional billing and overlapping surgeries. The list goes on and on.

E. C. Ellison (✉)
Department of Surgery, The Ohio State University, Columbus, OH, USA
e-mail: Christopher.Ellison@osumc.edu

© Springer Nature Switzerland AG 2019
A. R. Ferreres (ed.), *Surgical Ethics*, https://doi.org/10.1007/978-3-030-05964-4_5

In order to develop a strong ethical organization and avoid ethical breaches, it is essential that the modern surgical leader considers the ethical dimension in each and every decision. Achieving this goal is the ultimate ethical challenge for leaders [8]. It is essential that the modern surgical leader possess the personal qualities of a virtuous leader, to know her or his duty and adhere to their personal values when navigating the complex social and business environment in which we live and work. Per Emerson's quote being self-reliant and confident in your purpose and duty is critical. This alone is insufficient as the leader must connect with the organization they lead and the followers, who are so dependent on a trusting and just culture.

The fictional scenario in Case 1 frames some of the ethical challenges for surgical leaders and serves as an introduction to some of the concepts that will be outlined in this chapter.

Case Study 1

The following fictional scenario sets the stage for this chapter and highlights the ethical challenge of making decisions and the leadership role when there is tension between organizational decisions based on business performance and ethical decisions based on the best interest of the society. The friction between our social contract with the public and our organizational contract with specific performance objectives (quality, safety, patient satisfaction, LOS, readmission, value, and financial performance) is where ethical considerations become crucial for every leader.

Dr. Ann Leader, a general surgeon by training, is the CEO of a 5-hospital not-for-profit health system called Integrity Health with nearly 2000 beds and performing 100,000 surgical procedures annually. It is the leader in orthopedic procedures in the region and in particular hip and knee replacement performing 6000 procedures annually. A local competing for-profit health system, Profit Health, has been increasing the volume of joint replacements annually including robotic applications and has approached several of the busy orthopedic surgeons at Integrity Health to move their practices to Profit Health. The chief quality officer informed Dr. Leader of a high incidence of MRSA infection of implanted joints on the orthopedic service. The infection rate had increased to 8%. There have been no deaths. The hospital epidemiologist has determined a process issue in the sterilization of equipment was responsible for the infection and recommended stopping the procedures until the process error could be corrected and estimated this could take 8–12 weeks. The chief of orthopedics believes it was a sampling error and is adamant that the orthopedic service continues to run at the same volume without an interruption of service.

How would you handle this circumstance? What are the ethical dimensions you would consider as part of the decision-making process? What are the business decisions you would consider? How would you further investigate this incident? What is your internal and external communication plan? After reading this chapter, please return and reflect on the above questions.

Author's Comments

As the CEO in this situation and the moral fiduciary of Integrity Health, a decision must be made to protect the patients from further infection even though there would be loss of orthopedic procedure revenue. In addition, there is the potential that the program's future could be jeopardized as this incident may lead current surgeons to move their practices to Profit Health. The CEO must have the courage and wisdom to make a decision in the best interest of the public while at the same time having the guarded optimism that the organization will be able to reestablish trust with the public and recover. The CEO must oversee an objective study of the incident with internal or external experts and develop a corrective action plan. In addition the CEO must decide on how to communicate this information to the public and consult with the legal team on how to best to respond to lawsuits from infected patients or their families.

In order to be a successful leader in general, it is essential to consider the ethical dimension and how it interfaces with your organizational or departmental responsibilities. This is likewise important to your organization and followers for which you have the responsibility to care for and to lead in order to help them succeed. Whether a surgeon or another medical or nonmedical professional, leaders are often presented with situations that require them to consider several ethical dimensions prior to making a decision. This is a dynamic process and must be fluid responding to the organizational needs based on multiple factors. In these situations, an apparently innocuous decision may turn into a very dicey ethical issue. To best navigate today's complex world in health care, the challenge to the physician executive is to be an ethical leader and internalize the essential elements therein, foster the development of an ethical organization, and consider ethics in each and every decision. Developing a moral imagination will assist the leader in seeing ethical dilemmas and the ethical dimension in what seem to be routine decisions. Although surgeons will be leaders in many different situations, this chapter will focus the surgeon as a leader of larger organizations. However, the principles discussed can be applied to all leadership roles, regardless of the scope or size of the organization. Finally, bioethical dilemmas and considerations represent an important and distinct group of challenges that the leader in health care must be sensitive too. These will be covered in other chapters.

The primary objectives of this chapter are to provide the following:

- Definition of leadership and followers
- Definition of ethics
- Discuss why ethics is fundamental to leadership
- Definition of ethical leadership
- Creation of an ethical culture
- Examination of bad leadership
- The major ethical challenges of leadership
  - Power
  - Privilege
  - Information
  - Consistency
  - Loyalty
  - Responsibility
- Ethical decision-making
- Case studies in ethical and unethical leadership

Leaders and Followers

In general terms, leadership is the exercise of influence and power in a group context in order to achieve common goals. As Johnson points out, leaders are found wherever humans associate with one another [9]. These associations can be part of societal construct such as a village, county, or state legislature; a local parent-teacher organization; a societal movement; an organization such as a corporation, hospital, branch of the armed forces, or an industrial plant; or a team whether it be the premier soccer league, a health-care team, or a military unit. Leaders are the agents of change engaged in setting the vision, the goals, and the culture for the organization. Furthermore the leader has the major responsibility of furthering the needs and desires of their followers as well as themselves and their organization or service unit. A leader may be at the top of an organization/unit or the leader of subunit or division. It is no different in health care where the leaders may be presidents, CEOS, deans, chairs, division chiefs, section leads, or, on a smaller, but equally important, level, the leader of "the code blue team," a consult service, or an operating room team.

Johnson observed that the definition of leadership is incomplete without distinguishing between leading and following [9]. The leader usually is given the credit for both, success and failure. For example, a department chair in surgery will usually get credit for improving the mortality index, surgical admissions, and surgical volume but criticized and accountable for correcting a high readmission rate or suboptimal financial performance. In fact we know that in either circumstance the outcome is the result of efforts of many department members who are followers. Leaders and followers work together in complementary roles. The leader takes on a greater degree of responsibility for the direction of a group. The followers implement the plans and do the day-to-day work. Leaders through their actions affect followers' lives either negatively or positively. Hence, the leader has a moral responsibility to the followers as discussed below.

Leaders are sometimes followers and followers are sometimes leaders. In other words, we switch roles depending on the circumstances. For example, a division chief of vascular surgery is responsible for the clinical outcomes, patient satisfaction, and career development of his or her followers but in an MBA course may be a follower. Likewise a staff surgeon who is a follower may be a leader in the role as the coach of a soccer team or the president of the local parent-teacher association (PTA).

Although we often think leadership is related to a position, it is not a prerequisite to leading. All of us have the potential and will be given the opportunity to lead. For example, the junior resident during the day and at night may be the leader of the code blue team. There were many people who assumed leadership roles in the recent hurricanes (Harvey, Irma, and Maria) that impacted the USA and Caribbean in the fall of 2017 and the Las Vegas shootings in which Stephen Paddock killed 56 people and injured 489 on Sunday, October 1, 2017 [10]. They were put into devastating situations and rose up to the task of leadership. The leader-follower continuum is dynamic and changes upon the situational circumstances.

The Surgeon as a Leader

All surgeons will at times be leaders of teams. However, the size and complexity of the team may vary. In some circumstances the surgical leader is overseeing daily patient care. For example, the trauma surgeon is the leader of the team responding to a trauma alert, in the operating room the surgeon is responsible for the coordination of care of the surgical patient, and on rounds the chief resident is the leader of a smaller team of students and postgraduate trainees providing perioperative care. One can even reason that each surgeon-patient relationship involves some element of leadership from the point of view of the surgeon taking a role as a fiduciary of information and trust for patients as they go through the treatment process. In other circumstances the surgical leader has a larger scope of responsibility: for example, a clinical department, division, research program, college of medicine, or even a large health-care system. In these circumstances, the surgeon will have responsibility for diverse programs, operations, organizational outcomes and metrics in patient care, research, education, and financial performance. Regardless of the scope, size, or complexity of the organization, the principles of ethical leadership are fundamental to the success of the leader, his or her followers, and the organization or unit.

The Distinctive Attributes of the Surgical Leader

One could assume that a surgical leader is similar to all other leaders and offers no distinctive attributes. I argue against this assumption based on the fact that the features of surgical training which emphasize duty and moral obligations prepare the surgeon for ethical leadership. The surgeon-patient relationship is unique and bound by duty and trust. Based on training and experience, the surgeon brings a unique skill set, surgical ethical base, and thought process concerning leadership. Ferreres argues that surgery is a moral practice and as such the surgeon becomes a moral fiduciary agent [11]. The surgeons' role as a moral fiduciary is at the core of surgical ethics. By the very nature of the surgeon-patient relationship, the bond is characterized by trust. As a moral fiduciary, the surgeon is bound to focus on the patient's best interest, protect and promote that interest, and only secondarily be concerned with her or his own interests. By the nature of his or her training, the surgeon develops a core moral character built upon surgical ethics. Hence, the surgeon as a leader by training and experience has been exposed to the personal values and responsibility necessary to be an ethical leader. Albeit human nature and the drive for personal gain can derail the ethical core, so the surgeon and other leaders need to be very attentive to the ethical dimension in every aspect of leadership.

Ethics

The word ethics is derived from ethos, a Greek word meaning character, conduct, and/or customs. Ethics is concerned about what morals and values are found appropriate by individual members or groups of people in society. Ethics helps define the situational right and good vs. wrong and bad. Regarding leadership, ethics refers to a leader's character as well as what they do in word, action, or behavior.

Ethics Is Paramount to Leadership

Why are ethical considerations so critical for the leader? Ethics is paramount to leadership, because of the relationship between leaders and followers. Leaders through their actions affect followers' lives either negatively or positively [12]. The nature of influence depends on the leaders' character and behavior. As leaders have more power, hence they have greater responsibility with respect to their interactions and potential impact on their followers. Leaders influence followers in the pursuit and achievement of common goals. In these situations, leaders need to treat their followers as individuals and with respect and dignity. Furthermore leaders are instrumental in establishing organizational values and ethical climate which are determined by their own personal values. The success of an organization, particularly those in health care, is directly related to its satisfying the obligations of

the social contract between that entity and the people that it serves to provide access to and deliver safe, innovative, effective, value-based, and high-quality patient care. Such a leader must set the moral tone of the organization as the followers expect direction in challenging situations. These situations require that the leader have an ethical and moral character and the ability to consider moral dilemmas in the decision-making process. Ethical leadership is a dynamic process, balancing metric-driven performance with moral duty. As will be seen later in this chapter, there are many examples of decision-making seemingly devoid of ethical consideration which leads to an organizational chaos or crisis.

Ethical Leadership

The practice of ethical leadership is a part process involving both personal moral behavior and moral influence. It is a dynamic and ongoing process. A leader can be categorized as ethical or unethical based on his or her conduct and the consequences of his or her decisions or actions, duty, or character or virtue-based assessment. There is a widespread misconception that ethical leadership is not effective leadership. However, case studies have repeatedly shown that ethical leaders are frequently more and not less effective than unethical counterparts [13]. Ethical leaders were found to be more promotable and effective, more satisfied and engaged, more likely to create a culture of trust, and engage in socially responsible behavior.

Are there methods to assess ethical leadership? Yes. Three theories are commonly used to assess a leader's actions. Theories related to the consequences of ones' actions are called the *teleological approach*. *Telos* is a Greek word for purposes or ends. This approach uses the outcomes of a leader's actions, behavior, or conduct to define the leader's actions as ethical or unethical. The *deontological approach* (*deos* is a Greek word for duty) focuses on defining any action or behavior as being inherently good. The *virtue-based approach* focuses on the leader's character.

Teleological Assessment

Three categories of teleological decision assessment are employed: *ethical egoism, utilitarianism*, and *altruism. Ethical egoism* assesses the decisions and actions of the leader on the basis of achieving the greatest good for the leader. *Utilitarianism* assesses a leader's action on the basis of achieving the greatest good for the largest number of people. *Altruism* assesses the action of the leader on the dimension of demonstrating concern for the interest of others over the interests of the leader.

Ethical Egoism

Ethical egoism seems contrary to positive ethical leadership. Making a decision solely on the basis of achieving the greatest good for the leader seems at odds with meeting the leader's responsibility for her or his followers. Repeated decision-making using this model without consideration of how the decision could impact the organization may in fact be unethical or not, depending on the outcomes for the organization and followers.

Utilitarianism

Utilitarianism is based on the theory that ethical choices should be based on their consequences and is, in essence, attempting to do the greatest good for the greatest number of people. This process was formalized in the eighteenth and nineteenth centuries by English philosophers Jeremy Bentham (1748–1832) and John Stuart Mill (1806–1873) [14]. As summarized by Johnson, "there are four steps to conducting a utilitarian analysis: (1) clearly identify the action or issue under consideration, (2) specify all those who may be affected by the action, not limited to those immediately involved in the action" (patients, nursing staff, physicians, etc.), "(3) determine the good and bad consequences for those affected, (4) add up the good and bad consequences" [9, p. 147]. The action is morally right if the good outweighs the bad. Johnson advises the following when using this mode of decision-making: use your experience, assess the outcomes of prior decisions, screen off personal conflicts, and recognize the centrality of measuring impact [9]. He further advises the following cautions: outcomes are difficult to analyze, acknowledge potential unanticipated consequences, and those other decision-makers may make different decisions. Finally, be cognizant of unbridled utilitarianism [15].

Altruism

Altruism is based on placing the interests of others ahead of the leader's interests. Advocates of this approach to decision-making emphasize that love of neighbor is the ultimate ethical standard. Leaders focused on benefiting themselves will rely on personal achievements and rely on command and control as well as coercive paradigms to manage followers. In contrast the altruistic leader will pursue organizational goals, rely on followers, and give power away. Kanungo and Mendonca classify altruistic leaders in four groups based on the leader's focus and altruistic behaviors: individual-focused leader, group-focused leader, organizational-focused leader, and societal-focused leader (Table 1) [16]. An essential element of this decision-making style is self-sacrifice with leaders postponing or giving up personal benefits and perks and sharing with employees. Self-sacrificial behavior has a powerful impact on followers whether they are employees, staff, or faculty. This style is central in benevolent paternalistic leadership in which the leader sets a moral

Table 1 Altruistic behaviors

Individual-focused leaders provide
    Training
    Technical assistance
    Mentorship
Group-focused leaders provide
    Team building
    Participative decision-making
    Minority advancement programs
Organizational-focused leaders provide
    Commitment and Loyalty
    Protection of organizational resources
    Protection of whistle-blowers
Societal-focused leaders provide
    Contributions to promote social welfare
    Reduce pollution
    Ensure product safety
    Maintain and improve customer satisfaction
From: Kanungo and Mendonca [16]

example through selflessness, self-discipline, and a strong work ethic. Johnson notes successful application of this style requires the leader to consistently put the needs of followers before those of themselves, act as a role model, and use compassion as an important decision-making guide. He cautions the following: if is not possible to satisfy demands, the extent of the leader's duty may be blurred, many who express altruism fail to demonstrate such in word and action, and altruism may take on various forms [17].

Deontological Assessment

The founding basis of this approach is to assess the actions of leaders as doing what is ethically right. The argument is that whether an action is ethical depends on its outcome as well as whether the action is itself inherently good. In contrast to the standpoint, Immanuel Kant (1724–1804) argued that people should do what is morally right no matter what the consequences. Duty should play a role in ethical deliberations. Kant offers two powerful decision-making tools. First ask yourself if you would want everyone else to make the same decision. If not, take a pause and re-evaluate the decision. Second, always respect the dignity of others. Respecting the freedom of others to choose for themselves is fundamental to ethical leadership. To achieve this, leader can share information while avoiding deception and coercion. In applying this type of leadership, Johnson advises: meet your duty, reflect on whether you would want others to make the same decision, and consistently show respect for followers [18]. He cautions that every rule has exceptions, alternative moral obligations may conflict, ethical guidelines are at times impractical, and doing what is right in every stressful situation may be difficult [19].

Virtue-Based Approach

These theories are based on the leader's character. Interest in virtue ethics is prominent in both Eastern and Western thought. The Chinese phi-

losopher Confucius (551–479 BCE) emphasized that virtues are critical for maintaining relationships and for fulfilling organizational and familial duties [20, 21]. Aristotle taught that a person could be helped to become more virtuous. In other words virtuous behavior can be learned. He also espoused that more attention should be given to advising a person what to be rather than telling them what to do. Aristotle's virtues of an ethical person include generosity, courage, temperance, sociability, self-control, honesty, fairness, modesty, and justice. Furthermore, Velasquez opined that organizational managers should learn and demonstrate the following virtues: perseverance, public-spiritedness, integrity, truthfulness, fidelity, benevolence, and humility [22, 23].

Johnson points out that virtue ethics, until recently, was not popular with scholars being superseded by ethical theories as noted above. However, virtue ethics has regained interest and support [9, p.70–71]. Modern philosophers and psychologists have identified the following character strengths as important for today's leaders: courage, temperance, wisdom, justice, optimism, humility, compassion, and most importantly in my opinion integrity.

A person must exhibit courage to function as an ethical leader. Moral action can at times be risky and the ethical leader must have the strength of character necessary to model the ethical behavior despite the risk. These leaders create an ethical culture regardless of opposition from superiors or followers. Johnson states "They refuse to set their values aside to go along with the group, to keep silent when customers may be hurt or to lie to investors" [24].

The temperate leader practices moderation and controls destructive impulses. There are many intemperate leaders. They cannot control their emotions. They have a tendency to set unrealistic goals for the organization and followers. Frequently, they fail to live within their budgets, have inflated salaries, and try to control everything that goes on in their organization. Temperate leadership requires introspection. The temperate leader must recognize his or her destructive impulses [25].

Johnson identified the attributes of the wise leader: The wise leader (1) is exceptional at reasoning and employing a broad knowledge base, (2) seeks input from other thought leaders to make complex decisions, (3) demonstrates high emotional intelligence and respect of cultural differences, (4) is highly collaborative, (5) is authentically engaged with the organization and adaptable to changing circumstances, (6) is reflective with a keen sense own character, and (7) aspires to improve themselves, their followers, and organizations [25]. To practice wisdom, Johnson recommends the leader to be curious, inquisitive, collaborative, and analytical and think about the long-term impact of [25].

Justice is a particularly important trait for leaders, and fairness is central for every ethical decision. The leader should work to correct injustice and inequality caused by others.

Optimism is an equally important attribute of leaders. In this sense we are not speaking of unbridled optimism but rather realistic optimism where a leader grows from hardships and has the potential to help followers learn from setbacks and grow.

Humility is comprised of three components. The humble leader is self-aware and through introspection can assess his or her strengths and weaknesses. Owing to the recognition of one's own weaknesses, the humble leader is open to new ideas and information. The humble leader acknowledges there is a power greater than oneself. The humble leader will not develop an inflated view of oneself and their self-importance. They have the capacity to appreciate the worth and contributions of others [25].

Compassion is important to the ethical leader. It is an essential element of altruism. An orientation toward others rather than oneself is an important dimension that differentiates ethical from unethical leaders. An ethical leader recognizes that they serve the group rather than oneself.

Integrity is at the core of leadership. Leaders with this virtue live their values and keep their promises. There is nothing that undermines a leader more than lack of integrity. Establishing organizational trust is essential to achieving high-performance levels and employee engagement.

People who work in trusting environments are more productive and enjoy better working relationships [26].

Creating an Ethical Culture

The leader is the ethics officer for an organization [27]. The CEO is the chief ethics officer. Followers look up to leaders to be ethical role models. The leader, because of a position of authority, is legitimate. An ethical leadership challenge is to build one's credibility by living the values they talk about – the leader needs to walk the talk. Ethical leaders make sure that moral messages are not replaced by obsession with performance and profits but rather work to make the ethical goals and the performance goals complimentary. This builds an ethical climate.

Victor and Cullen classify ethical climates or cultures as follows [9, p. 330, 28]. The *instrumental climate* follows the principle of ethical egoism. Decisions are made for the selfish interests of the leader and the internal leadership group and not for the benefit of the followers and organization. In *caring climates* the number one consideration is concern of care for others. In *law and order climates*, decision-making is driven by criteria such as professional codes of conduct. In *rules climates* decisions are based on rules or policies developed by the organization. The final climate type is the *independent climate* in which members of the organization have wide decision-making authority.

In applying this scheme to health-care organizations and perhaps other types of organizations, I hypothesize that each organization may have a dominate climate but within an organization sub-climates might exist. Each organizational climate or sub-climate has unique ethical challenges. In assessing organizations, Johnson citing Victor and Cullen [28] notes those with instrumental climates are more prone to immoral decisions and behavior [9, p. 330]. These types of organizations tend to be in the for-profit sector and have an authoritarian-based leadership style. Caring climates promote follower loyalty. Rules-based cultures encourage ethical behavior but not organizational engagement. A major challenge for the leader is to understand the dominant culture and climate of an organization and the sub-climates in order to provide harmonious guidance.

Johnson points out that there is "no one-size-fits-all approach" [29]. The leader needs to identify the principles and attributes of moral organizations and apply them to create a dominant ethical culture. Johnson observes that the sine qua non of high-performing ethical organizations includes "risk recognition, zero tolerance of for destructive behaviors, justice, integrity, trust, process focus, structural reinforcement and organizational citizenship" [29].

Codes of ethics are useful tools to establish and maintain an ethical culture. Common elements include addressing conflicts of interest, maintenance of records of fund and assets, information management (includes HIPAA rules), outside relationships, employment practices (discrimination, sexual harassment, human resource issues), and other practices (health and safety, technology, innovation, use of company assets for personal benefit, institutional review boards (IRB), compliance programs).

Formal ethics training is important in onboarding new staff as well as reinforcing the current ethical position of the organization. Training should emphasize the moral danger signs; reduce destructive behaviors; promote trust and integrity; reinforce organization mission, vision, and values; and provide insight into ethical guidelines. Such training improves adhering to compliance guidelines which are critical for organizational integrity in the health-care sector.

Examining Bad Leadership

A leader is a person who has an unusual degree of power to create conditions under which other people live and work. Parker Palmer observes that the impact of leadership can be illuminating or the leader can cast an oppressive shadow [30]. The leader must take responsibility for her or his own character and work to evolve in a positive sense lest the act of leadership create more harm than good. Facing and reflecting on the dark side of leadership lessens the potential abuse of authority and can help the leader be more effective.

There are many ways to classify bad leadership. According to Kellerman, bad leaders can be ineffective, unethical, or both [31]. She identifies seven types of undesirable leaders: incompetent (lack academic or emotional intelligence and are careless), rigid (unwilling to accept new ideas), intemperate, callous, corrupt, insular (cordons off the leadership group from the organization), and evil. The Bond scholars identify seven clusters of destructive leadership behaviors:

- Cluster 1: Makes poor decisions, makes decisions without adequate information, and fails to prioritize.
- Cluster 2: Lacks critical skills, poor at negotiation, and cannot motivate followers.
- Cluster 3: Makes good decisions but is overly controlling and micromanages subordinates and managers.
- Cluster 4: Cannot deal with conflict, plays favorites, and behaves inconsistently.
- Cluster 5: Is not that bad or good and tends to be poor at seeking information and coordinating a team.
- Cluster 6: The leader isolates self from the rest of the organization.
- Cluster 7: Creates a situation of significant misery and despair by engaging in bullying behavior and lying [32].

Einarsen et al. found that 61% of respondents in a survey of Norwegian organizations reported ongoing destructive behavior in leaders [33]. Leaders could be classified as tyrannical, supportive but disloyal, derailed, and "laissez-faire," whereas constructive leaders cared about followers, were supportive, and used resources wisely.

A number of psychologists believe that unethical leadership is related to three similar personality traits: narcissism, machiavellianism, and psychopathy. Narcissists are self-absorbed and self-confident. A narcissistic tendency can be a positive trait of leaders as they have charisma and display confidence with bold decision-making. Unregulated this type of personality can lead to their demand for special privileges and abuse of power. Niccolo Machiavelli, an Italian philosopher, thought that a leader should maintain a virtuous image but use whatever means necessary to achieve their ends. These leaders can be highly skilled at manipulation and tend to be self-promoters. Psychopaths have a total lack of conscience and hence are very different than the aforementioned traits that according to psychologists can experience remorse and guilt to varying degrees. The psychopath is extroverted, energetic, and charming. Once they assume a leadership role, they tend to manipulate and subvert in order to secure additional power and privilege. All three types can undermine the ethical foundations of an organization. All would likely create a culture and climate devoid of trust and collaboration.

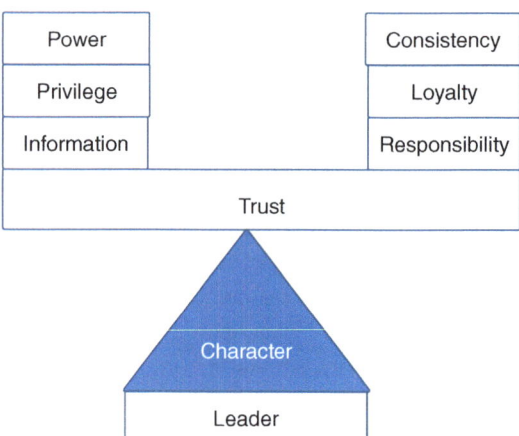

Fig. 1 For effective ethical leadership, the leader must exercise dynamic and balanced management of the six major challenges in order to create a culture of trust. To do so the leader must rely on her or his character and meet the duties of leadership. Failure to do so can lead to organizational imbalance and dysfunction as signaled by the cardinal symptoms, pressure to maintain numbers, lack of engagement with followers who display fear and silence, a bigger-than-life CEO, a weak board, unchecked conflicts of interest, and arrogant innovation. (Published with permission of E. Christopher Ellison MD FACS)

The Six Major Challenges of Leadership

Johnson identified six burdens of leadership [9]. These are truly the primary ethical challenges that the leader must manage in order to keep an organization balanced. The challenge of ethical leadership is the dynamic management of Johnson's burdens: power, privilege, information, consistency, loyalty, and responsibility. The illustration in Fig. 1 emphasizes that the leader must keep a delicate balance in managing these challenges. Failure to reach equipoise can lead to an organization that may be in ethical danger. The symptoms of such, according to Johnson, consist of several factors, of which the following are pertinent to health care in my assessment: pressure to maintain numbers, lack of engagement with followers displaying fear and silence, a bigger-than-life CEO, a weak board, unchecked conflicts of interest, and arrogant innovation ([9, p 332–333]).

Management of Power

Power is the fundamental foundation of influence. Power can be classified as coercive (punishment-based), reward-based, legitimate power (based on position), expert power (based on characteristics of the individual, and referent power (based on admiration or affinity for the leader) [34].

The effective and ethical leader modifies the application of his or her power. The abuse of power is more frequent than one thinks. In Europe 3–4% of employees report being the victim of bullying at least weekly, and 10–15% state that they have been the target of psychological aggression in the prior 6 months [35] (in the USA up to 90% of employees report being disrespected by the boss [36]). In this latter study, the authors noted that destructive leaders exhibited the following behaviors: deceit, constraint, coercion, selfishness, cruelty, disregard, and deification [36]. The followers of such leaders suffer low self-esteem and are less productive and, of course, less satisfied [36].

The more unchecked and unregulated a leader's power is, more risk there is for abuse. "Power Corrupts and absolute power corrupts absolutely" (Lord Acton). In health care there are few leaders that have completely unchecked power and control. In most organizations there is a governance structure such as boards or a hierarchical structure that modulates a leader's authority. The CEO of a medical center reports to a board. The medical school dean reports to the provost or university president and ultimately to the board of trustees. The chair of a department reports to the dean. It is important to appreciate that most leaders in large health systems, colleges of medicine, or departments in fact function as middle managers. They have a direct report to a superior and governing body and also have responsibility to the faculty and staff in their unit. For example, the chair of a department reports to the dean and must satisfy the needs of the faculty to be successful. Still unmodulated power is a risk for the organization. For example, a leader who isolates oneself from the organizational community and other leaders and not engaging that community will be viewed as more dictatorial than one who develops a leadership team representing various components of the organization and to engage them in the decision-making process. In a department of surgery, this leadership team may consist of vice chairs, division chiefs, as well as at large faculty. The inclusion of the latter is critically important and signals that the leader wants the input of the organization and that the leader realizes that collaboration is critical for success.

Despite the checks and balances, the style of the leader is critically important. The Bard Scholars Cluster 3 or micromanaging is one of the more serious dysfunctional leadership characteristics in surgery and medicine in general. In these circumstances the leader quickly disenfranchises the leadership team and sends the message that the team cannot be trusted and that he or she is the only one that can make decisions of any significant magnitude. The effective leader needs to develop confidence in the team and provide them some defined autonomy to oversee their areas of responsibility. For the CEO these would include hospital directors, for a dean department chairs, and for department chairs, division chiefs and faculty.

The crucial decision for the leader is to decide what types of power he or she should use, in what situations, and for what purposes, as well as how much power to delegate. Delegating power is highly motivating and very effective at promoting collaboration and organizational trust as well as increasing employee effectiveness and engagement. Clearly with delegation of authority, there may be some risks depending on the level of experience of the person to whom the authority is delegated. Hence there is the need to develop accountability measures and metrics to assess outcomes.

Management of Privilege

"With leadership comes great privilege." Much of the privilege is in terms of compensation. Over the past four decades, the salaries of chief executives in the USA have increased $ 15.2 million (including salary, bonuses, stock, and stock options) with an adjusted inflation rate of over 900% [37]. These large salaries are more common in the for-profit companies but occur in not-for-profit health-care organizations to a lesser extent as well. Johnson stated "In one year the compensation of the top 20 nonprofit hospital CEOS jumped 29.6% including major increases for Ascension executive Anthony Tersigni (who earned $7.1 million) and Ronald Peterson of Johns Hopkins Health System (who earned $1.7 million)" [9, p. 13, 38]. Leaders can potentially abuse perks as well because of their positions of influence. On the contrary, leaders can also send a powerful positive message if they contribute time and money to needy countries or organizations. With great privilege comes greater responsibility. Some of the world richest leaders have pledged to give away the vast majority of their wealth (Warren Buffett, Bill and Melinda Gates, Mark Zuckerberg, etc.) [9].

Regarding privilege and compensation, organizations can moderate the negative impact by establishing processes for setting formulaic compensation and awarding incentives. In the best of circumstances, compensation of the leader should be determined by a third party such as a board of trustees or directors or compensation committee.

Management of Information

A leader has more access to information than others in an organization. The leader is a fiduciary of information and must consider when and how much information to share. The leadership challenge is to meet the societal responsibilities while managing a potentially critical situation. If the information could negatively impact the organization, the leader may be tempted to cover up or alter the information. Wagner reported in a VA scandal audit that managers at the Veterans Administration disguised long wait times for veterans seeking medical care [39].

Other challenges that surround information sharing include whether to release information or not, the timing, privacy issues, and public reporting of medical outcomes such as infection rates or mortality. In most organizations a communication team is in place to advise the release of

Case Study 2
In late December of 2004, elevator workers at Duke Health Raleigh Hospital and Durham Regional Hospital drained hydraulic fluid into empty soap containers and capped them without changing the labels. Not long afterward, medical staff complained that some of their surgical tools felt slick [40].

Within a few hours of learning of the event, Dr. Victor Dzau, chancellor for health affairs and president and CEO of the Duke University Health System, reported in an interview that leadership "gathered a crisis team to assess the situation. The atmosphere was somber. We asked for facts. Were patients harmed? How could this have happened? Have we fixed the problem? Is this a systemic problem? Are any other hospitals affected? At that point, we did not know the circuitous chain of events that led to a mix-up that victimized both our health system and our patients (as we later learned, elevator company workers had drained used hydraulic fluid into empty detergent containers, which were then returned to the detergent company and subsequently redistributed to several hospitals as detergent). But soon I realized that everyone in the room felt the same way that

I did. Our first concern was for our patients. This was not the time to cast blame, but to jump into action."

According to the article, "The staff had discovered that some surgical instruments had been inadvertently washed in hydraulic fluid instead of detergent at our two community hospitals, Duke Health Raleigh and Durham Regional. The fluid was accidentally substituted for detergent in the multistage, high-heat cleaning process. The instruments were rinsed of the fluid in a washing machine and later sterilized with steam and dry heat. The leadership team agreed that we must immediately inform the 3800 affected patients and their physicians, and release new information as we learned more. Our discussion centered on how to deliver an accurate message while avoiding unnecessary anxiety or confusion. Our first letters went out on Jan. 6.

The leadership appointed an oversight team and designated the chief medical officer of each hospital as the primary patient contact, asking them to respond to patients and work with their physicians to provide appropriate care. Since the exposure to the instruments occurred over a defined period, the months of November and December, we were able to immediately examine the postoperative infection rate compared to normal rates. The infection control physicians put into place a vigilant surveillance program. There proved to be no out-of-the-ordinary spike in infections.

To assure that the sterilization procedure had not been compromised, we quickly sought an outside expert, Dr. William Rutala, a UNC professor and director of the Statewide Program in Infection Control and Epidemiology at the UNC School of Medicine. We received his final report and conclusions June 15. Using the actual used hydraulic fluid, Dr. Rutala found that replacing cleaning detergent with the fluid did not alter the effectiveness of the high-heat sterilization process. We also recognized at the outset that we should address any potential exposure to chemicals in hydraulic fluid (although we felt confident that any exposure would be slight, since the instruments were rinsed and sterilized before use). In January we retained world-renowned RTI International in Research Triangle Park to conduct a chemical analysis of the used hydraulic fluid, to determine what quantities of the fluid were left on the instruments, and to help us understand any risk to patients. The results showed that a miniscule amount of fluid was detected on the instruments that were determined not to be harmful to patients."

Do you think this was handled well? Is there anything else you would have done differently? Was the response fast enough? Was the detail in the communication appropriate? Were the ethical obligations met?

Author's Comment

In this example, the leadership successfully managed the situation by keeping the patients first and foremost, gathering the facts, and studying the impact of the hydraulic fluid on infection as well as any chemical exposure impact on the patients by creating a hotline and also offered patients the opportunity to be evaluated at an independent Duke clinic specializing in environmental medicine. Key to the success of information management by this group was collecting the factual and accurate information as quickly as possible while reassuring the patients. It takes time to collect and assess the information, and reporting inaccurate or speculative information would have created loss of trust between Duke Health and the patients. The goal was to allay their patients' fears and directly answer their questions.

newsworthy information and to assist with data collection for publicly reported information.

According to Johnson [9, p 15], unethical leaders have several deviant behaviors relative to

information: they "deny knowledge of the information they possess, hide the truth, fail to reveal conflicts of interest, withhold information that followers need, use information solely for personal gain, violate the privacy rights of followers, release information to the incorrect people, and prevent followers from releasing information that they are ethically bound to release."

The scenario below (Case Study 2) is a true example of information management around a potentially dangerous situation.

In Case Studies 3 and 4, the reader will find two other examples of sharing information. In making the decision to release or not release information that may affect the health of others, there is really only one choice and that is to release the information. Compare and contrast how the organizations in Cases 3 and 4 shared information with the public. In these situations maintaining public trust is critical and information sharing allows a company or hospital to meet its social contract.

Consistent Decision-Making

Leaders, whether in business or medicine, are human and are prone to develop closer relations with some of her or his followers than others. This may be people that the leader hired, people who are part of the leadership group, or people that share common interests such as the arts, golf, or perhaps their children go to the same school. Whatever the reason, there could be a tendency for the leader to treat this group of people more favorably than others unless they are cognizant that such relationships could lead to inconsistent decision-making. The in-group may have higher levels of trust or belong to the same clinical department or product line and share in the financial rewards of decisions that may favorably affect that unit. The out-group may not be as aligned with the leader. The relationships are not as supportive or trustworthy. When this group seeks access to resources, the leader, who is unaware of the obligation to be consistent and fair, will usually respond more cautiously or in the negative as opposed to an affirmative decision in the in-group.

The challenge for the leader, surgical or otherwise, is to be aware of these natural tendencies and try to control them in order to maintain consistent and equitable decision-making. In some instances, it is helpful to develop defined decision-making processes with use of arm's length third-party committees to distribute organizational assets such as capital or space (office, research, or clinical). Proponents of leader-member exchange theory encourage leaders to develop closer relationships with as many followers as possible including faculty, physicians, and other professional staff [41]. The process of engaging the organization may involve daily walk rounds, video blogs, and town hall meetings. The most effective is meeting as many people one on one as possible. In some large institutions, this may be difficult and assigning some of this to the leadership team can help. Remember there is no substitute for a handshake and a conversation no matter how short the duration.

Loyalty

The effective and ethical leader is loyal to her or his organization, customers, and followers and expects the leadership team and followers to be loyal as well. Customers in health care can include patients, students, residents, and fellows in training. Followers include physicians, nursing staff, OR personnel, and other health system and department employees. In addition, the leader must be true to the governing board and loyal to their interest. The obligations may extend to larger organizations such as a university or college or hospital system as well as the community in which the organization is located.

In addition, the leader must be loyal to the followers as noted above. Tension and friction not infrequently develops between the leader's fiduciary responsibility for organizational financial performance and loyalty to the followers. During times of financial stress and lower operating margins, leaders and boards need to focus on the bottom line. This may result in efforts to control expenses by reductions in staff and layoffs. In health care this is a particularly challenging issue

as patient care can be negatively impacted by high patient to nurse ratios. Reductions in OR staff can reduce the number of operating room suites which are available on a daily basis. Reductions in the ED can impact the number of patients seen and increase the hours on diversion. The patient staff ratios in ambulatory sites can lead to fewer patients being seen. Hence workforce cuts while helping the bottom line through expense reduction can lead to access issues for patients, and there is the theoretical concern for quality issues as well. Fortunately, in health care the staff is highly motivated and tends to put patients' needs before their own. Hence, in these situations health-care team may work overtime and fill in the gaps to provide the needed coverage. However this can lead to burn out and fatigue and may lead to turnover. The leader needs to be sensitive to these issues and communicate her or his sensitivity to the staff. Although in times of economic downturn layoffs may be necessary, less stressful for the employees could be sought. This might include other forms of expense reduction, including enhanced efficiency measures and focus on revenue cycle management. In addition most organizations have a back log of new hires that are pending. Prior to pursuing widespread workforce cuts, leaders could consider a hiring freeze. Every new position not filled saves a layoff. The key to loyal management is showing concern and free and open communication with full transparency.

The leader also has a loyal obligation to customers and patients. Decisions regarding organizational management and fiscal responsibility must be balanced by the needs of the customers. This is noted above in the staffing needs to maintain patient access and quality. Equally important is the need to deploy capital to keep up the infrastructure of the organization. If the facilities deteriorate, it may impact the patient confidence in the health-care organization and potentially quality of care as well as education and research programs. It seems reasonable that if an organization or department is having financial difficulties, the leadership team may need to cut back on capital expenditures to maintain the bottom line and maintain payroll. These decisions could be weighed against the impacts of lack of capital investment. If capital investments are being curtailed to maintain a bonus pool for senior executives or physicians or to build increased reserves, the decision may need to be reconsidered in that negative impact may far exceed the perceived gains. These dilemmas between performance, personal gain, ethical obligation, and duty are at the heart of ethical leadership.

Responsibility

Faculty, hospital, and departmental staff are responsible for their actions within the scope of their assignments. On the other hand, depending on the specific role, leaders may have a much broader scope of responsibility for performance of entire units including the department, service line, and health system. The leader is held accountable for the performance of these units as well as the individual actions of followers as prescribed by the code of conduct or human resource requirements and rules. In addition, the leader also has responsibility to the customers and patients.

A responsible leader makes reasonable efforts to prevent unacceptable employee-employee interactions and unethical behavior. Many of these occurrences are guided by institutional policies. For example, do you think a chair of surgery is accountable for unprofessional conduct of a physician on the trauma service? In this circumstance the scope of accountability is very broad. It is probably unreasonable that the chair is personally responsible for the unprofessional interaction, but he or she is responsible to the organization to be certain the faculty is aware of institutional policies guiding employee interactions and that appropriate corrective measures are taken to avoid future instances. Another example is when a faculty member publishes falsified scientific data. Is the chair or dean personally accountable for this occurrence? Many would argue no and some yes. However, the leader has an obligation to the organization and the scientific community to establish and maintain a culture of research integrity. This means having educational and compliance programs in place to ensure the faculty are aware of the personal ethical obligations for performing research. The individual faculty member is responsible for his or her actions. Contrarily, if the leader is aware of professional misconduct or research misconduct and takes no corrective action, then he or she is accountable for not meeting the obligations set forth by the organization. The leader is responsible for her or his actions or inaction set forth in the organizational policies and procedures. Other examples could be cited in the area of Medicare fraud and responsible billing practices. The leader is responsible for assuring that compliance programs are in place, that physician billing activity is monitored, and that corrective action plans are in place and used to intervene on individual physician billing practices. It follows that he or she is therefore accountable to the organization if there are lapses in such programs, yet may have little or no accountability for an individual physician's non-compliance. However, should the leader not take action when a physician is non-compliant, particularly after repeated infractions, then could the leader possibly share in the accountability for such actions?

The leader has a responsibility to patients to assure them a safe environment, with competent physicians and nursing staff. It is essential that the ethical leader abide by the medical staff bylaws and organizational policies concerning credentialing. The wise leader will defer final recommendations concerning privileges and credentialing the physician leadership of the medical staff. To act alone is unwise and fraught with personal risk. Likewise the leader has the organizational responsibility to assure that the each faculty and staff member is compliant with the recommendations and processes of the institutional review board that oversees human research. Unethical leaders would support circumventing these processes for a variety of conflicted issues involving research programs.

Ethical Decision-Making

The leader will face ethical dilemmas. Neuroscientists point out that the process of making ethical decisions is a complex cognitive process that involves logic and reason as well as intuition and emotion.

As cited by Johnson [9, p.177], Powers and Vogel identified six factors that underscore organizational ethical decision-making [42].

1. Moral imagination
2. Moral identification and sorting out important issues, priorities, and competing values
3. Moral evaluation: using analytical skills to evaluate options
4. Moral tolerance: realizing and accepting that there will be moral disagreement and ambiguity
5. Moral integration: educating the leadership team to anticipate possible ethical dilemmas in day-to-day decisions and incorporating this into the organizational culture
6. Moral obligation: developing a sense of duty regarding ethical decision in order motivate moral decision-making

According to Johnson, Rest has proposed the most accepted model of moral decision-making. He opines that ethical action is the result of four psychological processes: moral sensitivity or recognition, moral judgment, moral focus and motivation, and moral character [9, p. 177, 43].

Moral Sensitivity

Moral sensitivity is defined as being aware of ethical issues. Being insensitive to ethical situations can lead to moral breeches. As an example, Johnson cites the decision of the safety committee at Ford Motor to not repair the trouble-prone fuel tank on the Pinto automobiles as members apparently saw no problem saving money rather than lives. The fix would have cost $11 per vehicle. The Pinto was forcibly recalled by the National Highway Traffic Safety Administration in 1978 [9]. Another example of moral insensitivity is reflected in Case Study 3 concerning a meningitis outbreak related to compounding of steroids for spine injections, affecting over 800 patients and leading to 76 deaths. An example of heightened ethical sensitivity is demonstrated in

Case Study 4 on the Tylenol poisonings (1982) caused by potassium cyanide placed in Tylenol Extra Strength capsules by a worker in a Chicago drug store. The response of the CEO of Johnson & Johnson, James Burke, was to order the removal of all bottles of Tylenol Extra Strength (31 million) from the market at a cost of $ 100 million and only replaced them when they had developed a tamper-proof container.

Being sensitive to one's emotional intuition or gut reaction can heighten ethical sensitivity and moral imagination. This is at the core of ethical decision-making. If you do not think about the ethical ramifications of a decision, you will miss opportunities to make morally correct decisions. Another method to help frame the dilemma is the newspaper test. Imagine how the headlines would read in your local newspaper should there be an ethical breach driven by financial benefit for your hospital or department. If it feels bad, it will be bad and likely worse. In these situations it may be best to hit the pause button and rethink the ethical ramifications of the decision.

Moral Judgment

Moral judgment is the decision about which decision is right or wrong. Developing moral judgment is a complex cognitive process that involves progression from self-centered views of morality to broader definitions. Most leaders rely on rules and regulations when considering ethical decisions. Creating and sustaining an ethical culture facilitates ethical decision-making. Sound ethical decisions require education of the leadership, developing and encouraging a broader perspective and understanding moral principles. It is also essential to consider emotional and ethical blind spots and biases. Without acknowledging these, efforts to think and act ethically will be overwhelmed and ineffectual.

Moral Focus

Moral focus consists in developing a sense of ownership and motivation to make ethical decisions. Such focus is derailed by self-interest and hypocrisy. Incentives and positive emotions are essential to developing mature motivations. Disincentives will undermine any attempt to create a safe ethical environment and culture. One way to help foster this environment is the development of an anonymous ethics hotline. The followers need to understand that leadership takes ethical considerations seriously and want to know about any perceived violations. Having a confidential and no retaliation hotline can strengthen an ethical culture in an institution. People who witness ethical breeches should be free to bring them to the attention of the leadership without fear of retaliation affecting job and possible promotions. This will enhance all of the dimensions of an ethical culture including moral decision-making.

Moral Character

Moral character is required of the leader. Earlier in this chapter, we outlined the essential elements of moral character including courage,

---

Case Study 3: Meningitis Outbreak [46]

A New England Compounding Center meningitis outbreak which began in September 2012 sickened over 800 individuals and resulted in the death of 76. In September 2012, the Centers for Disease Control and Prevention, in collaboration with state and local health departments and the Food and Drug Administration (FDA), began investigating a multistate outbreak of fungal meningitis and other infections among patients who had received contaminated steroid injections from the New England Compounding Center (NECC) in Framingham, Massachusetts. The NECC was classified as a compounding pharmacy. Such pharmacies are authorized to combine, mix, or alter ingredients to create specific formulations of drugs to meet the

specific needs of individual patients and only in response to individual prescriptions. In October 2012, an investigation of the NECC revealed the company had been in violation of its state license because it had been functioning as a drug manufacturer, producing drugs for broad use rather than filling individual prescriptions. In December, federal prosecutors charged 14 former NECC employees, including President Barry Cadden and pharmacist Glenn Chin, with a host of criminal offenses. It alleged that from 2006 to 2012, NECC knowingly sent out drugs that were mislabeled and unsanitary or contaminated. The incident resulted in numerous lawsuits against NECC. In May 2015, a $200 million settlement plan was approved that set aside funds for victims of the outbreak and their families.

Study Questions
What type of leadership was at work in NECC?
If you were an employee at NECC, how would you have felt?
If you were a board member of the company, what actions would you have taken?
If you were a physician that ordered the injectable for NECC and administered it to a patient, what would be your communication plan with the involved patients?
What is your ethical obligation to a patient who contracted meningitis after you administered the injectable?

Case Study 4: The Tylenol Poisonings [47, 48]

Following the Tylenol killings in Chicago in 1982, Johnson & Johnson found itself in a precarious position. One of its leading products had killed seven people in a relatively small area. The response of Johnson & Johnson to the Tylenol poisonings in

1982 has received accolades as an example of how leaders should respond to a crisis impacting the public.

On September 29, 1982, 12-year-old Mary Kellerman died from taking one Tylenol Extra Strength capsule after waking up sick. Later that morning, Adam Janus took Tylenol and died shortly thereafter. To cope with their grief, Adam's brother, Stanley, and sister-in-law, Theresa, ingested Tylenol from Adam's bathroom, both dying within 48 hours. Tampered Tylenol would claim the lives of Mary Reiner, Paula Prince, and Mary McFarland, bringing the total death toll to seven people in the Chicago area.

Two firefighters called in to report a possible connection between the deaths and Tylenol, launching an immediate investigation. Laboratory tests showed that the victims had unknowingly ingested Tylenol Extra Strength capsules laced with potassium cyanide (KCN). KCN is an odorless, colorless substance that looks much like granulated salt or sugar. However, KCN can cause death within 45 minutes of exposure, commonly from cardiac arrest. Most of the victims had a cardiac arrest. Later reports said that the killer took several bottles of Tylenol off various drug store shelves, emptied the acetaminophen (the active ingredient in Tylenol) out of some capsules, filled them with KCN, and replaced the bottles back on the shelves. The killer used 65 mg in each tainted capsule. This was over 10,000 times the needed dose to kill a human being [49].

The company had never established a permanent public relations department other than an advertising and marketing division. Johnson & Johnson made several key decisions in response to this crisis. On October 5, 1982, 7 days after the first reported death, Johnson & Johnson issued a nationwide recall of all Tylenol Extra Strength capsules.

This included over 31 million bottles at an estimated retail value of over $100 million [48]. Their market share collapsed practically overnight from 35% down to 8%. By making this decision, Johnson & Johnson showed that they were not willing to take any risks with the public's safety, even if it cost them millions of dollars.

These incidents led to reforms in the packaging of over-the-counter medications and to federal anti-tampering laws. The actions of J&J to reduce deaths and warn the public of poisoning risks have been lauded as an exemplary public relations response to such a crisis.

In the days following the first death, J&J set up two 1-800 hotlines to deal with the massive panic and interest. One was set up for the general public for answering questions and fielding concerns of Tylenol users. The other was for news organizations that used a daily pre-recorded message with updated statements from the company. To more efficiently reach the public, CEO James Burke worked with the national television news shows such as 60 minutes and talk shows like The Phil Donahue Show. Using television as a communication vehicle allowed the public to put a face to the problem. This enabled J&J to gain credibility by shortening the distance J&J was putting between the company and the crisis.

In the weeks that followed, Johnson & Johnson continued making the good decisions working toward recovery. They worked closely with the Chicago Police Department, FBI, and FDA to find suspects. Customers were able to exchange Tylenol capsules for caplets, which are more difficult to tamper with. On November 11, 1982, Tylenol was reintroduced with a new, triple-sealed package. Tylenol became the first product in the industry to use a tamper-resistant packaging. The new sealing design made it nearly impossible for someone to tamper with the contents without being noticed.

Study Questions

What were the ethical decisions that J&J made?

Complete a teleological and deontological assessment of J&J's response to the incident.

Complete a utilitarian analysis. Was the J&J decision morally right?

What virtues were displayed by Johnson & Johnson CEO James Burke?

Compare and contrast this case with the case of NECC.

integrity, humility, optimism, justice, and compassion. These natural or learned virtues help the leader be attentive to the needs of others rather than personal priorities. The leader should be self-reliant on her or his ethical duty and not be persuaded by others to move from that point. Leaders driven by duty make and carry out ethical decisions based on a sense of loyalty and responsibility to followers and customers.

Ethicists Rushworth Kidder and Laura Nash have described ethical decision paradigms that will give the leader a process for making moral decisions [9, 44, 45]. Please see Table 2 and Table 3.

Table 2 Checkpoints in ethical decision-making (Kidder [44])

Recognize that there is a problem
Determine who the actor is
Gather the relevant facts
Test for right versus wrong issues
Test for right versus right issues
Apply relevant ethical standards (justice, fairness, utilitarianism, altruism)
Look for alternatives if values are irreconcilable
Make the decision
Revisit and reflect on the decision

Adapted from Kidder [44]: Fireside as cited in Johnson CE. Meeting the ethical challenges of leadership. 6th ed. Thousand Oaks, CA: SAGE; (2018). pp. 189–190

Table 3 Questions to guide ethical decision-making (Nash [45])

Have you defined the problem accurately?
How would you define the problem from the other side of the issue?
How did the situation occur?
To whom and to what do you give your loyalties to a person and an organization
What is the intention in making the decision?
How does the intention compare with the likely results?
Who could be injured by the decision or action?
Can you engage the parties in discussion before making a decision?
Are you confident that the position you take will be valid over a long period of time?
Could you comfortably disclose your decision to your superior, board of directors, Dean, etc.?
What is the symbolic potential of the decision if understood? Misunderstood?
Under what conditions would there be exceptions to the decision?

Adapted from Nash [45], as cited in Johnson CE. Meeting the ethical challenges of leadership. 6th ed. Thousand Oaks, CA: SAGE; 2018

Key Summary Points
- Ethics is critical to leadership.
- Ethical leadership is defined by a leader's actions and character.
- Balanced management of power, privilege, information, consistency, loyalty, and responsibility is the key to ethical leadership.
- The process of ethical decision-making involves moral imagination and considering the ethical dimension in each and every decision.

Summary

This chapter starts with an overview of real-life occurrences that face leaders and the sequela. The challenge to the surgical leader or for that matter a leader in any profession or business is to avoid bad outcomes to such occurrences by building a strong ethical organization and applying the ethical leadership principles reviewed in this chapter. Ethics is fundamental to leadership.

In order to know good leadership, we reflected on and examined bad leadership and the need to avoid the urge to go to the dark side of leadership, not infrequently driven by organizational greed or personal gain. The ethical leader must trust that their character is good, be self-reliant and unwavering in their duty, and make decisions that meet the organization's social contract and benefit the followers and customers of the organization. The leader must also be accountable to the performance metrics of the organization and the board or higher authority. At times there will be friction between these two responsibilities. At times there will be terrible events that you must manage in a socially responsible manner and make decisions that will negatively impact your bottom line or the financial aspects of your health-care organization. Whether a health system, medical school, department, or division, maintaining the unit's public reputation and trust should be the most important outcome by which your success is measured. Complex situations in which these dilemmas occur are where a moral imagination and consideration of the ethical dimension in each and every decision is so important.

We reviewed the six major challenges of leadership which are the management of power, privilege, information, consistency, loyalty, and responsibility. In addition, we explored the ethical decision-making process. Finally, we examined several case studies that the reader could delve into to apply the principles touched on in this chapter.

The ethical challenges of leadership are a complex topic. For those who seek to study this topic more, I highly recommend, Craig E. Johnson's book *Meeting the Challenges of Ethical Leadership* [9] which served as a source for much of the information in this chapter.

References

1. Emerson RW. Self-Reliance in Essays 1st and 2nd Series 1920; London and Toronto : JM Dent and Sons LTD. New York, NY: EP Dutton and Co.
2. http://fortune.com/2017/10/20/equifax-breach-credit/. Accessed 13 Nov 17
3. http://money.cnn.com/2017/09/12/news/companies/equifax-pr-response/index.html?iid=EL. Accessed 13 Nov 17.
4. http://www.espnfc.com/blog/fifa/243/post/2630853/fifa-timeline-blatter-and-platini-banned-more-arrested/. Accessed 13 Nov 17.
5. https://www.si.com/college-basketball/2017/09/29/what-we-know-about-each-school-fbi-investigation. Accessed 13 Nov 17.
6. http://www.tampabay.com/news/business/corporate/should-the-former-wellcare-ceo-convicted-of-fraud-be-able-to-cash-in-stock/2279153. Accessed 13 Nov 17.
7. Stewart JB. Blind Eye: how the medical establishment let a doctor get away with murder. New York, NY: Simon and Shuster, Inc; 1999.
8. Bass BM. Bass and Stogdill's handbook of leadership. 3rd ed. New York, NY: Free Press; 1990.
9. Johnson CE. Meeting the ethical challenges of leadership. 6th ed. Thousand Oaks: SAGE; 2018.
10. https://www.cbsnews.com/news/las-vegas-shooting-victims-names-latest-list/. Accessed 6 Nov 2017.
11. Ferreres AR, Angelos P, Singer EA. Ethical issues in surgical care. Chicago: American College of Surgeons; 2017.
12. Yukl G. Leadership in organizations. 8th ed. Upper Saddle River: Pearson/Prentice Hall; 2012.
13. Brown ME, Trevino LK. Socialized charismatic leadership, values congruence and deviance in work groups. J Applied Psychology. 2006;91:954–62.
14. Troyer J. The classical utilitarians: Bentham and Mill. Indianapolis: Hackett; 2003.
15. CE J. Meeting the ethical challenges of leadership. 6th ed. Thousand Oaks: SAGE; 2018. p. 148.
16. Kanungo RN, Mendonca M. Ethical dimensions of leadership. Thousand Oaks: SAGE; 1996.
17. CE J. Meeting the ethical challenges of leadership. 6th ed. Thousand Oaks: SAGE; 2017. p. 162.
18. CE J. Meeting the ethical challenges of leadership. 6th ed. Thousand Oaks: SAGE; 2018. p. 150.
19. CE J. Meeting the ethical challenges of leadership. 6th ed. Thousand Oaks: SAGE; 2018. p. 151.
20. Provis C. Virtuous decision making for business ethics. J Bus Ethics. 2010;91:3–6.
21. Sim M. Remastering Morals with Aristotle and Confucius. Cambridge, UK: Cambridge University Press; 2007.
22. Velasquez MG. Business ethics : concepts and cases. 3rd ed. Englewood Cliffs: Prentice Hall; 1992.
23. Northouse PG, editor. Leadership: theory and practice. 6th ed. Thousand Oaks: Sage; 2013.
24. Johnson CE. Meeting the ethical challenges of leadership. 6th ed. Thousand Oaks: SAGE; 2018. p. 72.
25. Johnson CE. Meeting the ethical challenges of leadership. 6th ed. Thousand Oaks: SAGE; 2018. p. 75.
26. Dirks KT. The effects of interpersonal trust on work group performance. J Appl Psychol. 1999;84:445–55.
27. Resick CJ, Dickson MW, Smith DB. Leaders, values and organizational climate: examining leadership

strategies for establishing an organizational climate regrading ethics. J Bus Ethics. 2004;55:223–24128.

28. Victor B, Cullen JB. A theory and measure of ethical climate in organizations. Greenwich: JAJ Press; 1990.

29. Johnson CE. Meeting the ethical challenges of leadership. 6th ed. Thousand Oaks: SAGE; 2018. p. 330.

30. Palmer P. In: Spears LC, editor. Insights on leadership, service stewardship, and servant-leadership. New York, NY: John Wiley; 1996). Leading from within. p. 197–208.

31. Kellerman B. Bad leadership: what it is, how it happens, why it matters. Boston, MA: Harvard Business School Press; 2004.

32. Erickson A, Shaw JB, Agabe AZ. An empirical investigation of the antecedent behaviors and outcomes and bad leadership. J of Leadership Studies. 2007;1:26–43.

33. Einarsen S, Aaland MS, Skogstad A. Destructive leadership behavior: a definition and conceptual model. Leadership Quarterly. 2007;18:207–2016.

34. French RP, Raven B. The bases of social power. In: Cartwright D, editor. Studies in social power. Ann Arbor: University of Michigan Institute for Social Research; 1959. p. 150–67.

35. Einarsen S, Hoel H, Zaptf D, Cooper CC. The concept of bullying and harassment at work : the European tradition. In: Einarsen S, Hoel H, Zaptf D, Cooper CL, editors. Bullying and harassment in the workplace : developments in theory, research and practice. 2nd ed. Boca Raton: CRC Press; 2011. p. 3–40.

36. Hornstein HA. Brutal bosses and their prey. New York, NY: Riverhead; 1996.

37. For highest paid CEOs the party goes on, New York Times.com Krantz M, (2015 December 9).

38. Sandler M (2015) CEO pay soars for top not for profits. Modern Health Care.

39. Wagner D (2014) VA scandal audit : 120,000 veterans experience long waits for care. The Arizona Republic.

40. Opinion Medicine. 2005. https://today.duke.edu/2005/07/dzauoped.html/. Accessed 6 Nov 2017.

41. Schriesheim CA, Castor SL, Cogliser CC. Leadermember exchange (LMX) research : a comprehensive review of theory, measurement, and data-analytics practices. Leadership Quarterly. 1999;10:63–114.

42. Powers CW, Vogel D. Ethics in education of business managers. Hasting-on-Hudson, NY: Institute of society, Ethics, and the Life Sciences; 1980.

43. Rest JR. Moral development : advances in research and theory. New York, NY: Praeger; 1986.

44. Kidder RM. How good people make tough choices. Resolving the dilemmas of ethical living. New York, NY: Fireside; 1995.

45. Nash LL. Ethics without the sermon. In: Andrews KR, editor. Ethics in practice. Managing the moral corporation. Boston, MA: Harvard Business School Press; 1989. p. 243–57.

46. https://en.wikipedia.org/wiki/New_England_Compounding_Center_meningitis_outbreak. Accessed 31 Oct 2017 "Pharmacist at center of 2012 fungal meningitis outbreak sentenced to 9 years in prison". Fisher Broyles LLP. Retrieved 28 July 2017.

47. Market, H. How the Tylenol murders of 1982 changed the way we consumer medication. Health 2014. https://www.pbs.org/newshour/health/tylenol-murders-1982 Accessed 31 Oct 2017.

48. https://en.wikibooks.org/wiki/Professionalism/Johnson_%26_Johnson%27s_Response_to_the_1982_Tylenol_Poisonings. Accessed 12 Oct 2017.

49. Emsley J. Molecules of murder: criminal molecular and classic cases. Cambridge, UK: Royal Society of Chemistry; 2008. p. 174.

Suggested Reading

Ferreres AR, Angelos P, Singer EA (2017) Ethical issues in surgical care. Chicago, Illinois: American College of Surgeons. Dr. Ferreres and colleagues summarize many of the issues in ethical surgical care. This book is important for ethical surgical leadership. This includes a tool box for surgical ethics. Important considerations are that of ethical issues in surgical intervention, conflicts of interest, and relationships with industry and ethical issues in the mentor–mentee relationship.

Johnson CE (2018) Meeting the ethical challenges of leadership (6th ed.) Thousand Oaks, Ca.: SAGE. Craig Johnson is an authority on the ethical challenges of leadership. This sixth edition text is a guide to navigating the issues in ethics as they occur in day-to-day leadership. The scope is all areas of leadership. Many case studies are presented to assist the reader in further developing their skills as an ethical leader.

https://www.rcseng.ac.uk/library-and-publications/college-publications/docs/surgical-leadership-guide/. The link above is to the guide to surgical leadership published by the Royal College of Surgeons of England. This covers all the essential elements of an effective surgical leader. It is highly recommended.

What Is Surgical Professionalism?

Steven M. Steinberg and Andrew L. Warshaw

Key Points
- Professionalism is highly valued by both patients and surgeons and is a bedrock of the doctor-patient relationship.
- Understanding professionalism requires recognition of unprofessional behaviors and actions.
- Keeping a professional stance requires facing the challenges from circumstances in surgical practice and in the evolving health-care environment.

Professionalism is and has been, inarguably, the most important characteristic valued by both patients and surgeons. If a surgeon is not perceived as behaving professionally, a patient's trust in the surgeon is much more likely to be called into question. This chapter deals with the topic of professionalism in the practice of surgery, challenges to surgeons' professionalism,

S. M. Steinberg (✉)
Division of Trauma, Critical Care and Burn,
The Ohio State University, Columbus, OH, USA
e-mail: steven.steinberg@osumc.edu

A. L. Warshaw
Massachusetts General Hospital and Partners
HealthCare, Massachusetts General Hospital,
Harvard Medical School, Boston, MA, USA
e-mail: awarshaw@partners.org

and our suggestions on meeting those challenges.

Even before modern surgeons evolved from our European predecessors, professionalism has been important for those practicing the art of surgery. Guy de Chauliac [1], sometimes referred to as the father of surgery, completed his seven-volume work titled *Chirurgia Magna* in the 1360s, and it remained the preeminent European text on surgery for five centuries. Earlier in his career, he wrote what has been translated to be "What a Surgeon Ought to Be." It is worth reading in its entirety:

The conditions necessary for the surgeon are four: First, he should be learned; second, he should be expert; third, he must be ingenious; and fourth, he should be able to adapt himself. It is required for the first, that the surgeon should know not only the principles of surgery, but also those of medicine in theory and practice; for the second, that he should have seen others operate; for the third, that he should be ingenious, of good judgment and memory to recognize conditions; and for the fourth, that he be adaptable and able to accommodate himself to circumstances. Let the surgeon be bold in all sure things, and fearful in dangerous things; let him avoid all faulty treatments and practices. He ought to be gracious to the sick, considerate to his associates, cautious in his prognostications. Let him be modest, dignified, gentle, pitiful, and merciful; not covetous nor an extortionist of money; but rather let his reward be according to his work, to the means of

A. R. Ferreres (ed.), *Surgical Ethics*, https://doi.org/10.1007/978-3-030-05964-4_6

the patient, to the quality of the issue, and to his own dignity.

It is easy to recognize many of the modern-day precepts of professionalism in Guy de Chauliac's assertion. If one of the preeminent surgeons of the day actually spelled out these characteristics in detail, the implication is that there must have been practitioners who did not follow these principles. One can infer that the perception of our professionalism was of great importance to leading surgeons, especially to distinguish themselves from the common tradesman of the day.

Barber surgeons were first recognized in medieval times and made their first appearance in medieval European monasteries at approximately 1000 AD to assist monks in maintaining their required tonsure. These early surgeons would also perform other minor procedures such as bloodletting and tooth pulling. Over the next few hundred years, their surgical repertoire expanded as very few physicians performed operative procedures. They were, however, viewed as technicians and were looked down upon by the better-educated physicians of the day. Subsequently surgery and surgeons in Europe took diverse paths in development.

In Paris, the surgeons' organization was established in 1210 AD when the College of St. Cosme was founded. The College's members were divided into two groups, one of whom wore long robes and the other short. Only the long-robed, who were considered to be physicians, could perform surgery. A three-way conflict between physicians, surgeons, and barbers ensued over who would control surgery. When anatomy was introduced into the curriculum, it created even more confusion as only physicians could supervise the dissections, while the surgeons actually wielded the knives. In 1516, the conflict ended when the surgeons yielded authority to the physicians, at least in part because of a shared distaste of the barbers.

Italy took a different tack. Because the medical schools in Salerno, Bologna, and Padua trained physicians to operate, no division between physicians and surgeons developed. The medical school in Florence kept physicians and surgeons as separate professions but, in 1349, instituted the Florentine statute that gave the barbers secondary status compared to physicians.

In England, barbers and surgeons were members of separate guilds. The Fellowship of Surgeons was founded in 1368, and the Company of Barbers was established in 1376. King Henry VIII merged them in 1540 into the United Barbers-Surgeons Company. This fact in itself may indicate the low esteem of surgeons at that time! Nonetheless, at that time, surgeons on the mainland like Ambroise Paré were elevating the work of surgeons above that of the common barber, and eventually barbers were banned from performing operative procedures except for bloodletting and tooth pulling. In 1745, due to the separation in practices of barbers and surgeons that had evolved, the surgeons left the United Barbers-Surgeons Company and established the Company of Surgeons. In 1800, a royal charter was granted to this Company, which became the Royal College of Surgeons of England. Similar colleges arose in Scotland and Ireland as well as other parts of the British Empire. A common thread was the evolution of surgery from a trade to a profession and, with it, the codification of professional values.

The *Merriam-Webster* dictionary defines profession as "a calling requiring specialize knowledge and often long and intensive academic preparation" and a professional as "characterized by or conforming to the technical or ethical standards of a profession; exhibiting a courteous, conscientious, and generally businesslike manner in the workplace." Souba summarized the meaning of profession as "a collegial discipline that regulates itself by means of mandatory, systemic training. It has a base in a body of technical and specialized knowledge that it both teaches and advances; it sets and enforces its own standards; and it has a service orientation, rather than a profit orientation, enshrined in a code of ethics. To put it more succinctly, a profession has cognitive, collegial, and moral attributes." These qualities are well expressed in the familiar sentence from the Hippocratic oath: "With purity and with holiness I will pass my life and practice my Art." [2]

It is important to distinguish our profession from a vocation. The most commonly held definition of vocation is simply the work in which a person is employed, a means of making a living. However, vocation can also be defined as a "calling," which implies that it may be more than just a job. While physicians and surgeons certainly are employed to care for the sick and physicians in the United States are well rewarded with a substantial income for their work, most did not become physicians for pecuniary purposes. Medical school applicants commonly indicate that they wish to become a doctor for altruistic and humanistic reasons. Indicating that passion, William Osler stated "You are in this profession as a calling, not as a business; as a calling which extracts from you at every turn self-sacrifice, devotion, love and tenderness to your fellow man. We must work in the missionary spirit with a breath of charity that raises you far above the petty jealousies of life." [3] It seems very clear that the standards of a profession are much higher than just those that define a "job." It is those standards that elevate professions above mere vocations. It is also those standards that are being continuously challenged.

Nonetheless, there are forces at play that threaten to disrupt our professionalism, some internal to medicine and surgery and others external. Possibly the most important force is the change in our society and more specifically the rate of change that has created stresses. Challenges to our professionalism and professional standing include a philosophical and attitudinal change from paternalism to patient autonomy, the medical malpractice crisis, the rise of entrepreneurship in medicine with its potential for conflict of interest, the commercialization of health care, the commoditization of surgical care, the introduction of new technology, and more overt competition among health providers and systems. We will address each of these and offer possible approaches to avoid the loss of professionalism in surgery. We must be clear that we do not believe that the changes in our health-care environment and practice are necessarily bad for professionalism. None of these changes are to be feared and many have a positive flip side. It is simply important to understand them in order to adapt effectively.

Paternalism Versus Patient Autonomy

The rise of patient autonomy in medical decision-making is a relatively recent occurrence. Both of this chapter's authors' careers began at a time when it was more common for a patient to respond to a recommendation with "whatever you say. You are the doctor" rather than a request for more information in order that the patient could make a truly informed decision. The arguments in favor of paternalism have been twofold: the comprehension of the cognitive and technical aspects of surgical care is beyond the typical layperson, and the doctor believes he/she best understands what is in the patient's best interest. At its extreme, the paternalistic approach allows the patient very little input into decisions, and the physician proceeds with the treatment plan he/she thinks is best. Even the American Medical Association, in its 1847 Code of Ethics, states: "The obedience of a patient to the prescriptions of his physician should be prompt and implicit. He should never permit his own crude opinions as to their fitness, to influence his attention to them." [4] There is no requirement for the physician to explore the patient's circumstances, culture, or preferences. After the appalling examples of the human experiments by the Nazis during World War II and the Tuskegee experiments in which African-American men were observed instead of treated for syphilis so that the natural history of the disease could be determined, strict guidelines have been developed for human experimentation and informed consent. These discussions have helped fuel the rise of patient autonomy as an ethical principle. Autonomy calls for respect and observance of the decisions of the patient regardless of how faulty, deficient, or dangerous those positions might seem, as long as the patient is competent.

Some physicians have had a difficult time adjusting to the change in these ethical doctrines, and some have resisted them. The outcome may be a degradation of the doctor-patient relationship such that the physician presents the patient with a menu of choices, but without explaining the pros and cons of each or offering alternatives. This incomplete presentation leaves the patient with autonomy on personal decisions but without a proper foundation. The surgeon at this extreme becomes a technician rather than an advisor. More recently shared decision-making, in which the surgeon seeks to understand the factors driving the patient's views and to present the risks and benefits germane to that individual, has gained increasing acceptance and implementation. Atul Gawande, in his book *Being Mortal: Medicine and What Matters in the End*, illustrates the issue very well in a series of examples that share a common thread; it is only after the doctor and patient truly have an honest conversation about what is important to the patient that the doctor can help construct a treatment plan that will assist the patient in achieving the goals and endpoints that embody the patient's wishes [5]. For example, does the patient prefer to undergo a risky operation for a small chance of cure? A mastectomy vs a lesser excision and radiation? Prostatectomy vs observation for low-grade cancer? In the 2016 AMA Code of Ethics, the message has changed completely in comparison with the 1847 version. It now states "The health and well-being of patients depends on a collaborative effort between patient and physician in a mutually respectful alliance." [6]

Payment Model

For most of the history of health care, the most care delivered by surgeons has been via a direct pay-for-procedure model. Social anthropologists have pointed out that unit-based payment can influence surgeons, unconsciously or consciously, to offer operations that may be of unproven value or even unnecessary. As Albert Jonsen mentioned, the conflict between altruism and self-interest is the one that characterizes the relationship between the physician and the patient (New Engl J Med 1983; 308: 1531–1535). Clearly, the provision of care that may not benefit a patient but stands to benefit the surgeon, whether intentional or not, subtracts from the professionalism of the surgeon. Any surgeon can think of procedures that were previously thought to be "mandatory" that now, after further study, have been shown to be of doubtful benefit. In general surgery, inguinal hernia repair is a good example. In the past, the mere presence of an inguinal hernia was believed to be an indication for inguinal herniorrhaphy. We now know that many patients, particularly elderly patients, who have no or relatively minor symptoms can be safely observed and many will never require an operation at all. The role of appendectomy in both children and adults has also been called into question, and there are a number of studies that indicate that in uncomplicated acute appendicitis, the majority of patients will recover without surgery. Yet some surgeons – albeit in the belief they are delivering the right care – may be unconsciously influenced by their own financial benefit to operate.

In order to reduce the delivery of care that is unnecessary or of unproven benefit, as well as to reduce overall cost of health care, new payment models are being investigated. In the United States, accountable care organizations (ACOs) are being tested. As opposed to providing payment for specific services, ACOs pay for populations of patients such that the economic benefits to doctors are greater for assuring health rather than treating illness. It is postulated that changing the payment model will reduce the amount of unnecessary care being provided. However, a word of caution is that changing the payment model in this way may eliminate one challenge to physician professionalism but may introduce a different, and just as negative, challenge. If we are financially incentivized to avoid unnecessary care, the unintended outcome may be to withhold indicated care. At the end of the day, professionalism means that the surgeon is obliged always to place the health and welfare of the patient first.

The Rise of Technology in Medicine and Surgery

Technologic advancements in both diagnostics and therapeutics in surgery have been occurring at a rapid pace [7]. Minimally invasive surgery, first in the form of laparoscopic surgery and more recently with robotic surgery, first burst upon the scene in the 1980s. Minimally invasive techniques such as these tempt both surgeons and patients unwisely to lower the bar for deciding on an elective operation. Computed tomography (CT) has been utilized commonly – and perhaps excessively – so that a patient admitted to the surgical service *without* a CT scan is almost a rarity. CT, magnetic resonance imaging (MRI), and other types of imaging often precede and even replace the performance of a complete history and physical examination. This technology, as wondrous as it is, has come at the expense of the "laying on of hands," the performance of a good history and physical examination, and the consequent development of the personal doctor-patient relationship that is the bedrock of the surgical professional.

Even the simple interposition of technology in between the physician and patient in the notification of test results challenges our professionalism and the doctor-patient relationship. Many health systems have developed protocols to notify patients of many of their test results by automatic email, without the requirement that a physician or any other sort of provider deliver the results and be available to explain them to the patient. We have assumed that the patient always will want a more rapid result, but do they? [8] Friedman suggests, and we agree, that what patients want is a "physician who actually cares" enough to explain results of test, their implications, and next steps [9].

The explosion of social media on the Internet has created its own challenges to professionalism. The posting of patient pictures or other identifiable protected health information online for any reason and the posting of inflammatory or derogatory information about people or institutions constitute unacceptable behavior. The permanent inability to retrieve or erase those postings has created a serious liability to one's reputation and career. A precept of medicine that dates to Hippocrates is the oath to keep patients' information as "holy secrets."

In a 2014 study, Langenfeld and colleagues reported on the use of social media by 996 surgical residents. Just under one-third had a publicly identifiable Facebook page [10]. They assessed each of the residents' Facebook pages for unprofessional posts such as depicting binge drinking, sexually suggestive photographs, and Health Insurance Portability and Accountability Act (HIPAA) violations of protected health information. Of the Facebook posts of those 319 residents, 12% contained blatantly unprofessional content, 14% were found to have potentially unprofessional content, and there was no difference in unprofessional behavior based on gender of resident posting the item. These data indicate that fully one-fourth of the next generation of surgeons have either behaved unprofessionally or gave that perception. In 2015, the same research group performed a similar study on the faculty of the abovementioned residents [11]. Of the 758 general surgery faculty identified, just under 26% had identifiable Facebook pages. Ten percent had potentially unprofessional postings, and 5% had clearly unprofessional postings. In the case of the faculty, clearly unprofessional postings were made only by the male surgeons, and those behaviors were more common in young male surgeons in their first 5 years of practice. In taking these two studies together, it is clear that age is inversely related to posting either potentially or clearly unprofessional material on Facebook. What is not so obvious is whether older age is associated with a lower risk of posting such materials because one learns to avoid inappropriate postings over time or whether younger surgeons are more likely to use social media, its use being associated with unprofessional behaviors. In the first case, there is hope that social media in the course of time may have a diminishing role in promoting unprofessional behavior with proper education of its users. The maxim of "don't send an email and don't post anything online that you wouldn't want your parents or your patients to see as the headline in tomorrow's newspaper" continues to be true.

Commercialization of Health Care

As health care has become a profitable commodity, many surgeons, and particularly those who are employed by health systems, have become viewed as high-priced technicians whose primary job is to produce relative value units (RVUs) and contribute to the health system's bottom line. For small and rural hospitals, the general surgeon is viewed as a vehicle for financial survival. The external pressure to produce clinical activity and increase the number of patients and operative procedures may introduce a conscious or subconscious motivation by administrators to require more cases, perhaps with questionable justification. While this behavior is clearly unconscionable and unprofessional, the pressure is real, especially for the employed surgeon who is receiving regular reports on personal productivity which determines compensation.

In all fairness, commercialization of health care has not been all bad. When viewed at a system level or higher, larger health-care systems may be better able to provide the right care, including subspecialty and multidisciplinary care, at the right time and at lower cost than can individual practitioners. Furthermore, larger systems have the capacity to measure and improve outcomes of care, including both quality and cost.

Appropriate Limitations to Practice and Responsibility to Keep Up To Date

A key tenet of professionalism is the obligation for self-improvement. Without continuous life-long learning, the surgeon of tomorrow will practice just the medicine of today – no better. A corollary is the voluntary confinement of one's practice to one's current expertise, not the broad content of remote training. Failure to recognize the sometimes insidious reduction in proficiency

as we age provides another trap if the surgeon continues to provide care for conditions better treated by other providers with greater skills.

Surgical practice is changing very rapidly as new technology and new knowledge emerge. What has been accepted truth is constantly being enlarged or discarded. Professionalism demands constant improvement through self-study, using tools such as the Surgical Education and Self-Assessment Program (SESAP) and Continuing Medical Education conferences. The assumption of current competence must be proven by specialty-specific Maintenance of Certification (MOC). We have a duty to refer patients whose disease process falls outside our current expertise to other surgeons who can better manage the patient's clinical problems.

Fear of Malpractice Litigation

The fear of litigation appears to have had a significant impact on surgeons' perception of professionalism, but the level of concern about medical malpractice liability has waxed and waned. Whereas for many years that concern ranked in the top four in the annual survey of the American College of Surgeons' Board of Governors, in 2015 medical liability fell to the 10th position of importance behind other problems such as the burden of the electronic medical record, process of MOC, and funding of graduate medical education. Nonetheless, the fear of litigation has been shown to drive some surgeons to perform unnecessary tests to cover all potential eventualities in case of litigation, expediency replacing professional judgment.

Failure of Systems-Based Practice

In a 2007 study performed by the American College of Surgeons, unprofessional practices fostered many malpractice claims [12]. Of 460 closed claims that were reviewed, 229 were attributed, at least in part, to technical errors, but the surgeon's technical ability was judged to be inadequate in only 11%. More commonly, unprofessional judgment related to technical competence contributed to the bad outcome for the patient: performing procedures outside of one's usual scope of practice; failure to refer cases that required skill outside the surgeon's abilities; failure to consult intraoperatively: failure to refer patients to a tertiary care center for problems that exceeded the abilities of the provider or institution to meet the patient's needs; failure of attending surgeons to properly supervise trainees; failure of trainees to ask for needed attending supervision; failure of aging surgeons to recognize their declining cognitive or technical skills and consequently to limit their scope of practice; and failure to introduce new technology into the practice. Other behavioral failures that contributed to bad patient outcomes and liability claims included failure to communicate with the patient and/or family (34%), pursue an abnormal symptom or test result (25%), recognize a postoperative problem (25%), or adequately assess a surgical problem preoperatively (19%). In 78% of the examined closed claims, at least one behavioral failure was identified [13].

Self-Regulation: The Code of Silence

Doctors are often hesitant to report or confront colleagues with their failings and errors. The tenet of self-regulation of a profession requires corrective action when there is recognition of a failure or deficiency in the practice of surgery. To meet professional responsibility, proper actions include communication with the relevant erring colleague, patient, family, and, when appropriate, the colleague's supervisor, institution, or even authorities.

As previously mentioned, medical errors or perception of negligence can lead to malpractice litigation, but professionalism dictates that a surgeon called to testify must be fair both to the patient and the defendant surgeon. An expert witness should be prepared to speak the truth for either side. Justice cannot be served if surgeons will only speak for the defense but decline to criticize another surgeon. When a surgeon is found guilty, that should be stated in order to protect the patient's rights. Professionalism requires that we act on behalf of the patient as well as our colleagues. According to Antiel and colleagues, the reasons for the reluctance of doctors to testify against other doctors, a "code of silence," fall into four main categories: error as a contested concept, the glass house effect, fear of retaliation, and the diffusion of responsibility. Ultimately, remaining silent is a selfish, self-protective, and unprofessional behavior [14].

Competition

Competition for patients, either between individuals or groups, can escalate to unprofessional behaviors. Competition between individual surgeons can lead to unsubstantiated injurious claims of substandard care. Even the peer review process has been used unfairly to defame a competitor with the goal of compromising that surgeon's standing and practice. Turf battles between subspecialties that perform the same or similar procedures (e.g., thyroidectomy by either endocrine surgeons or otolaryngologists; spine surgery by either neurosurgeons or orthopedic surgeons) have sometimes led to public competition with overt advertising or, worse, derogatory and unfounded claims of substandard quality of care or difference in outcome.

Conflict of Interest

Real and perceived conflicts of interest have been identified as major ethical and professionalism problems. The topic has been well studied, and it has been shown that even small inducements, such as ballpoint pens and coffee mugs, may influence physician prescribing and choices of treatment. Consequently, many institutions have restrictive policies related to pharmaceutical and device manufacturers and their attempt to bias physician opinion and institutional purchasing decisions. Some prohibit industry representatives from even entering the institution. Essentially all academic institutions and educational programs require physicians to disclose any relationship they have with relevant commercial interests. Decisions impacting patient care must not be determined by any consideration other than the welfare of the patient.

Solutions

A good starting place for understanding professional behavior is in the American College of Surgeons Fellowship Pledge [15], which states:

> Recognizing that the American College of Surgeons seeks to exemplify and develop the highest traditions of our ancient profession, I hereby pledge myself, as a condition of Fellowship in the College, to live in strict accordance with the College's principles and regulations.
>
> I pledge to pursue the practice of surgery with honesty and to place the welfare and the rights of my patient above all else. I promise to deal with each patient as I would wish to be dealt with if I were in the patient's position, and I will respect the patient's autonomy and individuality.
>
> I further pledge to affirm and support the social contract of the surgical profession with my community and society.
>
> I will take no part in any arrangement or improper financial dealings that induce referral, treatment, or withholding of treatment for reasons other than the patient's welfare.
>
> Upon my honor, I declare that I will advance my knowledge and skills, will respect my colleagues, and will seek their counsel when in doubt about my own abilities. In turn, I will willingly help my colleagues when requested.
>
> I recognize the interdependency of all health care professionals and will treat each with respect and consideration.

These precepts are further detailed in the 2003 American College of Surgeons Code of Professional Conduct [16]:

> *As Fellows of the American College of Surgeons, we treasure the trust that our patients have placed in us because trust is integral to the practice of surgery. During the continuum of pre-, intra-, and postoperative care, we accept the following responsibilities:*
>
> - *Serve as effective advocates of our patients' needs*
> - *Disclose therapeutic options, including their risks and benefits*

- *Disclose and resolve any conflict of interest that might influence decisions regarding care*
- *Be sensitive and respectful of patients, understanding their vulnerability during the perioperative period*
- *Fully disclose adverse events and medical errors*
- *Acknowledge patients' psychological, social, cultural, and spiritual needs*
- *Encompass within our surgical care the special needs of terminally ill patients*
- *Acknowledge and support the needs of patients' families*
- *Respect the knowledge, dignity, and perspective of other health care professionals*

Our profession also is accountable to our communities and to society. In return for their trust, as Fellows of the American College of Surgeons, we accept the following responsibilities:

- *Provide the highest quality surgical care*
- *Abide by the values of honesty, confidentiality, and altruism*
- *Participate in lifelong learning*
- *Maintain competence throughout our surgical careers*
- *Participate in self-regulation by setting, maintaining, and enforcing practice standards*
- *Improve care by evaluating its processes and outcomes*
- *Inform the public about subjects within our expertise*
- *Advocate for strategies to improve individual and public health through communication with government, health care organizations, and industry*
- *Work with society to establish just, effective, and efficient distribution of health care resources*
- *Provide necessary surgical care without regard to gender, race, disability, religion, social status, or ability to pay*
- *Participate in educational programs addressing professionalism*

This code of professional conduct emphasizes the following four aspects of professionalism: (1) a competent surgeon is more than a competent technician; (2) while ethical practice and professionalism are closely related, professionalism also incorporates surgeons' relationship with both patients and society; (3) unprofessional behavior must have consequences; and (4) professional organizations are responsible for fostering professionalism in their membership.

The Accreditation Council for Graduate Medical Education includes professionalism as one of their six key competencies along with patient care, medical knowledge, practice-based learning and improvement, interpersonal and communication skills, and systems-based practice. These six key competencies have filtered up and down the medical education chain of command to the point that they are expected of any medical student, resident or fellow, or practicing physician. Our medical students and trainees are routinely evaluated on these competencies. They have rightfully become an explicit part of our health-care culture.

From a practical perspective, what can an individual surgeon do to maintain his/her professionalism? Other than working to comply with the abovementioned precepts and code of behavior, there are two simple strategies. First, keep the doctor-patient relationship sacrosanct. From the first to the last encounter, treat patients as you would like to be treated – with respect, care, and consideration of their goals – and as an equal partner in their health care. Communicate with the patient the way you would want your doctor to communicate with you. Most of us expect the truth. Do not substitute technology for that relationship; a CT scan may help you make a diagnosis, but only you can explain the findings to the patient and discuss the implications of the findings. While health care has become big business in the United States and around the world, avoid the temptation to think of your patients in fiscal terms. Understand the health-care goals of your patients, and recommend only the pathway that will best help them achieve those goals. Second, treat everyone the same, regardless of who they are and whatever their social standing or resources. Our biggest reward as surgeons comes from the satisfaction of helping a patient maintain or regain health. That sensation is truly priceless.

In conclusion, while there are many external forces that compete with or impede our aspirations to behave as professionals, the choices that must be made to get us there are really our own. Being a professional requires continuous self-improvement; self-regulation keyed to our individual current competence, contributing to the equitable regulation of our colleagues and our surgical practices; and always putting the welfare of the patient before our own.

Concluding Remarks

- Changes in health-care delivery, the doctor-patient relationship, and society have created stresses that are challenging professional behavior.
- Being a professional requires continuous self-assessment and self-improvement.
- Being a professional requires self-regulation keyed to current competence.
- Being a professional requires the equitable self-regulation of the form, function, and practice of surgery.
- Principally, being a professional requires putting the welfare of patients first, treating them with care and respect as partners in understanding and achieving their goals.

References

1. Watters DA. Guy de Chauliac: pre-eminent surgeon of the middle ages. ANZ J Surg. 2013;83:730–4. https://doi.org/10.1111/ans.12349.
2. Adams F. The genuine works of Hippocrates, translated from the Greek. London: Balliere, Tindall and Cox; 1939.
3. Hinohara S, Niki H, editors. Osler's "Way of Life" and other addresses, with commentary and annotations. Durham: Duke University Press; 2001.
4. American Medical Association (1847). Code of Medical Ethics of the American Medical Association. https://www.ama-assn.org/sites/default/files/media-browser/public/ethics/1847code_0.pdf. Accessed 23 Oct 2017.
5. Gawande A. Being mortal: medicine and what matters in the end. New York: Henry Holt and Co.; 2014.
6. American Medical Association (2016). Code of Medical Ethics of the American Medical Association.

https://www.ama-assn.org/delivering-care/patient-rights. Accessed 23 Oct 2017.

7. Angelos P. Ethics and surgical innovation: challenges to the professionalism of surgeons. Int J Surg. 2013;11:S2–5. https://doi.org/10.1016/S1743-9191(13)60003-5.

8. Choudhry A, Hong J, Chong K, et al. Patients' preferences for biopsy result notification in an era of electronic messaging methods. JAMA Dermatol. 2015;151:513–21. https://doi.org/10.1001/jamadermatol.2014.5634.

9. Friedman EM. You've Got Mail. JAMA. 2016;315:2275–6. https://doi.org/10.1001/jama.2016.1757.

10. Langenfeld SJ, Cook G, Sudbeck C, et al. An assessment of unprofessional behavior among surgical residents on Facebook: a warning of the dangers of social media. J Surg Ed. 2014;71:e28–32. https://doi.org/10.1016/j.jsurg.2014.05.013.

11. Langenfeld SJ, Sudbeck C, Luers T, et al. The glass houses of attending surgeons: an assessment of unprofessional behavior on Facebook among practicing surgeons. J Surg Ed. 2015;72:e280–5. https://doi.org/10.1016/j.jsurg.2015.07.007.

12. Griffen FD, Stephens LS, Alexander JB, et al. The American College of Surgeons' closed claims study: new insights for improving care. J Am Coll Surg. 2007;204:561–9. https://doi.org/10.1016/j.jamcollsurg.2007.01.013.

13. Griffen FD, Stephens LS, Alexander JB, et al. Violations of behavioral practices revealed in closed claims reviews. Ann Surg. 2008;248:468–74. https://doi.org/10.1097/SLA.0b013e318185e196.

14. Antiel RM, Blinman TA, Rentea RM, et al. When a surgical colleague makes an error. Pediatrics. 2016;137(3):e20153828. https://doi.org/10.1542/peds.2015-3828.

15. American College of Surgeons (2017). American College of Surgeons Fellowship Pledge. https://www.facs.org/member-services/join/fellows/fellowreq. Accessed 10 Dec 2017.

16. American College of Surgeons (2003). American College of Surgeons Code of Professional Conduct. https://www.facs.org/about-acs/statements/stonprin#code. Accessed 10 Dec 2017.

Suggested Literature

Cullen MJ, Konia MR, Borman-Shoap EC, et al. Not All Unprofessional Behaviors Are Equal: The Creation of a Checklist of Bad Behaviors. Med Teach. 2016;39:85–91. https://doi.org/10.1080/0142159X.2016.1231917. A categorization of unprofessional behaviors and rank order of dimensions of professionalism.

Hochberg MS, Berman RS, Pachter HL. Professionalism in Surgery: Crucial Skills for Attendings and Residents. Adv Surg. 2017;51:229–49. https://doi.org/10.1016/j.yasu.2017.03.018. A set of bullet lists of skills that every practicing surgeon should consider and work to master.

McGinnis LS. Presidential Address: Professionalism in the 21st Century. Bull Am Coll Surg. 2009;94:8–18.. President McGinnis's address at the 2009 Convocation of the American College of Surgeons that highlights professionalism in surgery, provides past and current examples of highly professional surgeons, and acts as a call to action for the fellows of the American College of Surgeons.

Souba WW, Steinberg SM. Professionalism in surgery. ACS surgery: principles and practice. Elements of contemporary practice. New York: WebMD; 2008. p. 1–4.

Ethics in Academic Surgery

Charles W. Kimbrough and Timothy M. Pawlik

Key Points

- The multiple roles of academic surgeons can generate dual loyalties, where responsibilities to patients may conflict with responsibilities to research.
- Human subjects research must satisfy ethical conditions including voluntary informed consent, scientific value, and favorable risk-benefit ratios.
- Researchers have an ethical responsibility to maintain research integrity and avoid misconduct including falsification, fabrication, and plagiarism.
- All real or perceived conflicts of interest must be appropriately disclosed.
- Good mentorship and the cultivation of an ethical character are crucial to provide a moral compass in academic surgery.

C. W. Kimbrough
Department of Surgery, The Ohio State University, Wexner Medical Center, Columbus, OH, USA

T. M. Pawlik (✉)
Department of Surgery, The Urban Meyer III and Shelley Meyer Chair for Cancer Research, The Ohio State University, Wexner Medical Center, Columbus, OH, USA
e-mail: Tim.Pawlik@osumc.edu

Introduction

Academic surgery represents a challenging but rewarding career path, which is at the forefront of exciting new innovations and a rapidly evolving knowledge base. The pace of technological advance and discovery over the last century has been unlike any time in human history. In just over 60 years since Watson and Crick described DNA, we are on the verge of manipulating the human genome, and insights into molecular biology have led to impressive developments in precision therapy. Technical advances have widened the scope of what can safely be achieved in the operating room, all while minimizing the invasiveness of procedures. Nonetheless, many times these developments raise as many questions as answers. Important questions regarding what we ought to do for our patients and the role we have in society underscore many of these scientific and technical advances. At its core, surgery is still an exercise in applied ethics. To this point, advances in medicine often bring unanticipated ethical challenges without clear answers or precedent to guide behavior. Nonetheless, academic surgeons have the obligation to practice socially responsible science [1]. As a profession, surgeon-scientists are afforded a privileged place in society. Through prolonged study and training, we possess highly specialized knowledge and a skill set that affords us the ability to determine our own professional codes and standards. In return, the public places great trust in us to use

© Springer Nature Switzerland AG 2019
A. R. Ferreres (ed.), *Surgical Ethics*, https://doi.org/10.1007/978-3-030-05964-4_7

our skills and knowledge to the best benefit of society.

There exists a rich tradition and history on medicine's social responsibilities, with considerable focus on the physician-patient relationship. Some of the earliest recorded histories of medicine reveal a preoccupation with ethics, as both the Egyptians and Babylonians prescribed rules for physician behavior [2]. Most medical students still take some version of the Hippocratic Oath. Interestingly, most modern literature on the ethics of academic surgery has been published only over the last several decades. In turn, concepts of medical ethics have evolved over time with many discussions now focused on the core principles of bioethics that include autonomy, nonmaleficence, beneficence, and justice [3]. These principles can assist academic surgeons who have many roles, including clinician, scientist, administrator, and educator. Competing goals among these roles can introduce a number of ethical tensions and dual loyalties [4]. For example, as surgeons, we have a fiduciary duty to our individual patients, yet as researchers we have a duty to future patients and society as a whole. One can easily imagine similar ethical tensions emerging among other roles assumed by the academic surgeon. For example, the pressure to be academically productive ("publish or perish") may contribute to misconduct such as plagiarism, fraud, exploitation of subordinates, or any number of conflicts of interest [5]. Furthermore, administrative responsibilities lead to duties to the healthcare system, which must be balanced against the needs of individual patients.

For surgeons trained in evaluating objective data and making decisive judgments, the "subjectivity" and nuances of ethical reasoning to deliberate trade-offs and balance competing demands can represent a different challenge for the clinician. Ethical dilemmas may arise with no clear answer, and each side can be supported by equally valid arguments. Learning how to approach and manage these dilemmas is, however, critical for any surgeon-scientist. Even though certain ethical problems may present mutually exclusive options, most ethical dilemmas can still be approached in a systematic fashion (see Box 1). A solid understanding of existing precedents and guidelines is necessary to resolve many ethical challenges that arise in academic surgery. In this chapter, we explore specific facets of academic surgery, identify where ethical conflicts can arise, and discuss prior precedents, guidelines, laws, and other resources that can help guide ethical decision-making.

Box 1 A Method for Ethical Decision-Making
Many theories have been proposed to tackle ethical dilemmas, although there is no clearly superior approach. Shamoo and Resnik describe a practical method to approach ethical problems. While this particular approach does not need to be strictly applied and may vary depending on the particular circumstances, these steps can serve as a useful framework for approaching ethical problems.

1. Define the problem, question, or issue.
2. Gather relevant information.
3. Explore the viable options.
4. Apply ethical principles, institutional policies, or other rules or guidelines to the different options.
5. Resolve conflicts among principles, policies, rules, or guidelines.
6. Make a decision and take action.

Adapted from: Shamoo and Resnik [24].

Human Subjects Research

Historical Development

Research on human subjects can represent a fundamental conflict of interest for surgeon-scientists. As clinicians, surgeons have a primary responsibility to place the needs of each particular patient above all else. However, as scientists we seek to expand the overall understanding of disease and improve therapies to benefit society as a whole. The nineteenth-century story of Dr. William Beaumont provides a representative example of these dual loyalties. While a young military surgeon, Dr. Beaumont treated and saved the fur trapper Alexis St. Martin following a shotgun wound to the abdomen. When St. Martin developed a chronic gastrocutaneous fistula, Beaumont used St. Martin to conduct a series of experiments on human digestion. Beaumont greatly advanced the understanding of gastric physiology, ultimately publishing his work [6]. However, St. Martin became increasingly frustrated by Beaumont's experiments and made multiple attempts to dissolve their relationship. Similar examples from the nineteenth century stirred debate regarding the objectification of human patients for scientific research.

In the twentieth century, technological advances and the rise of the modern state amplified the tension between medical research and patients' rights. In particular, the topic of human subjects research was brought into focus by the war crimes that occurred under the Nazi regime in the 1930–1940s. Many of these atrocities were sanctioned by official government policies and occurred despite vigorous ethical debates at the time [7]. The subsequent Nuremberg trials led jurists to articulate a series of principles that protected human subjects involved in research. Many consider the Nuremberg Code as the beginning of modern bioethics. The Nuremberg Code's principal points emphasized that "voluntary consent of the human subject is absolutely essential" and any experiment should have a favorable risk-benefit ratio with "fruitful results for the good of society" [8]. In 1964 the Declaration of Helsinki further developed the ethics of human subjects research and addressed perceived shortcomings of the Nuremberg Code. Overall, the Declaration of Helsinki emphasizes "the health of the patient is the first consideration" and research "can never take precedence over the rights and interests of individual research subjects" [9]. The Declaration of Helsinki further outlines specific principles that promote the protection of patients, a favorable risk-benefit ratio in research, and independent review of research protocols.

Despite the principles outlined by the Nuremberg Code, a climate of unethical research persisted after World War II. To this point, in 1966 Dr. Henry Beecher published a landmark review of research studies that abused patient's rights. Egregious examples included the artificial induction of viral hepatitis in mentally challenged children and the injection of live cancer cells into patients without their knowledge [10]. A few years later, the revelation of the infamous 40-year-long Tuskegee Syphilis study led Congress to establish the National Commission for the Protection of Human Subjects of Biomedical and Behavioral Research. The Commission was given the task of developing guidelines for ethical research and published their results as the Belmont Report in 1979 [11]. The report clearly defined the boundary between clinical practice and research and conceptually linked ethical principles to facets of human subjects research in order to protect vulnerable populations. Specifically, the Belmont Report suggested respect for persons should guide informed consent, beneficence should drive risk-benefit analyses, and justice should underlie patient selection for research [12].

The principles outlined in the Belmont Report were converted into official federal policy by the Department of Health and Human Services through 45 CFR 46, known as the Common Rule. The Common Rule defines research as any systematic investigation designed to contribute to general knowledge and includes all human subjects research. In order to ensure that research on human subjects remains ethical, the Common Rule established external review of research protocols through institutional review boards (IRBs). IRBs consist of members of the research institution as well as the community and can only approve research if the following conditions are met: (i) risks to subjects are minimized, (ii) risks to subjects are reasonable in light of potential benefits, (iii) subjects are selected in an equitable manner, (iv) informed consent is obtained, (v) provisions exist for safety monitoring, (vi) data confidentiality will maintain subject privacy, and (vii) protections exist for vulnerable populations. The Common Rule outlines additional regulations that cover informed consent and documentation requirements. Researchers have an ethical duty not to proceed with any human subjects research without IRB approval. Once a study is underway, IRBs have the responsibility to monitor its progress and may suspend or modify the protocol if necessary. In turn, researchers are obligated to report any adverse or unanticipated events to the IRB.

Ethical Requirements for Clinical Research

Although the Common Rule establishes regulatory requirements for human subjects research, ethical research must satisfy more than a legal minimum. While IRB review and the Common Rule establish necessary conditions for research studies, bare adherence to these regulations may not be sufficient. Research studies occur within a variety of different backgrounds, where case by case differences challenge rote application of regulations. This distinction has led several authors to propose a generalized approach to clinical research ethics. For instance, Emanuel and colleagues proposed several ethical requirements for clinical research. In this conceptual model, seven key requirements establish a framework for evaluating the ethics of clinical studies (see Box 2). These requirements attempt to ensure that research subjects are not merely used, but respected, and that research projects contribute to the social good [13]. Similar to the Belmont Report, these requirements can be linked back to broad principles including respect for autonomy, justice, beneficence, and nonmaleficence. By anchoring research requirements to ethical principles, these guidelines provide a universal framework to approach the ethical dimensions of clinical research.

3. Fair subject selection – Vulnerable populations should not be targeted for risky research, nor should beneficial research favor privileged groups.
4. Favorable risk-benefit ratio – Risk to individual subjects must be minimized, while benefits are enhanced; the potential overall gain for society must outweigh a study's risks.
5. Independent review – Independent bodies not affiliated with the research must be available to review, approve, emend, or terminate a study.
6. Informed consent – Subjects must be informed of the research and give voluntary consent; consent may be withdrawn at any time.
7. Respect for enrolled subjects – Subjects should have appropriate privacy protections, the right to withdraw from a study, access to information regarding any results or new risks, and their well-being monitored.

Adapted from: Emanuel et al. [13].

Appropriate application of such requirements demands a necessary level of expertise. Researchers must not only be skilled in the appropriate research methodology and statistical analysis (see Data Management); surgeons should also understand the ethical dimensions to study design such as subject selection and risk-benefit ratios. For instance, randomized controlled clinical trials represent the gold standard for clinical research but require a significant knowledge of both methodology and ethics. For a clinical trial to be valid, it must be designed to address a state of "clinical equipoise." As defined by Freedman, clinical equipoise exists when there is "honest professional disagreement among expert clinicians about the preferred treatment" [14]. There are several implications to this definition. Once sufficient evidence accumulates to favor one treatment over another, clinical equipoise is disturbed, and it is no longer ethical to

> **Box 2 Seven Ethical Requirements for Clinical Research**
> 1. Value – Research must enhance health or knowledge.
> 2. Validity – Studies must adhere to a rigorous and scientifically valid methodology.

withhold the proven treatment. This determination can be heavily influenced by trial design and may require sophisticated analysis, often by independent review boards. In addition, patients must be clearly informed of the lack of professional consensus and be given the option to participate or decline. As Freeman notes, "We do not conscript patients to serve as subjects" [14].

Informed Consent

Informed consent remains a cornerstone of human subjects research. Just as research ethics evolved over the last 50 years, the process of informed consent has undergone similar revision. Traditional paternalistic conceptions of the physician-patient relationship have given way to increased emphasis on patient autonomy, transparency, and shared decision-making [15]. The overall process of informed consent should recognize the inherent worth of patients as moral agents and respects their right to independently decide the best course of action. In the research setting, subjects may need a significant amount of information to make a truly informed decision. Many patients are being asked to participate in clinical trials or studies at the edge of medical knowledge, involving issues that are difficult or unresolved even for medical experts. While many subjects are happy to simply "follow the doctor's orders," others may have a profound distrust of the medical community stemming from publicized examples of research abuses. For the surgeon-scientist, informed consent therefore becomes a critical mechanism to facilitate communication and instill trust. While documentation of consent receives heavy emphasis, it is the process of open communication, trust, and understanding between the researcher and subject that forms the true basis of informed consent.

From a practical standpoint, informed consent consists of three general steps: disclosure, subject understanding, and subject decision-making free of coercion. Disclosure entails the provision of information to potential subjects on the risks, benefits, and alternatives of participation in a study. Researchers must understand that subjects may be naïve to many of the medical details and present information in an easy-to-understand format incorporating lay terminology, images, or similar strategies. Recent updates to the Common Rule require that consent forms provide information that a "reasonable person" would want to make an informed decision and that forms are organized to facilitate easy understanding of why one may or may not want to participate in research [16]. Despite these measures, subjects can proceed through the informed consent process without truly becoming "informed." For this reason, it is necessary to gauge a potential subject's understanding of the information that has been presented. Often this can be performed by asking patients probing questions or having them explain the study details back to you in their own words. During this exchange, the researcher can help clarify any misunderstanding or address additional concerns that a subject may have. It is not only important to assess the subject's comprehension of the technical details but also how they believe any potential outcomes will impact their life. Only after a person demonstrates understanding can they make a truly informed decision. Even then, this may be a process that requires consultation with family, friends, or repeat conversation with their surgeon.

Surgical Innovation

In addition to research on human subjects, surgeons are uniquely positioned to innovate during the course of routine operations and patient care. Surgery has a rich tradition of innovation and as Riskin and colleagues note "most surgeons innovate on a daily basis, tailoring therapies and operations to the intrinsic uniqueness of every patient and their disease" [17]. This constant tinkering to improve technical skill and operative technique has led to the incremental improvement of postoperative outcomes and patient care. Many times this is a conservative, gradual process with small adjustments in technique that pose minimal additional risk to patients. However, rapidly advancing technology offers the opportunity for more radical innovations through novel, but unproven, approaches. Such advances may place patients at unacceptable levels of risk for little or no benefit, and some may actually harm patients. Currently, there is little regulation that applies to surgical innovation, and the boundary between innovation and research is not always clear.

In order to help surgeons decide when innovation crosses over into research, the Society of University Surgeons (SUS) developed guidelines that distinguish between the two [18]. The distinction is critical as it affects not only the level of oversight required but also determines when informed consent is necessary for patient participation. Unlike true innovation, variation is defined as a minor modification that is generally unplanned and involves only a slight shift in technique. Variations do not increase the risk to the patient or extend the time of anesthesia. In contrast, any planned, systematic investigation that leads to generalizable knowledge is considered research. An example of research may be a randomized trial of carotid endarterectomy versus carotid stenting. For projects that qualify as research under this definition, all the Common Rule regulations and IRB oversight discussed above are mandatory. Innovation is much more difficult to define. According to the SUS, an innovation is new or modified procedure that differs from accepted local practice, with outcomes that have not been described, and that may pose risk to the patient [18]. An alternate definition is any surgical procedure that has not been described in a North American surgical text. All planned innovations must be disclosed to patients prior to surgery during the informed consent process or postoperatively if an unplanned innovation occurs during surgery.

Research Integrity and Publication

While ethical abuses against human subjects have been heavily publicized over the last half-century, recently there has been increased awareness of fraud and misconduct related to research data and the publication of study results. Egregious examples of fraud emerged in the 1980s, highlighted by a physician who was accused of forging data that formed the basis of over 100 publications while he was employed at Emory and Harvard [19]. As a result, new federal regulations were developed that required institutions to investigate allegations of research misconduct [20]. The Office of Research Integrity (ORI) was established in 1992 to respond to these allegations, as well as promote integrity and responsible research practices. As part of this mission, the ORI has developed clear definitions and guidelines regarding research misconduct (See Box 3) [21].

Box 3 Federal Research Misconduct Policy

Research misconduct is defined as fabrication, falsification, or plagiarism in proposing, performing, or reviewing research or in reporting research results.

- Fabrication is making up data or results and recording or reporting them.
- Falsification is manipulating research materials, equipment, or processes or changing or omitting data or results such that the research is not accurately represented in the research record.
- Plagiarism is the appropriation of another person's ideas, processes, results, or words without giving appropriate credit.
- Research misconduct does not include differences of opinion.

From: Steneck N. ORI Introduction to the Responsible Conduct of Research. Department of Health and Human Services, Washington D.C.: US Government Printing Office, 2007.

Fabrication, falsification, and plagiarism are clear examples of misconduct and dishonesty in research. Whereas fabrication is the outright invention of data that is not supported by experimental results, falsification is the misrepresentation of data or results obtained from experiments.

Falsification may include the omission of data points or the selective use of data to obtain anticipated results. Perhaps the most overt form of misconduct is plagiarism, where one takes the work of another and passes it off as their own. Self-plagiarism can occur when someone duplicates their own work without proper citation. All three can have serious consequences for the researcher, including public embarrassment or censure, loss of funding, or "academic suicide" with all loss of credibility. From a broader perspective, repeated episodes of misconduct erode the public trust in scientific research.

In addition to fabrication, falsification, and plagiarism as outlined by the ORI, a wider range of scientific misbehavior has been described. In one survey of over 3000 scientists, one-third admitted that they had engaged in a variety of questionable research practices [22]. These included overlooking flawed data or methodologies, nondisclosure of conflicts of interest, or changing study design or results in response to pressure from a funding source. The exact prevalence of research misconduct is unknown, with survey estimates ranging anywhere from less than 2% to as high as 72% [23]. The higher estimates suggest that misconduct results from more than just a "few bad apples." In fact, structural pressures inherent to the research environment itself may be conducive to academic misconduct. Many of the tangible rewards in academics – status and prestige, promotion and tenure, or grants and funding – are tied to one's publication records contributing to an environment described as "publish or perish" [24]. Often the quantity of publications receives just as much emphasis as quality, which rewards slicing projects into the "minimal publishable units" or duplicating content. Many of these papers may have little to add to existing scientific knowledge. In addition, the overemphasis placed on positive results by medical journals may encourage data mining or "massaging" data to arrive at the "right" results. For example, 1 review of 74 randomized controlled trials published in highly regarded journals found selective reporting of trial results. There were multiple discrepancies between protocol-specified and published endpoints with positive unplanned findings significantly outnumbering negative results [25].

Even in the absence of intentional fraud, many research findings may be misleading simply due to inappropriate methodology or statistical naiveté [26]. The increasing complexity of studies and the requisite sophistication of analyses introduce a significant potential for error. However, perhaps as a consequence of "publish or perish," many studies do not adhere to good research practices. In fact, it has been suggested that the majority of published research findings are false and that up to 85% of research resources are wasted [27]. Low statistical power, incorrect choice of statistical tests, and poor understanding of pretest probability all contribute to the increased chance of false-positive results and misinformed conclusions. These phenomena may be even more pronounced among small studies, as well as reports with small effect sizes and those testing multiple hypotheses [28]. In fact, many research results have not been reproducible or proven robust over time, including those published in well-regarded, high-impact journals [29]. Flawed studies can propagate misinformation through the medical literature, which ultimately may divert future resources to misguided follow-up studies, obscure good results, or possibly harm patients. Beyond just producing credible results, researchers therefore have an ethical responsibility to ensure that sound methodology and analysis underlie all research projects.

Good Research Practices

Increased recognition of research misconduct and questionable methodologies has led to critical reevaluation of the research environment. Multiple reforms have been proposed to improve the quality and integrity of biomedical research [1, 27, 29–31]. Reform proposals range from improvements in study design and execution to changing the underlying research culture. From a process standpoint, good research is based upon the principles of the scientific method, whereby thoughtful hypotheses are tested by empirical data gathered through well-designed experiments. Transparent, verifiable, and reproducible data are critical to this process. Increased use of collaborative research, registration of trials and protocols, sharing of raw data, and standardized definitions may all help improve the reproducibility and transparency of results [27]. Early involvement of a statistician with more appropriate use of statistical methods can improve the credibility of data analyses. Journals have recognized the importance of many of these measures, and several have introduced new recommendations to encourage the use of more stringent research methodologies. For instance, Nature announced new author guidelines including the increased disclosure of raw data, more thorough description of methods, and possible auditing by independent statisticians [32].

Beyond questions around research methodology, other proposed reforms have sought to alter the underlying research culture. The competitive nature of some fields deters collaboration, and the priority given to those who publish first contributes to a "winner-takes-all" system [30]. In addition, hierarchal relationships may discourage discussion of questionable results or practices within a research laboratory. The result can be the suppression of open communication both within and between research labs. The cultivation of open research environments is critical to improve the quality of research. An open research environment allows ambiguous or dubious results to be discussed, and measures can be implemented expeditiously to correct wrong answers or faulty processes.

The current promotion and reward system is another aspect of research culture that may incentivize poor-quality research. Promotion, funding, or other rewards can often be unduly influenced by the quantity of research output. This may encourage projects that help build a CV, but have little impact on a researcher's field. Performance reviews can involve individuals with limited knowledge of an investigator's specific focus of research and are therefore poor judges of the research quality. Careful peer review of the scientific quality of an investigator's work may be one strategy to better capture the contributions of a

researcher to his/her field [30]. Aligning reward structures with a researcher's true impact may help decrease the overall incidence of research misconduct.

Authorship

Given the emphasis on publication, authorship can be a particularly contentious topic. Authorship has implications for career advancement and promotion; the volume of medical publications and the rising number of multi-author papers reflect the importance placed on getting author credits [33]. Several ethical conflicts related to authorship have been described. Researchers may have little involvement with the underlying research but become attached to publications as "honorary" authors. Honorary authors have made no significant contribution to the publication, but attempt to justify inclusion for any number of reasons (e.g., "I obtained the grant funding for this project" or "Several of the patients studied were from my practice"). On the other hand, junior researchers can be denied first authorship despite having performed the bulk of the data collection, analysis, and manuscript preparation. In order to help clarify any controversy surrounding authorship, the International Committee of Medical Journal Editors has established clear criteria on what constitutes authorship (see Box 4). Many journals now request that any article submissions conform to similar criteria and ask that each author details their specific contributions to the project [34].

In order to help avoid controversies related to authorship, clear discussion should be held at the beginning of any research project. The principal investigator or senior author needs to agree on clear expectations and guidelines with co-authors, particularly any junior or lead authors. The ultimate order of the authors generally reflects the importance of each individual's role in the paper. If necessary, author roles may need to be revisited and reordered during the course of a project – such contingency plans should be clearly discussed and established beforehand.

Box 4 ICMJE: Defining the Role of Authors and Contributors

The ICMJE recommends that authorship be based on the following four criteria:

- Substantial contributions to the conception or design of the work or the acquisition, analysis, or interpretation of data for the work
- Drafting the work or revising it critically for important intellectual content
- Final approval of the version to be published
- Agreement to be accountable for all aspects of the work in ensuring that questions related to the accuracy or integrity of any part of the work are appropriately investigated and resolved

Adapted from: The International Committee of Medical Journal Editors. Recommendations for the Conduct, Reporting, Editing, and Publication of Scholarly Work in Medical Journals. Updated December 2016.

Conflicts of Interest

To maintain research integrity, ethical scientific research should strive to limit sources of bias. Conflict of interest is one particular form of bias that has attracted significant attention. While attention has generally focused on areas of inappropriate financial gain, conflicts of interest can occur related to industry sponsorship, publication/authorship, medical education, and promotion. In general, a conflict of interest arises anytime that professional judgment concerning a primary interest (e.g., valid research) tends to be unduly influenced by a secondary interest (e.g., industry sponsorship) [35]. As opposed to ethical conflicts, where both sides of an argument may have equal claim to validity, in a conflict of interest, only the primary interest has claim to authority. This does not mean that secondary interests

are illegitimate but that they should not dominate the primary interest. For instance, while industry sponsorship is often necessary and desirable to conduct trials, it should not impact the validity of the underlying research. Unfortunately, empirical evidence suggests industry funded clinical trials are "positive" up to 85% of the time, whereas approximately 50% of government funded trials have "positive" results [26].

Even when a true conflict of interest may not exist, the perception of a conflict of interest can degrade confidence in the integrity of a research project or surgeon-scientist. In a sense, perception is just as important as reality. Any researcher must therefore be cognizant of any and all potential conflict of interests, as well as know the appropriate steps to properly address them. The first step to manage a conflict of interest is full disclosure of any real or potential conflicts of interest. The Institute of Medicine (IOM) recommends that any institution that carries out medical research should develop a conflict of interest committee to manage these issues within their institution [36]. Individual institutions are responsible for outlining the rules of what constitutes a conflict of interest and how they must be disclosed [37]. While the IOM has provided recommendations for relationships that should be disclosed (see Box 5), it is incumbent on each surgeon-scientist to familiarize themselves with their home institution's policy. In general, disclosure should include any financial compensation for the specific investigator or any of his/her family members. Independent institutional review of any potential conflict of interest is necessary, as individual researchers may not be able to objectively assess when a conflict of interest exists. While many times researchers with a conflict of interest can continue their work with appropriate oversight, occasionally they may need to divert themselves from projects or industry relationships.

Box 5 Recommend Industry Relationships Requiring Disclosure
- Research grants and contracts
- Consulting agreements
- Participation in speakers bureaus

- Honoraria
- Intellectual property, including patents, royalties, and licensing fees
- Stocks, options, warrants, and other ownership
- Position with a company
- Technical advisory committees, scientific advisory boards, and marketing panels
- Company employee or officer, full or part time
- Authorship of publications prepared by others
- Expert witness for a plaintiff or defendant
- Other payments or financial relationships

From: Lo and Field [36].

In addition to the institutional oversight of conflicts of interest, the US government has introduced regulations aimed at increasing transparency in physician relationships with industry. Provisions of the 2010 Patient Protection and Affordable Care Act require pharmaceutical companies and medical device manufacturers to report all payments or transfers of value to physicians and teaching hospitals. More popularly known as the "Sunshine Act," payments as low as $10 per instance or $100 per year require disclosure [38]. Among other forms of compensation, transfers of value also include gifts, entertainment, travel, or meals. Data from the first 2 years of the program demonstrated payments to physicians totaling $3.4 billion in 2013, rising to $6.6 billion in 2014 [39]. The Centers for Medicare and Medicaid Services maintains a website that provides free public access to data regarding payments to individual physicians and teaching hospitals.

Mentorship

Perhaps no area of academic surgery has the potential to cultivate ethics as much as effective mentoring and education. Not only do mentors

teach the technical skills and provide the feedback trainees need to develop a research skill set, more importantly, they play a critical role in modeling ethical behavior. The word ethics derives from the Greek *ethos*, meaning character or habit. In this sense, professionalism and ethics do not result as much from a technical curriculum as they are acquired through long guidance under teachers who embody these moral qualities [40]. Although it is critical that trainees understand the guidelines, regulations, and rules that govern research, there is no guarantee that this knowledge will lead to ethical behavior. Instead, mentoring should instill a deep appreciation of fundamental ethical principles. For instance, Markman argues the fundamental principles of human subjects research that should be emphasized by mentors include (1) informed, voluntary consent, (2) scientific value of the study justifies the risks, and (3) any potential hazards to patients are minimized [41]. Many times adherence to these principles requires attitudes or values that cannot be taught in formal courses.

Nonetheless, the mentorship relationship can be subject to abuse. Ideally the mentor-trainee relationship is fiduciary – trainees rely on their mentors to look after their best interests. However, there exists a disparity in expertise, power, and experience in the mentor-trainee relationship that may leave the trainee vulnerable to exploitation [24]. Examples include overloading trainees with low-yield activities, "hijacking" a trainee's ideas or projects, or not giving sufficient time to truly be effective mentors [42]. On the other hand, trainees have a responsibility to be good mentees by taking a proactive role in the relationship, as opposed to passively expecting their mentor do everything for them. Trainees need to come to meetings prepared, make good use of their mentor's time, and have a clear goal or agenda in mind [43].

Conclusion: Ethics as a Continuing Endeavor

The development of research ethics and contemporary precedent can help guide surgeon-scientists. It is crucial to understand, however, that these guidelines are not absolute given that the rapid growth of scientific knowledge will continue to demand evolution of ethical codes. In reality, regulations, guidelines, and recommendations need to serve as the starting point for continued conversation over ethical behavior. As Dr. Lawrence McCullough notes "Medical Ethics [can be] understood as a moral response to scientific and technological change...[such a] moral response is required to address scientific and technological changes that are unprecedented and therefore threaten to outstrip society's moral capacities" [44]. In other words, scientific discovery reshapes our codes of ethics at the very same time that we seek these principles for guidance. Knowledge threatens to outpace wisdom. To be successful, simply following guidelines will not be enough – we should strive to develop ethical character [5]. A focus on character shifts the emphasis away from action prescribed externally by codes to action guided internally by a sense of what one ought to do. A foundation of ethics built in this manner would be able to respond morally to novel situations without any pre-existing guidelines or rules or in situations where prior guidelines no longer seem to apply.

The unique nature of academic surgery will continue to challenge surgeons with ethical dilemmas. Conflicting loyalties among the different roles assumed by surgeon-scientists require a balancing act between our responsibilities to patients and society as a whole. A solid understanding of ethical codes and precedent is critical to inform our moral reasoning as we engage these problems. More than this, we must also challenge ourselves to develop the character necessary to provide moral orientation and direction when new ethical dilemmas arise.

Box 6 Concluding Remarks

- Rapid technological advances and the dual roles assumed by surgeon-scientists have introduced multiple ethical challenges to the practice of academic surgery.
- Learning how to manage such challenges is a crucial skill that every academic surgeon must develop.
- Although understanding ethical precedent and guidelines is critical to manage ethical dilemmas, one must strive to develop an ethical character.

References

1. Resnik DB, Elliott KC. The ethical challenges of socially responsible science. Account Res. 2016;23(1):31–46.
2. American College of Physicians Ethics Manual. Part I: history of medical ethics, the physician and the patient, the physician's relationship to other physicians, the physician and society. Ad Hoc Committee on Medical Ethics, American College of Physicians. Ann Intern Med. 1984;101(1):129–37.
3. Beauchamp TL, Childress JF. Principles of biomedical ethics. Oxford: Oxford University Press; 2012.
4. Pawlik TM, Schwarze ML. Ethics in surgical research. Success in Academic Surgery, Vol. 1. Springer lLink; 2017. p. 43–57.
5. Pawlik TM, Platteborze N, Souba WW. Ethics and surgical research: what should guide our behavior? J Surg Res. 1999;87(2):263–9.
6. Beaumont W. Experiments and observations of the gastric juice and the physiology of digestion. Plattsburgh: FP Allen; 1833.
7. Moreno JD, Schmidt U, Joffe S. The Nuremberg code 70 years later. JAMA. 2017;318(9):795–6.
8. The nuremberg code. JAMA. 2017;276(20):1691–1.
9. World Medical Association Declaration of Helsinki. Ethical principles for medical research involving human subjects. JAMA. 2017;310(20):2191–4.
10. Beecher HK. Ethics and clinical research. N Engl J Med. 1966;274(24):1354–60.
11. National Commission for the Protection of Human Subjects of Biomedical and Behavioral Research. The Belmont report : ethical principles and guidelines for the protection of human subjects of research. Washington, D.C.: US Government Printing Office; 1979.
12. Friesen P, Kearns L, Redman B, et al. Rethinking the Belmont report? Am J Bioeth. 2017;17(7):15–21.
13. Emanuel EJ, Wendler D, Grady C. What makes clinical research ethical? JAMA. 2000;283(20):2701–11.
14. Freedman B. Equipoise and the ethics of clinical research. N Engl J Med. 1987;317(3):141–5.

15. Childers R, Lipsett PA, Pawlik TM. Informed consent and the surgeon. J Am Coll Surg. 2009;208(4):627–34.
16. Menikoff J, Kaneshiro J, Pritchard I. The common rule, updated. N Engl J Med. 2017;376(7):613–5.
17. Riskin DJ, Longaker MT, Gertner M, et al. Innovation in surgery: a historical perspective. Ann Surg. 2006;244(5):686–93.
18. Biffl WL, Spain DA, Reitsma AM, et al. Responsible development and application of surgical innovations: a position statement of the Society of University Surgeons. J Am Coll Surg. 2008;206(6):1204–9.
19. Stewart WW, Feder N. The integrity of the scientific literature. Nature. 1987;325(6101):207–14.
20. Benos DJ, Fabres J, Farmer J, et al. Ethics and scientific publication. Adv Physiol Educ. 2005;29(2):59–74.
21. Steneck N. ORI introduction to the responsible conduct of research. Department of Health and Human Services. In: US Government Printing Office. Washington D.C; 2007.
22. Martinson BC, Anderson MS, de Vries R. Scientists behaving badly. Nature. 2005;435(7043):737–8.
23. Sarwar U, Nicolaou M. Fraud and deceit in medical research. J Res Med Sci. 2012;17(11):1077–81.
24. Shamoo AE, Resnik DB. Responsible conduct of research. New York: Oxford University Press; 2015.
25. Raghav KP, Mahajan S, Yao JC, et al. From protocols to publications: a study in selective reporting of outcomes in randomized trials in oncology. J Clin Oncol. 2015;33(31):3583–90.
26. Steen RG. Misinformation in the medical literature: what role do error and fraud play? J Med Ethics. 2011;37(8):498–503.
27. Ioannidis JP. How to make more published research true. PLoS Med. 2014;11(10):e1001747.
28. Ioannidis JP. Why most published research findings are false. PLoS Med. 2005;2(8):e124.
29. Begley CG, Ioannidis JP. Reproducibility in science: improving the standard for basic and preclinical research. Circ Res. 2015;116(1):116–26.
30. Casadevall A, Fang FC. Reforming science: methodological and cultural reforms. Infect Immun, Vol. 80. United States; 2012. p. 891–896.
31. Casadevall A, Ellis LM, Davies EW, et al. A framework for improving the quality of research in the biological sciences. MBio. 2016;7(4):e01256-16
32. Announcement: reducing our irreproducibility. Nature. 2012;492:34–6.
33. Kornhaber RA, McLean LM, Baber RJ. Ongoing ethical issues concerning authorship in biomedical journals: an integrative review. Int J Nanomedicine. 2015;10:4837–46.
34. Lundberg GD, Glass RM. What does authorship mean in a peer-reviewed medical journal? JAMA. 1996;276(1):75.
35. Thompson DF. Understanding financial conflicts of interest. N Engl J Med. 1993;329(8):573–6.
36. Lo B, Field MJ. Conflict of interest in medical research, education, and practice. Institute of Medicine. Washington D.C.: The National Academies Press; 2009.
37. Freischlag JA. Academic medical centers write their own rules. J Vasc Surg. 2011;54(3 Suppl):19s–21s.
38. Rosenthal MB, Mello MM. Sunlight as disinfectant-new rules on disclosure of industry payments to physicians. N Engl J Med. 2013;368(22):2052–4.
39. Agrawal S, Brown D. The physician payments sunshine act--two years of the open payments program. N Engl J Med. 2016;374(10):906–9.
40. Kinghorn WA. Medical education as moral formation: an Aristotelian account of medical professionalism. Perspect Biol Med. 2010;53(1):87–105.
41. Markman M. Mentoring in the ethics of clinical research: an ongoing need. Curr Oncol Rep. 2007;9(4):235–6.
42. Chopra V, Edelson DP, Saint S. A PIECE OF MY MIND. Mentorship malpractice. JAMA. 2016;315(14):1453–4.
43. Zerzan JT, Hess R, Schur E, et al. Making the most of mentors: a guide for mentees. Acad Med. 2009;84(1):140–4.
44. McCullough LB. Laying medicine open: understanding major turning points in the history of medical ethics. Kennedy Inst Ethics J. 1999;9(1):7–23.

Suggested Reading

Beecher HK. Ethics and clinical research. N Engl J Med. 1966;274(24):1354–60. Landmark report detailing multiple examples of research misconduct and violations of humans subjects rights.

Emanuel EJ, Wendler D, Grady C. What makes clinical research ethical? JAMA. 2000;283(20):2701–11. Discussion of 7 proposed requirements that comprise a framework for evaluating the ethics of clinical research studies.

Pawlik TM, Platteborze N, Souba WW. Ethics and surgical research: what should guide our behavior? J Surg Res. 1999;87(2):263–9. Discussion of the principles that anchor ethical behavior in surgical research, with the proposal that character should serve as a moral guide.

Shamoo AE, Resnik DB. Responsible conduct of research. New York: Oxford University Press; 2015. Comprehensive review of the ethical issues surrounding biomedical research. Covers multiple topics and is well supplemented with case studies.

The nuremberg code. JAMA. 2017;276(20):1691–1. Considered among the founding documents of modern bioethics, the 10-point Nuremberg code outlines protections including informed consent and argues any research must benefit society.

Ethical Issues of the Mentor-Mentee Relationship

Alberto R. Ferreres

"To hold the one who has taught me this art as equal
to my parents and to live my life in partnership with him"

The Hippocratic Oath [1]

Key Points
- Surgical mentoring of young trainees and faculty enables them to nurture, hone, and acquire knowledge, experience, and judgment in order to progress in the surgical environment.
- The role of the mentor is a fiduciary one, and the relationship should be founded upon trust and loyalty.
- A mentor is not synonym of trainer, role model, faculty, or boss.
- Both sides should acknowledge the fact that the mentoring relationship may come to an end and is no longer needed.
- This relationship may be established and developed at different stages of academic and nonacademic life.

The term mentor has a Greek origin, *men* refers to the "person who thinks," while *tor* is the male

A. R. Ferreres (✉)
Department of Surgery, University of Buenos Aires, Buenos Aires, Argentina

Department of Surgery, University of Washington, Seattle, WA, USA
e-mail: aferre17@uw.edu

and *trix* the female. So the meaning of mentor refers to "the man who thinks."

The Oxford English Dictionary provides the following meanings of the noun mentor:

1. Allusively, one who fulfills the office which the supposed Mentor fulfilled toward Telemachus
2. An experienced and trusted advisor

So, it is mandatory to dive into Homer's *The Odyssey* to get an insight of the term and its connotations. The origin of this term was linked to the character of Mentor, mentioned in Homer's work. In Greek mythology [2], Mentor was the son of Alcimus and Asopia and is considered the one who supposedly took care of Telemachus, Odysseus' son, during the latter's long trip. Nonetheless a careful review of this epic poem will prove that this assumption is not true [3]. These quotations provide further details:

- "Mentor rose to speak. Mentor was an old friend of Odysseus, to whom the King had entrusted his whole household when he sailed, with orders to defer to the aged Laertes (Odysseus' father) and keep everything intact" (:43)
- "Remember your old friend and the good turns I have done you in the past. Why, you and I were boys together" (:333, Odysseus speaks)
- "Odysseus has come home. And he has killed the rogues who turned his whole house inside

out, ate up his wealth, and bullied his son" (:341, words of an old servant)

- "I have been told that a whole crowd of young galants are courting your mother and running riot in your house as uninvited guest" (:56, Nestor of Gerenia)
- "Is it not enough that all this time, under pretext of your suit, you have been robbing me of my best, while I was too young to understand? I tell you, now that I am old enough to learn from others what has happened and to feel my own strength at last, I will not last until I have let hell loose upon you" (: 45, Telemachus' statement to Mentor)

The only opportunities when Mentor's role was according the expectations were when goddess Pallas Athena took his disguise, as the goddess of war and wisdom. So it is clear that the demeanor of Mentor in Homer's *The Odyssey* was not ethical neither beneficial to Telemachus.

The current idea of what a mentor is and deserves to be may be found in *The adventures of Telemachus, the son of Ulysses* published in 1699 by Francois Fenelon, a French bishop and theologian, who was tutor to Louis XIV's grandson.

Some of the quotes from his book may illustrate the nature of a true and loyal Mentor [4]:

- "…I have been seeking my father all over the sea, in the company of this man (Mentor) who was to me another father" (:54)
- "Oh happy Telemachus! You will never be bewildered as I have been bewildered, while you have such a guide and instructor! Mentor, you are the master!" (:136)
- "Forget not, my son, the pains I took when you were a child, to make you as wise and as valiant as your father" (:160)
- "Mentor regulated the whole course of the life of Telemachus in order to raise him to the highest pitch of glory" (:215)

His ideas were very influential at his time, as proved by his successors: Marquis Caraccioli (1719–1803) and Jean Jacques Rousseau (1712–1778). In the book *Emile, or On Education* by the latter, the main character is given a copy of Fenelon's *The adventures of Telemachus* in the initial stages.

The first use of the term mentor in the English language can be attributed to Lord Chesterfield (1694–1773) in a letter to his son dated March 8, 1850, where he wrote: "There are resolutions which you must form and steadily execute for yourself, whenever you lose the friendly care and assistance of your mentor." But the concept had been developed since earlier times and was related to the guilds of craftsmen in England during the Middle Ages. The elders and owners transferred their knowledge and abilities to their pupils or apprentices, who were required to work for about 12 years or until they were 21 years old. This methodology of training and developing human resources was changed due to the advent of the Industrial Revolution.

In the surgical arena, the traditional method of training was the apprenticeship model of one to one (master and pupil), as previously described. But this paradigm shifted after William Halsted, surgeon in chief at Johns Hopkins, introduced the surgical residency, a system where the trainees spent 5 or more years living in a hospital (hence, the term of residents) learning clinical and surgical skills, as well as performing research under the guidance of faculty and attendings. This model has been largely replicated and represents the standard of surgical training all over the world.

Along the world history, religion has also adopted mentorship for the trainees: some examples include the guru-disciple relationship in Buddhism and Hinduism, as well as the master-disciple system in rabbinical Judaism and different aspects of Christianism.

The modern use of the terms mentor and mentorship should be traced to the North American business arena and the social movements starting in the 1960s. The importance of mentorship was highlighted by the group of Yale's social scientists led by Levinson; he considered the mentor as the key factor in order to achieve the dream of another individual [5]. Barondess in his Presidential Address when inducted as President of New York Medical Society introduced the term mentor in the medical field. He made special reference to the relationship between Bill Dickey and Yogi Berra, both New York Yankees' catchers and how the latter mentioned "Bill is learning me his experience." This phrase illustrates in a very simple fashion the core of this relationship [6]. Edward Copeland III' s Presidential Address at the American College of Surgeons in 2006 was titled "The role of a mentor in creating a surgical way of life" [7].

According to the Oxford English Dictionary, a mentor is an experienced and trusted advisor. In other words, a mentor is someone who sees more talent and ability within you than you see in yourself and helps bring it out of you. In the medical and surgical arena, mentoring is the process whereby an experienced, highly regarded, empathic person (the mentor) guides another individual (the mentee) in the development and reexamination of their own ideas, learning and achieving professional and personal development; the mentor, who often but not necessarily works in the same organization or field as the mentee, achieves this by listening and talking in confidence to the mentee.

So, what does the mentor do on behalf of his or her mentee? The mentor provides knowledge, expertise, and guidance to achieve the promotion of the mentee to a higher level and also to hone the personal qualities of the latter. But there is a difference between a mentor and a trainer, a mentor is not necessarily the one who teaches and trains a new generation of surgeons but the one who provides tutorship and guidance to progress the academic and surgical ladder. So it is desirable for the mentor to be altruistic, available, , accountable, active, and appreciative.

An assessment throughout the lives of successful public figures demonstrates that most of them recognized the role of a mentor at some stage of their lives. But there are also many examples of failed mentor-mentee relationships. According to G. Steiner, there are three main scenarios or relationship structures within a mentor and his or her mentee [8]:

• Masters who destroyed their mentees
• Mentees who have betrayed and destroyed their mentors

• The interchange or "eros" of mutual confidence, which should be the goal of very mentor-mentee relationship in order to be successful and beneficial

Surgical ethics overflies this relationship in multiple ways: there is a moral responsibility on each side of this dyadic relationship which represents an essential component of surgical academics and lies at the core of professionalism. Both sides of this important relationship should bear in mind their ethical duties and obligations in order to achieve a successful and profitable relationship and in this way prevent abuses of power and other inequities, which will put an end to this nurturing link. The principles of biomedical ethics developed by Beauchamp and Childress represent a very useful framework and a very valuable tool when confronted to ethical dilemmas and conflicts in everyday surgical life [9]. The ethical principles include beneficence, nonmaleficence, justice, and autonomy.

Beneficence: it refers to the virtue of acting for the benefit of others. David Hume considered that the duty to benefit others arises from social interactions and is grounded on reciprocity, which consists in the act or practice of making an appropriate and usually proportional return. This principle is undoubtedly linked to the fiduciary duty of the mentor toward the mentee and vice versa. Both sides should behave in an altruistic manner, keeping in high regard the interest, benefit, and welfare of the other counterpart.

Nonmaleficence: it reflects and means the avoidance of harm to the other side. Many mentor-mentee relationships have been characterized by hurting the other participant of the relationship. It is grounded on the "Primum non Nocere" dictum, which should not be attributed to Hippocrates but to Parisian pathologist Auguste Francois Chomel (1788–1858) [10].

Justice: justice should be considered as the fair, equitable, and appropriate treatment of what is due or owed to persons, which consists in the moral duty to act on the basis of fairness and equality. John Rawls introduced the liberty and the different principles in his concept of justice as fairness [11]. In a practical sense, it highlights the right to be treated equally.

Autonomy: the concept of autonomy overviews the decision-making of both mentor and mentee, as participants of this dyadic relationship. The word autonomy derives from the Greek autos (self) and nomos (rule, governance) and originally referred to the self-rule of independent city states in ancient Greece. There are two essential conditions to the autonomy of both parties: liberty, meaning freedom from external coercion and agency, the capacity to act intentionally.

Both mentor and mentee should exercise an autonomous behavior, and their relationship should be an example of self-decision and free will. Both mentor and mentee should choose being in this relationship and profit from each other. The ideas of Immanuel Kant and John Stuart Mill have had a strong impact in the approaches to autonomy, representing the moral imperative and the utilitarian approach [12].

The foundation of the mentor-mentee relationship is grounded in the fiduciary bond between both sides. This fiduciary relationship entails the mentor's duty to act in the mentee's best interest and not his own one. The fiduciary role is a key element for the success of this relationship. Being a fiduciary should be understood as being "a person holding the character of a trastee, in respect to the trust, confidence, good faith and candor which it requires and having the duty to act primarily for another person's benefit in matters connected with such undertaking" [13]. Another way to explain the fiduciary role is the fact that the mentor surgeon must be beneficent, zealous to serve the interests of the mentee, and the duty of good faith should be especially strong. In matters related to the fiduciary relationship, it is elemental that a fiduciary owes a duty of undivided and undiluted royalty to those whose interests the fiduciary is to protect. There is a sensitive and inflexible rule of fidelity as well as a duty of loyalty, requiring avoidance of conflict of interest between both sides. Being a fiduciary means that the mentor role should be characterized by:

- Placing the mentee's welfare above your own
- Treat each mentee as you would wish to be treated

- Value each individual
- Do unto others as you would have them undo unto you

The ethical practice of mentorship involves the undertaking of principles, values, and virtues. The adherence to the ethical principles in a strict fashion will keep this relationship between the mentor and his or her mentee in due way and will also help preventing the first two scenarios described by Steiner [8]. So, this relationship will comply with three must-do actions:

- Respect each other's rights and dignity
- Act honestly and promoting fairness and equity in the relationship
- Advocate the welfare, benefit, trustworthiness, reliability, and loyalty

Mentors should focus on the beneficent aspects of their support toward the mentee, and probably this ethical principle is predominant and of the higher impact in the relationship, and the second one should be to avoid harm to the mentee, who is usually in a weaker situation [14].

Some of the ethical issues involve some of the elements of the mentor-mentee relationship, and care should be taken to prevent future conflicts among both parties. The goal of this relationship should be very clear from the first moment to both sides; having a clear and deep understanding of the reasons of this bond will help to prevent any confusion which may derive in unethical conducts as well as abuse of power. Communication should focus on the goals to achieve and objectives and characterized by openness, sincerity, and honesty. Trust is of paramount importance for the success of the mentor-mentee relationship, but it takes a time to build it. When trust does not flourish, the relationship must terminate, to prevent eventual harm. It should be understood that the improvement and development of this relationship are a process with different stages.

The ethical principles, above mentioned, represent a roadmap for both parties in order to achieve progress in this relationship and prevent its failure. The misuse of power throughout this relationship represents a particular concern.

These power inequities may be acknowledged or not by either sides; strict adherence to four ethical principles should be useful to set appropiate boundaries and prevent this situation. Integrity is another moral value that should trascend this bond. An ethical mentor-mentee relationship needs to avoid eventual profit or benefits, as for example the mentee getting an underserved position or appointment. Mentors should promote fairness and equity not only to their mentors but to other members of the academic faculty and also refrain from requesting specific favors to their mentees. In summary, the mentor should be a master of virtues and a moral role model.

- How ethical issues can be undertaken under this relationship in order to prevent its failure?
  - Underneath the ethical mentor-mentee relationship
- Honesty: the general fiduciary commitment to protect and promote the interests of the mentees if SE is to guide the clinical judgment and practice of surgeons
- Ethical concept of the mentor as a fiduciary: a person holding the character of a trustee, in respect to the trust and confidence involved in it and scrupulous good to act primarily for another's benefit in matters connected to that understanding
- (Brody BA, Engelhardt HT. The major moral considerations. 1987)
- Fidelity: faithfulness to a person, cause, or belief demonstrated by continuing loyalty and support, by the mentor and the mentee
- Fidelity: trustworthy, consistent, reliable, responsible, and fiduciary relationship
- Success: reciprocity, mutual respect, clear expectations, personal connection, and shared values
- Value: greater productivity, more rapid promotion, academic retention, personal profit, and adequate balance
- Failure: poor communication, lack of commitment, personality differences, competition, conflicts of interests, and mentor's lack of experience
- Abuse of power
- Misuse of power: abuse, sexual harassment, academic exploitation, Wrongful management of power, and inappropriate boundaries

Concluding Remarks

- The concepts and features in the term Mentor do not origin from Homer's *The Odyssey* but from Fenelon's *The Adventures of Telemachus*.
- The surgical mentor-mentee relationship should be ruled by strict ethical principles.
- The role of the mentor is a fiduciary one: to seek the benefit of the mentee and not his or her own benefit.
- The values to be followed are honesty, integrity, fidelity, royalty, and fairness.
- It is very important to prevent power inequities as well as the misuse and/or the abuse of power from either side.

References

1. Edelstein L. The Hippocratic oath. Baltimore: The Johns Hopkins Press; 1943.
2. Homer. The odyssey. New York: Penguin; 1997.
3. Roberts A. Homer's mentor. Duties fulfilled or misconstrued. Hist Educ Soc Bull. 1999;64:313–29. Available in www.nichols.us/homers_mentor.pdf. Accessed 13 Jan 2018-08-05.
4. Fenelon F. The adventures of Telemachus, the son of Ulysses. Whitefish: Kessinger Legacy Reprints; 2010.
5. Levinson DS, Darrow N, Klein EB, Levinson M. Seasons of a man's life. New York: Random House; 1978.
6. Barondess JA. President's address: a brief history of mentoring. Trans Am Clin Climatol Assoc. 1995;106:1–24.
7. Copeland EM III. The role of a mentor in creating a surgical way of life. Bull Am Coll Surg. 2006;12:9–13.
8. Steiner G. Lessons of the masters. Charles Eliot Norton lectures 2001–2002. New York: Open Road; 2005.
9. Beauchamp TL, Childress JF. Principles of biomedical ethics. 4th ed. New York: Oxford University Press; 1994.
10. Smith CM. Origin and uses of "Primum non Nocere": above all, do no harm! J Clin Pharmacol. 2005;45:371–7.
11. Rawls J. A theory of justice (revised edition). Cambridge: Harvard University Press; 1999.
12. Ferreres AR. Ethical debate: the ethics of not performing extended lymphadenectomy in patients with gastrointestinal cancer. World J Surg. 2013;37:1821–8.
13. Black HC. Black's law dictionary. St Paul: West Publishing; 1970.
14. Ferreres AR, Pellegrini CA. Ethics of the mentor-mentee relationship (chapter 15). In: Ferreres AR, Angelos P, Singer EA, editors. Ethical issues in surgical care. Chicago: American College of Surgeons. p. 2017.

Surgical Ethics and the Surgical Societies: What Are We Doing?

Richard I. Whyte

Introduction

The focus of this chapter, where the subject of Ethics fits in the role of professional surgical societies, is best addressed firstly through a definition of terms, an exploration into the role of surgical societies (and professional organizations in general), and then through an empirical view of how several surgical societies incorporate ethics into their mission and practice. One can then ask the question, what should surgical societies do with respect to ethical issues and, conversely, what ethical issues or conflicts should surgical societies avoid?

Ethics is considered as the study of standards of conduct and moral judgment. Incipient to this is the term "moral"—a term that assigns rightness or wrongness to certain actions, thoughts, or concepts. Common morality refers to a set of tenets that are applicable to life in general and are not relative to specific cultures or groups [1]. These would include the following: do not kill, do not cause pain or suffering to others, prevent evil or harm from occurring, rescue persons in danger, tell the truth, nurture the young and dependent, keep promises, do not steal, do not punish the innocent, and obey just laws. Variations in these are consistent across many cultures and eras and appear to apply to the human condition in general.

The topic of "biomedical ethics" is a somewhat narrower term, but its current usage refers to a broad range of ethical issues related to clinical medicine, medical research, biology, life sciences, reproductive technology, and even social conditions. The core principles of biomedical ethics have been well described by Beauchamps and Childress in their now standard textbook on the subject, *Principles of Biomedical Ethics* (now in its eighth edition) [2]. In this work, the authors describe four pillars of biomedical ethics: respect for autonomy, beneficence, nonmaleficence, and justice, and frame much of their subsequent discussion on this framework. While these principles form a strong basis for examining surgical ethics, Fox has lamented that biomedical ethics has, historically, not dealt with larger, societal issues that relate to health—issues such as poverty, economics, and racial and gender disparities—and how they affect both health and specific public policy determinations [3].

Within biomedical ethics is the subject of professional ethics—at least as it is related to the practice of medicine. At a general level, professional ethics may incorporate the rules of common morality but are generally standards of conduct applicable to a certain profession. While some of these standards are unwritten, others may be found in written policies sponsored by organizations such as professional societies, State Medical Boards, hospital policies and

R. I. Whyte (✉)
Harvard Medical School, Boston, MA, USA
e-mail: rwhyte@bidmc.harvard.edu

© Springer Nature Switzerland AG 2019
A. R. Ferreres (ed.), *Surgical Ethics*, https://doi.org/10.1007/978-3-030-05964-4_9

bylaws, professional accreditation organizations, and educational institutions.

On a broad sense, surgical ethics would include all topics that cover the standards and the moral judgments of surgeons in their professional role. The list of topics is extensive, and many of the topics are covered in this book, but at any given time, some topics are more germane. Currently, some of the "hot topics" include consent for surgery, end-of-life care, futile care, the role of trainees in the operating room, the surgeon's role in social media, regulation of surgical innovation, global surgery, disclosure of surgical errors, gender reassignment surgery, and issues related to medical malpractice. Others, such as the practice of concurrent, or overlapping, operations, demonstrate a fairly short, but intense, level of interest and can be correlated with coverage in the lay press.

Professional Surgical Societies

To uncover the role of ethics in professional surgical societies, the role of such organizations needs to be defined. While it is difficult to generalize, most medical professional surgical organizations are nonprofit organizations whose purpose is to further the issues of the profession, the interests of the individuals in the profession, and the public. Rothman pointed out that professional medical associations play a vital role in medical education, the development and publication of practice guidelines, and that they define ethical norms for their members through codes of conduct for professional behavior [4].

Surgical societies are a diverse set of organizations. Some are based as primarily national organizations. The American College of Surgeons (ACS) and Society of Thoracic Surgeons (STS) are two such organizations although each permits international membership as either standard membership for Canadians or international membership for others. Other surgical societies are international in nature, the European Society of Thoracic Surgery (ESTS), for example, while others are regional (Southern Thoracic Surgical Association (STSA), Pacific Coast Surgical

Society) or even local (Boston Surgical Society, San Francisco Surgical Society). Some surgical societies are specialty specific, the American Pediatric Surgical Society, for example, while others are far more broad-based, the American College of Surgeons as an example. Some such organizations are focused on a primarily academic membership (American Surgical Association (ASA), Society of University Surgeons (SUS)), while others design their membership criteria to attract as many practitioners in the field as possible (the Society of Thoracic Surgeons as an example). Others, such as the North American Spine Society, are focused on a particular type of surgery yet include surgeons of different specialties (both neurosurgery and orthopedics in this example) and actively solicit membership of "affiliated" practitioners such as nurses, physical therapists, nurse practitioners and chiropractors. There are also some surgical societies based on specific academic affiliation— the Coller Society of the University of Michigan, for example. Finally, one could consider quasi-regulatory bodies such as the American Board of Surgery as a "professional surgical society" since it is composed primarily of members of the profession although its role is distinctly different from the other societies listed above.

Given the diversity of surgical professional societies, it should not be surprising that their missions are also quite diverse. While the purpose of this chapter is not to categorize all surgical societies, the specific interests and goals of the societies generally reflect their membership, and their missions are often apparent from published mission statements. For example, the mission of the American College of Surgeons is "improving the care of the surgical patient and to safeguarding standards of care in an optimal and ethical practice environment" [5]. Similarly, the mission of the Society of Thoracic Surgeons is to "enhance the ability of cardiothoracic surgeons to provide the highest quality patient care through education, research, and advocacy" [6]. In a similar vein, the goal of the North American Spine Society is to utilize "education, research and advocacy to foster the highest quality, ethical, value- and evidence-based spine care for patient"

[7]. While the motto of the American Association for Thoracic Surgery is "A Century of Modeling Excellence," its bylaws indicate that its purpose is "to associate persons interested in, and carry on activities related to, the science and practice of thoracic surgery, the cure of thoracic disease and the related sciences; to encourage and stimulate investigation and study that will increase the knowledge of intrathoracic physiology, pathology and therapy, and to correlate and disseminate such knowledge; to hold scientific meetings featuring free discussion of problems and developments relating to thoracic surgery, and to sponsor a journal for the publication of scientific papers presented at such meetings and other suitable articles" [8].

On a more general level, the ethical underpinnings of professional medical associations have been examined relatively infrequently. Pellegrino and Relman pointed out that "the history of professional medical associations reflects a constant tension between self-interest and ethical ideals that has never been resolved" [9]. They went on to warn against focusing too much attention on the economic concerns, avoiding excessive revenue generation from for-profit business ventures (unrelated to the central purposes of the organization should be avoided), avoiding co-branding with commercial entities, and ensuring editorial independence of owned scientific journals. Rothman went on to amplify the concerns about interactions between medical professional organizations and commercial entities [10]. More recently, Minkoff and Ecker examined the differences between guild interests and professional obligations: the former being more concerned with the interests of the members of the guild, while the latter putting emphasis on responsibilities to patients [11].

Ethics in Surgical Professional Organizations

Given the diversity of missions, it should not be surprising that the role of ethics in surgical societies also varies widely. As noted above, some surgical societies feature ethics in some fashion in their very mission statement. Others have ethics committees, and others combine ethics, professional standards, and discipline. Some include ethics as an educational opportunity, while some have set up special task forces to address specific ethical issues relevant to their society's membership. The Royal Australasian College of Surgeons, for example, has focused considerable efforts into combating bullying, discrimination, and sexual harassment in surgery in Australasia and the American Academy of Otolaryngology-Head and Neck Surgery has addressed issues related to expert witness testimony [12, 13].

While some professional organizations have strayed into controversial ethical and political waters, Vogelstein has argued against doing so [14]. Interestingly, Rothman has argued exactly the opposite—suggesting that medical professional organizations expand their advocacy efforts [10].

From an ethical point of view, this approach represents the so-called political ethics, representing the relationship between professional associations and their community.

As to specific surgical societies, the situation of the American College of Surgeons (ACS) comes first. This organization of roughly 25,000 surgeons in the United States and Canada as well as international members incorporates the concept of an ethical practice in its mission statement—a fact that suggests that the organization places a high level of importance on sound ethical practices. The ACS has a Committee on Ethics whose primary responsibility is to coordinate presentations related to surgical ethics at the organization's Annual Congress. The Committee sponsors an Ethics Colloquium, a session consisting of peer-reviewed abstracts, and it sponsors an annual named lectureship devoted to some topic related to surgical ethics. Furthermore, the ACS supports ongoing education in surgical ethics through financial support of fellows to participate in the University of Chicago MacLean Center's Ethics Fellowship. The ACS Committee on Ethics has a focus on ethics, as opposed to professional standards and discipline—which is addressed by a separate committee—and has recently sponsored presentations and discussion covering issues such as the role of residents (trainees) in medical missions, new concepts in informed consent, and surgical futility and has recently published a textbook, *Ethical Issues in Surgical Care* [15]. Notably, the ACS Committee on Ethics is not involved in professional advocacy, discipline, or societal and professional standards as the organization has separate committees that address these areas. Furthermore, the ACS publishes its professional standards, not through its ethics committee, which as noted above focuses more on ethics education, but through its Statements of Principles [16]. This series of statements describes the College's code of professional conduct and addresses a wide range of issues ranging from responsibilities of the surgeon, to commitment to education and research, and to specific topical issues such as simultaneous or overlapping operations.

The Society of Thoracic Surgeons also has an ethics committee—however its title is "Standards and Ethics Committee." This committee serves to "represent the Society, under the direction of the Board of Directors, in matters relating to standards of conduct in the specialty and in matters pertaining to medical ethics and discipline which involve members of the Society" [17]. While the committee's deliberations are confidential, the STS has published a number of statements related to professional standards that are based, at least in part, on the recommendations of the committee, for example, the organization's Code of Ethics, and policy statements on "Ethical Standards for Cardiothoracic Surgeons Relating to Industry" and "Statement on the Physician Acting as an Expert Witness" [18, 19]. As part of its educational efforts, the STS sponsors a luncheon debate on a current ethical topic and, jointly with the American Association for Thoracic Surgery, sponsors the Cardiothoracic Ethics Forum which, in turn, sponsors an annual surgical ethics course.

While not wishing to focus solely on cardiothoracic surgical organizations—but using them as examples—the American Association for Thoracic Surgery has supported an annual ethics course as a supplement to its annual meeting, and both the Society of Thoracic Surgeons and the Southern Thoracic Surgical Association have included ethics debates as standard part of the programs of their annual meetings. Other surgical organizations have acted similarly. For example, the Society of American Gastrointestinal and Endoscopic Surgeons (SAGES) has incorporated ethics based talks, primarily focused on the introduction of new technology, into its annual meeting program.

Two additional examples include the American Pediatric Surgical Association and the Royal Australasian College of Surgeons. The American Pediatric Surgical Association has an ethics committee whose purpose is to "provide education that is relevant to the active practice of pediatric surgery" and which will "provide resources for resolution of ethical dilemmas and a forum for discussion of particular cases" [20]. The Royal Australasian College of Surgeons has adapted into what appears unique, or at least unusual, in that its ethics committee is registered Human Research Ethics Committee. Specifically, the Ethics Committee "reviews research proposals, ensuring the privacy, welfare and rights of those involved are upheld, offers independent advice to RACS via the Professional Development and Standards Board on ethical matters, and aims to develop policies and position statements relevant to the protection of those taking part in research projects" [21].

Controversies in How Surgical Societies Handle Ethical Issues

What, then, should surgical societies do with respect to ethical issues? While this, obviously, should be left to the individual society, the answer depends on the mission of the organization—whether it is primarily focused on education of its members, advocacy for a certain issue, or a guild-like role of advocating for its members—or a combination thereof. One need that crosses essentially all professional societies is that of internal assessment: is the organization appropriately balancing its multiple roles, or is it becoming too focused on one area—profit-making interest at the expense of education of its members, for example. While this may be done through informal ongoing discussions among leaders of the organization, it can also be accomplished through discussions of specific controversies germane to the organization.

Surgical societies periodically face "hot button" issues that affect significant numbers of their members—or even their patients. One recent example of this relates to the role of social media in the interactions between individual surgeons and their patients (and patients' families). While this has come up recently in several surgical organizations, how each organization manages the debate will be specific to the organization: one may deal with it in an ethics committee, another may deal with it using an ad hoc committee dedicated specifically to the matter, and yet another may deal with the issue through a committee dedicated to the organization's internal policies and standards [22, 23]. Furthermore, dissemination of the organization's recommendations may be published either internally or externally through the forms of web pages, white papers, journal editorials, or other means. One potential problem, however, is that different organizations may well come to different conclusions.

How should surgeons and surgical organizations handle situations where codes of conduct, or policies, conflict with each other? This situation can easily arise when an individual surgeon belongs to several different surgical societies and each has a different mission. One example of this is the issue of live surgical broadcasts as an educational activity. While there may be advantages to watching a live broadcast of an operation, it can be argued that it creates divided loyalties for the surgeon. While he or she has the obvious obligation of doing his or her utmost for the patient at hand, he or she also has an educational responsibility. Is it possible that the patient's operation could be compromised—perhaps unnecessarily prolonged—in order to make an educational point? Is it possible that the surgeon's attention is diverted away from the patient in order to answer a question from the audience? While the answer to both of these is obviously "yes," one would hope that the surgeon would do his/her best to avoid any compromise in patient care.

Nonetheless, the potential drawbacks of a live broadcast should be balanced against any benefits. Would a carefully edited recording not provide a more efficient and directed educational product? Is "real-world" decision-making lost in the presentation of an edited product? While reasonable people may disagree on such questions, it is clear that different surgical professional organizations have come down on opposite ends of the debate. Is one wrong and the other right? In such cases, the wording of each organization policies may be critical. Is the organization simply not sponsoring live surgical broadcasts or is it enjoining its members from participating in such activities? While the first is simple to parse for the individual member—the responsibility of sponsoring an activity lies with the organization, not the individual member—the latter is more problematic and may reflect subtle differences in priorities of the organizations, one where the focus is on peer education and the other where patient advocacy is paramount.

Concluding Remarks

- The role of ethics in surgical societies is a broad one and one that is determined by the mission of the organization, the structure of the organization, and the needs of its members.
- There is no single way to incorporate ethics into a surgical society, but as Pellegrino and Relman stated, "medicine is, in essence, a moral enterprise," and, as such, no surgical organization can completely escape the need to deal with ethics at some level [9].
- Whether the organization simply looks internally and assesses itself, whether it makes ethics a central part of its educational offerings, whether it takes on larger, more controversial, issues such as global medicine and the role of poverty in healthcare, or is somewhere in between will depend on the goals of the organization and the needs of its members.
- While surgical societies may address issues specifically related to surgery, they may take on more expanded roles—advocating for their

members patients and advocating for specific political goals—but they must understand that other surgical and medical societies will do the same and that one of the dangers of entering the political realm is that while the ethics of an issue may be clear from one perspective, even a moral perspective, the development of policies that can actually be enacted needs to be tempered by the political realities of the time.

References

1. Beauchamps TL, Childress JF. Principles of biomedical ethics. 7th ed. New York: Oxford University Press; 2013. p. 3.
2. Beauchamps TL, Childress JF. Principles of biomedical ethics. 7th ed. New York: Oxford University Press; 2013. p. 13.
3. Fox R. Is medical education asking too much of bioethics? Daedalus. 1999;128(4):1–25.
4. Rothman DJ, McDonald WJ, Berkowitz CD, et al. Professional medical associations and their relationships with industry: a proposal for controlling conflict of interest. JAMA. 2009;301(13):1367–72. https://doi.org/10.1001/jama.2009.407.
5. https://www.facs.org/about-acs.
6. https://www.sts.org/about-sts.
7. https://www.spine.org/WhoWeAre/Leadership Governance/AboutUs.
8. http://aats.org/aatsimis/AATS/Association/By-Laws_and_Policies/By-Laws/THE_BY-LAWS.aspx.
9. Pellegrino ED, Relman AS. Professional medical associations ethical and practical guidelines. JAMA. 1999;282(10):984–6. https://doi.org/10.1001/jama.282.10.984.
10. Rothman DJ. Medical professionalism — focusing on the real issues. N Engl J Med. 2000;342:1284–6.
11. Minkoff H, Ecker J. When guild interests and professional obligations collide. Obstet Gynecol. 2017;130(2):454–7. https://doi.org/10.1097/AOG.0000000000002138.
12. https://www.surgeons.org/news/racs-apologises/.
13. Svider PF, Eloy JA, Baredes S, Setzen M, Folbe AJ. Expert witness testimony guidelines: identifying areas for improvement. Otolaryngol Head Neck Surg. 2015;152(2):207–10. https://doi.org/10.1177/0194599814556721. Epub 2014 Nov 11.
14. Vogelstein E. Professional hubris and its consequences: why organizations of health-care professions should not adopt ethically controversial positions. Bioethics. 2016;30(4):234–43. https://doi.org/10.1111/bioe.12186. Epub 2015 Aug 26.
15. Ferreres AR, Angelos P, Singer EA, editors. Ethical issues in surgical care. Chicago: American College of Surgeons; 2017.
16. https://www.facs.org/about-acs/statements/stonprin.
17. https://www.sts.org/about-sts/governance-and-leadership/committees-council-operating-boards-workforces.
18. https://www.sts.org/about-sts/policies/cardiothoracic-surgical-organizations-standards-interactions-companies-0.
19. https://www.sts.org/about-sts/policies/statement-physician-acting-expert-witness.
20. http://www.eapsa.org/about-apsa/committees/?cid=ETHICS.
21. https://www.surgeons.org/about/governance-committees/committees/ethics-committee/.
22. Rouprêt M, Morgan TM, Bostrom PJ, et al. European Association of Urology recommendations on the appropriate use of social media. Eur Urol. 2014;66(4):628–32. https://doi.org/10.1016/j.eururo.2014.06.046. Epub 2014 Jul 16.
23. https://www.facs.org/about-acs/statements/106-social-media.

Ethical Issues in Surgical Research

Richard Jacobson, Laurel Mulder,
and John Alverdy

Key Points

- Historically, ethical guidelines were established "ad hoc" in response to unethical research behavior. Such guidelines did not offer a complete means of evaluating the ethics of study design.
- Emanuel et al. in 2001 proactively proposed seven requirements that are necessary and sufficient to evaluate the ethical conduct of human subject research.
- Intrinsic barriers make scientific validity, and thus ethical research conduct, more complex in the field of surgery compared to nonsurgical fields.
- The "IDEAL" recommendations for surgical innovation simplify and standardize the ethical conduct for the practice of novel surgical techniques.
- Special topics: communication during informed consent, sham surgery, and authorship.

Surgical research has been criticized for producing low-quality evidence in the form of case series rather than randomized controlled trials. Performance of randomized trials for surgical intervention is inherently more difficult than similar investigation in the medical world due to the nature of surgical disease and therapy. However, in the past decade, the stages of development of surgical technique have been formalized in ways that maximize scientific validity and thus ethics. Surgeons should be aware of what constitutes research versus personal variations in technique and the protocols for developing novel techniques.

Quality improvement (QI) and quality improvement research have increased dramatically in recent years. The line between quality improvement and human subject research is often ambiguous, and although risks to patients are frequently limited to privacy concerns, investigators should be aware of what defines pure QI versus human subject research and the ethical guidelines governing each.

R. Jacobson
University of Chicago, Department of Surgery, Chicago, IL, USA

Rush University Medical Center, Department of Surgery, Chicago, IL, USA

L. Mulder
Rush University Medical Center, Department of Surgery, Chicago, IL, USA

J. Alverdy (✉)
University of Chicago, Department of Surgery, Chicago, IL, USA
e-mail: jalverdy@surgery.bsd.uchicago.edu

© Springer Nature Switzerland AG 2019
A. R. Ferreres (ed.), *Surgical Ethics*, https://doi.org/10.1007/978-3-030-05964-4_10

The hierarchical nature of surgical departments often places investigators in difficult position when assigning authorship. Formalized guidelines are in place to aid in this process; however ultimately communication between all participants is paramount to prevent and avoid unethical and painful situations.

Surgical research, like clinical surgery, creates sustainable value in the societies it serves only if practiced ethically. Surgeons are uniquely positioned in the research community due to the history and culture of our profession. Surgical research is governed by the core ethical principles that apply to all biomedical investigation. However, nuances of surgical disease [1] and the surgeon-patient relationship [2], along with the need for research in surgical technique and devices, are the factors that warrant specific consideration of ethical conduct in surgical research.

Recognizing that it would be difficult to review or identify all ethical dilemmas that confront surgeons involved in research, this chapter will focus in the following aspects: (1) outlines common and critical issues, (2) provides the reader with an understanding of the principles guiding the ethical conduct of surgical research, and (3) explains the context in which these principles arose. Box 1 highlights the idea that we believe underlies all ethical analysis of research conduct.

Box 1 Overarching Theme of Ethical Principles Guiding Surgical Research
Human research subjects are contributors to the common good. They place themselves at risk in order to advance societal knowledge of biology and the treatment of disease in others. As such, investigators owe their subjects exhaustive self- and external examination of their methods with an appreciation of established ethical guidelines in mind and the recognition that their subjects are providing an invaluable resource to both society and the investigator.

Historical Perspective

The ethical conduct of research on human subjects in the modern era is grounded in the failures of the past. In our profession's early history, there are countless examples of unethical research practices including nontherapeutic vivisection and forced sterilization carried out on vulnerable populations under the "care" of surgeons [3]. Out of atrocities committed in Germany in the 1930s and 1940s rose the Nuremberg Code, which established the principles of voluntary informed consent and minimalization of risk to subjects [4]. Expanding on the Nuremberg Code, the Helsinki Declaration was developed in 1964 and subsequently revised by the World Medical Association to protect the rights of the individual research subject from violation in the name of any greater societal good [5]. More recently, the National Commission for the Protection of Human Subjects of Biomedical and Behavioral Research released the Belmont Report, which codified the concepts of respect for persons, beneficence, and justice in response to grossly unethical human experimentation that took place in the Tuskegee Syphilis Study [6]. These concepts were the foundation of the Common Rule established in 1991 by the American Department of Health and Human Services to govern baseline ethical requirements for study design and implementation in government-funded human subject research [7]. The Common Rule established institutional review boards (IRBs) to protect the rights of research subjects and provided additional protections for populations vulnerable to harm from unethical research practices including prisoners, children, and pregnant women. Updates to the Common Rule will go into effect in 2018, the most notable of which is the waiver of informed consent for biobanking of unidentified specimens [8]. These landmarks have been aptly described as the "property of all humanity [9]."

Unifying Principles of Ethical Research Conduct

An important distinction exists between clinical research and individualized patient care. The overarching goal of research is hypothesis testing and advancement of general knowledge. Patient care is the application of a previous hypothesis test with the primary goal of maximizing well-being to the individual [10]. This distinction allows that the ethical treatment of research subjects may differ from ethical treatment of patients in clinical practice. Biomedical research necessarily implies some degree of risk to individual subjects. If an intervention were known to improve outcomes in the absence of risk, it would be the standard of care, and investigation would be unnecessary. Research subjects volunteer for risk exposure and treatment in a fashion that may not maximize individual well-being, whereas treatment of a patient in clinical practice in any fashion other than one intended to maximize individual well-being would be unethical.

The potential for harm to subjects mandates scrutiny of the ethics of proposed research prior to the induction of a study. In 2001, Emanuel et al. published a series of requirements for evaluation of the ethics of clinical research studies. The authors recognized that the various academic and governmental guidelines in existence at the time left investigators with incomplete and often unclear instructions for ethical conduct of research. This was attributed to the reactionary, ad hoc establishment of ethical guidelines for research that persisted through the late twentieth century. Prior to these guidelines, no single set of rules was both necessary and sufficient to determine whether an investigation was ethical. The requirements described below were designed as a unified code of ethics for clinical research. They serve a parallel purpose to the accepted guiding principles of ethical clinical practice: beneficence, nonmaleficence, respect for autonomy, justice, respect for dignity, and veracity [11]. The requirements assume honesty and responsibility from all parties and apply to clinical research in all forms. The seven requirements correspond to the respective phases of investigation: development, implementation, and review. While they are not specific to surgery, it is our belief that these requirements are the most complete and applicable tool for evaluating the ethics of surgical research. The requirements as described in the original manuscript are summarized in Box 2

[12]. Requirements 1, 2, 3, 4, and 7 are considered necessary for the conduct of ethical research. Requirements 5 and 6 are in place to ensure enforcement of the necessary requirements, minimize conflicts of interest, and ensure subject autonomy.

> **Box 2 Summary of the Seven Requirements for Ethical Conduct of Clinical Research [12]**
>
> 1. *Social or scientific value*: This requirement applies to hypothesis development and requires familiarity with the current knowledge base and needed future directions of study in a field. Human research subjects are necessarily exposed to risk. To be considered ethical, the proposed research must include potential benefit to the well-being of patients or add to the scientific knowledge base. This prevents the exploitation of subjects and wasting of scarce resources on hypothesis tests without potential social or scientific benefit. A proposal to evaluate the relationship between eye color and severity of cholecystitis would violate this requirement.
> 2. *Scientific validity*: This requirement applies to methods of data collection and analysis and requires statistical expertise, familiarity with study populations, and pragmatism. Even if performed to test a potentially beneficial hypothesis, invalid data collection methods negate the value of the hypothesis test. This requirement also specifies that intentionally underpowered trials lack validity and thus true scientific value. Adherence to this principle prevents exploitation of subjects and wasting of resources. A proposal to evaluate operative versus nonoperative management of acute appendicitis with five patients per group would violate this requirement.
> 3. *Fair subject selection*: This requirement applies to decisions on recruitment

along with inclusion and exclusion criteria. It requires knowledge of epidemiology, vulnerable populations, and an understanding of the virtue of justice. Fair subject selection requires that selection of the study population maximizes validity and minimizes risk and that the population being studied stands to benefit from the research being conducted. Under this principle, a valuable and scientifically valid investigation conducted at a safety net hospital that in the end would disproportionately benefit the rich and powerful is considered unethical.

4. *Favorable risk-benefit ratio*: This requirement applies to study design and implementation. It requires extensive scientific knowledge in the field of study and social values. Favorable risk-benefit ratio requires a study protocol that minimizes risk and maximizes benefit to individual subjects and ensures that the benefits of participation outweigh the risks. Maximization of benefit to the individual is constrained to improvements in well-being. Extraneous benefits such as compensation do not weigh into this calculus. This requirement importantly assumes that the extent and probability of potential benefit and harm to the individual is clearly understood by the patient. The calculus allows that societal benefits may in some cases outweigh risk to the individual, as in the case of phase 1 safety trials and studies where risk to subjects is minimal. A study evaluating management of asymptomatic lipoma with chemoradiation versus radical excision violates this requirement.

5. *Independent review*: This requirement applies to study design and serves to prevent conflict of interest in investigators. Moral hazard exists for investigators in the development, implementation, and review of human subject research. IRBs and data and safety monitoring boards exist to minimize the impact of external pressures such as monetary gain and the desire to advance one's career on the completion of high-quality research. The review process also helps to ensure that principles of ethical research are followed.

6. *Informed consent*: This requirement applies to enrollment and requires scientific knowledge, communication skills, and respect for autonomy. The principle of informed consent requires that patients have full knowledge of the risks, benefits and alternatives to study participation, and full autonomy free from coercion in the decision to participate in research. There are two notable exceptions to these criteria: emergency situations where genuine clinical equipoise between two treatments exists and in the case of surrogate decision-makers practicing substitute judgment for patients unable to make decisions regarding participation in the trial. The responsibility of the surgeon-scientist involved is first to the research subject and second to the community intended to benefit from their research. Surgeons have a fiduciary responsibility to individual patients that supersedes their role as scientists. To minimize conflict of interest, persons other than the principle investigator of a study should perform the informed consent with potential subjects.

7. *Respect for potential and enrolled subjects*: This requirement applies to enrollment and accrual. It ensures the protection of privacy and subject well-being and requires scientific knowledge and communication between subjects and investigators. To fulfill this requirement, investigators must (1) respect rules regarding protected health infor-

mation, (2) keep subjects apprised of developments in their individual health and the study intervention, (3) allow subjections to freely withdraw from studies, (4) vigilantly monitor well-being of subjects throughout the study and care for them accordingly, and (5) disclose to subjects the results of the study and what knowledge or benefit their participation yielded.

Intrinsic Barriers to Ethical Surgical Research

The ideal investigation of an intervention would be a multicenter, double blind randomized controlled trial that includes relevant patient groups without undue risk to each individual subject. Aspects of surgical practice and culture frequently prohibit such investigations. Many current surgical procedures were passed along through generations of hierarchical training programs and "grandfathered" in to practice without rigorous scientific comparison to alternative approaches [13]. These intrinsic factors distinguish surgical from nonsurgical research and warrant consideration in modern study design.

First, variations in methods and surgical expertise add a layer of complexity to the design of surgical research. While nonsurgical investigations do involve a level of diagnostic skill and critical thinking, this requirement is compounded with the technical abilities of individual surgeon in surgical investigations. Standardization of delivery of a medical therapy is more straightforward than standardization of surgical management for research purposes. Varying availability of instruments and care pathways among institutions raises concerns about the internal validity of multicenter trials. In the past, the single-surgeon or single-center case series was the solution of choice to standardization of methods [14]. However, this method leads to poor generalizability of results thus decreasing external validity of the results. One surgeon's outcomes under study conditions may differ vastly from typical outcomes of the same procedure in the community at large.

Learning curves and individual expertise have a role in the implementation of surgical therapy that is absent in nonsurgical interventions. Outcomes early in the life cycle of an operation generally do not match those that can be expected when the procedure is established. For instance, outcome improvements along the learning curve for laparoscopic colectomy are well described in studies including surgeons at various levels of experience [15]. This too impacts internal validity, as the results of a trial comparing various interventional therapies necessarily depend upon the point in each therapy's life cycle at which the trial takes place.

Surgical intervention often represents a physiologic challenge to the patient that is not encountered in trials of medical therapy. This can be prohibitive to the development of randomized trials comparing operative to nonoperative treatment, in the form of channeling bias [16]. Channeling bias occurs when investigators rightly assign patients to one intervention or another if the alternative branch of the study is considered too risky. In practice, this leads to disparate risk profiles of the patient groups in a study and necessarily decreased internal validity.

Finally, the degree of personal responsibility that surgeons take for their outcomes impacts the degree of willingness to participate in investigations of novel surgical therapies. If surgical therapy fails, the surgeon rather than the therapy itself is more often held accountable. This equates to the unlikelihood of separating their roles as care providers from their roles as scientists for surgeons participating in a clinical investigation. If a pharmaceutical approach fails, blame frequently falls on the drug and the state of the science, not the treating physician.

Pathways to Ethical Innovation

With the above concerns in mind, McCulloch et al. have developed the IDEAL model, a strategy for practical and scientifically valid surgical innovation. This pathway couples innovation and continuous evaluation using a staged rollout of novel invasive procedures. This model is summarized in Box 3.

> **Box 3 Summary of the IDEAL Recommendations for Surgical Innovation [17]**
>
> *Stage 1 – Innovation*: This stage entails proof of concept for a novel intervention, performed by select innovators for the first time on humans, whether in a planned or unplanned fashion. The procedure is still in development and best practices are not established. Authors publish complete technical descriptions of the procedure and details of patient selection in case reports. Ethics committee approval is recommended but institution-dependent. Outcomes, particularly adverse events, should be reported to avoid replication of ineffective or dangerous methods.
>
> *Stage 2a – Development*: This stage involves development of a standardized technique. The procedure remains in its infancy and is performed again on a small number of patients by innovators and early adopters. At this point prospective evaluations of safety and efficacy are performed, and the initial learning curve is established. In the past, innovation at this phase has been published as retrospective case series. However, prospective, planned out study with IRB approval is both scientifically and ethically superior and should be performed whenever possible.
>
> *Stage 2b – Exploration*: Once technique is standardized, replication outside the original center is appropriate to determine scalability of the technique. Early adopters should perform rigorous tracking of patient-reported and clinical outcomes in the form of uncontrolled prospective case series. Initial response and complication rates are critical to power analyses in the planning of randomized trials to be performed in later stages.
>
> *Stage 3 – Assessment*: If initial outcomes are comparable to the established standard of care and true clinical equipoise exists, progression to a comparative trial is appropriate. Ideally, the new procedure is compared to the existing standard in a randomized controlled trial; however as discussed above, difficulties exist in designing such trials for surgical therapy. If an RCT is not feasible, alternative comparative methods such as controlled interrupted-time series studies and others are a viable, albeit less valid alternative.
>
> *Stage 4 – Long-term study*: Established procedures should proceed to long-term monitoring of outcomes and rare events. Monitoring is generally achieved through a registry compiled with administrative data. Longitudinal data collected at a single center is valuable; however between-center comparison can be fraught with complications related to risk adjustment.

Personal developments in the art of surgical technique are considered improvements in individual patient care. Parameters such as suture material and technique, laparoscopic port placement, and patient positioning are at the discretion of the operating surgeon and should reflect what he or she believes is in the best interest of the patient. Often in the usual discourse of patient care, alterations in technique come about that, if disseminated and generalized, would contribute to the existing knowledge base and help large groups of patients in the future. An example of this would include the critical view of safety in laparoscopic cholecystectomy [18]. It is our view that if and when an investigator has intent to publish or otherwise disseminate a technique, he or she is obliged to treat the use of the technique as research, and its investigation should proceed as outlined in the IDEAL recommendations.

Special Considerations

Quality Improvement

In recent years, surgeons and trainees have increasingly participated in systematic, data-guided analysis of healthcare processes to improve the quality of care. Quality improvement (QI) initiatives measure adherence to evidence-based guidelines for processes of care across disciplines. Initiatives to continuously review practice patterns and surgical outcomes represent an attempt to maintain clinical standards of care at the level of the most recent research-based guidelines.

QI projects often resemble both research and normal clinical practice [19]. This lack of distinction often creates an ethical gray area – are QI projects subject to the same ethical requirements as human subject research? The *Belmont Report* states that research is undertaken with the intent to develop generalizable knowledge, whereas practice is intended only to improve the well-being of an individual. Kass et al. argue that in current practice patterns, this distinction is nearly impossible, as the goal of delivering the best possible care to the individual patient is often inseparable from the continuous processes of institutional learning and improvement. Practice-derived data drives production of generalizable knowledge, and that knowledge is rapidly incorporated into clinical practice in an iterative cycle of analysis and implementation [20].

Entities identified as QI initiatives range from departmental morbidity and mortality conferences to the creation of vast administrative datasets such as the National Surgery Quality Improvement Program (NSQIP). As QI initiatives range widely in scope, scale, and purpose, the line between QI and human subject research is often blurred, and ethical oversight is difficult. This is further complicated by varying definitions among institutions and federal agencies. QI projects are generally subject to less oversight by institutional review boards (IRBs) than human subject research. Under the premise that continuous quality improvement is in the best interest of patients and designed to promote well-being with minimal risk, informed consent is frequently waived or considered part of a global consent for treatment during prospective QI studies. Studies that involve human subjects, but only through retrospective review of medical records, involve no physical risks to the subjects. These studies again involve minimal risk and undergo expedited IRB review without the requirement of informed consent. Studies that utilize protected health information (PHI) do involve significant risks to patient privacy, however, and this risk should be recognized and mitigated by the investigator with appropriate data security. Any study that involves more than minimal risk to the participant should be fully subject to the seven requirements for the ethical conduct of research as outlined above. It should proceed through IRB review and require informed consent on the part of the subject.

Communication During Informed Consent

Communication is the cornerstone of the informed consent process in clinical research. The burden of communication lies with the investigator and not with the patient [21]. It is necessary to explain the risks and potential benefits of participation in terms that the patient understands. This may include but is not limited to the use of illustrations, certified foreign language interpreters, and nonscientific terminology. If, after the informed consent discussion, the patient is unable to clearly explain the goals of the trial and the risks associated with participation, informed consent has not been obtained even if that patient signs a document attesting that it has. In informed consent proceedings in the clinical practice setting, the guiding principle is that of veracity, which requires complete honesty from the provider to his or her patients when conveying information about their condition and its progression and prognosis.

An informed consent for clinical research must include the attendant risks of participation along with the goals of the study and potential benefits. Participants' preconceived notions of clinical research and the specific innovations at use in the study must also be accounted for. Cultural and societal factors influence perceptions of clinical research and surgery at large but also specific interventions. Out of therapeutic optimism, patients may assume that because a surgical technology is new, it is superior to previous methods [22]. In the setting of surgical innovation, this is not necessarily true, and again, the burden of truth lies with the investigator. It follows that in discussing an investigative robotic approach to surgery with a patient, the surgeon would be justified in explaining to the patient that use of the robot offers increased range of motion through articulation inside the abdomen. However, if, for example, the surgeon fails to disclose that outcomes in robotic-assisted cholecystectomy are not superior to conventional laparoscopy (at least at the present time), they have committed an unethical lie of omission.

Sham Procedures

The relative paucity of randomized controlled trials in the surgical literature is well documented. This has been attributed to difficulties with study design, particularly in blinding patients and surgeons to the treatment group, and the ethics of sham surgery. Sham-controlled investigations historically have been utilized to dispute the utility of procedures that were "grandfathered" into practice based on basic pathophysiology and animal studies, but never rigorously tested. Most frequently, sham operations are used to remove ineffective procedures from general clinical practice. Such has been the case for arthroscopic procedures in degenerative meniscal tear [23], gastric "freezing" for duodenal ulcer [24], and internal mammary artery ligation for angina [25]. Investigators frequently encounter moral hazard in their approach to such studies, given that surgeons in a fee-for-service system may suffer financially if a procedure is disproven.

Sham surgery is often necessary to maximize scientific validity in investigations of invasive procedures. A sham operation controls for the placebo effect, which may be pronounced in surgical compared to medical therapy due to higher levels of therapeutic optimism [16]. However, unlike placebo controls in randomized trials of medical therapy, which are biologically inert substances that cause no harm to the subject, sham surgery puts the research subject at risk of pain and complications of anesthesia. Opponents criticize the idea that subjects are necessarily harmed without any reasonable expectation of improvement in their condition. However as discussed above, the goals of clinical research differ from individualized patient care, and a small amount of risk is acceptable if it is a necessary component of an ethically designed study. Phase 1 clinical trials regularly expose human subjects to untested pharmaceuticals with the potential to do harm so that the rest of society may benefit from drug safety data. Some authors pose that the risks of sham surgery are not categorically distinct from the risk to subjects in phase 1 trials [26]. It is understandable, however, that surgical investigators have misgivings about causing physical injury to subjects, a reality rarely confronted by researchers in nonsurgical fields.

Industry Relationships

The use of surgical instruments and materials manufactured by corporate entities in the healthcare industry necessitates a relationship between surgeons and the makers of their tools. Often, industry representatives have expertise in the use of instruments or materials that exceeds that of the operating surgeon, particularly early in the life of a device. The American College of Surgeons released guidelines for the presence and role of healthcare industry representatives in the operating room as it relates to individual patient care [27]. To our knowledge no similar set of guidelines exists to govern the role of industry in surgical research investigating the use of devices and materials. A meta-analysis of industry vs nonprofit-funded research showed a clear loss of clinical equipoise with bias toward the implementation of industry-sponsored products [28]. We recognize the importance of industry relationships to foster innovation and advances in clinical care and see a clear role for industry in hypothesis development and procurement of materials. However, we believe it is self-evident that profit motive has no place in the results and conclusions of a hypothesis test. Disclosure of industry relationships in the presentation and publication of scientific data has been significantly bolstered by recent requirements for public reporting of financial relationships between physicians and industry.

Authorship

Publication is essential in academia for career advancement and promotion. Publication also has social and financial implications. Ethical dilemmas arise when attempting to give appropriate credit to those who deserve it and avoiding listing those members who did not contribute in a meaningful manner.

Efforts to standardize criteria for authorship have been made to avoid ethical issues. The International Committee of Medical Journal Editors (ICMJE) defines the role of authors and contributors based on four criteria: (1) substantial contributions to the conception or design of the work or the acquisition, analysis, or interpretation of data for the work, (2) drafting the work or revising it critically for important intellectual content, (3) final approval of the version to be published, and (4) agreement to be accountable for all aspects of the work in ensuring that questions related to the accuracy or integrity of any part of the work are appropriately investigated and resolved. If there are other contributors that do not meet all four criteria, they should not be listed as authors, but should be acknowledged for their contribution to the work. By establishing these standardized criteria, there is less ambiguity in terms of who should be included as an author [29].

Pitfalls exist, however, when following this system. For example, in collaborative projects, no one author may fit all four criteria. These standards can be used as a guideline, but each case should be viewed independently. Because of this, authorship should be determined at the beginning of a research project. This allows for roles to be clearly defined prior to writing a manuscript [30]. It is equally important for journals and editors to have written authorship guidelines.

The concept of authorship applies widely across academics but also specifically to the surgeon-scientist. Often surgical residents are required to have research experience during their training. Resident research programs supported by faculty mentorship are essential during training to teach these principles early in an individual's career. The definition of "first" and "senior" authorship can vary greatly between fields. Authorship is unfortunately not always determined by contribution but instead unethically "gifted" to individuals based on seniority or honor [30]. Mentors should recognize how the power dynamics of mentor-mentees can be problematic in determining authorship [31]. This stresses the importance of having authorship discussions early in the research process. Principal investigators/attending surgeons should make clear what is expected of the junior researcher to achieve first authorship. Furthermore, mentors in this situation should be aware of their power and be certain not take advantage.

The gender gap is also a consideration as a part of the ethical debate of authorship. The number of women in medicine has increased drastically in recent years, but women are still underrepresented in academic surgery. This directly applies to women's involvement in publication and authorship as well. In female-dominated fields such as obstetrics and pediatrics, there has been an overall increase in women as first and senior authors. However, the numbers are still low in surgical journals. This may be related to the number of women in the field but it is clear that a gap still exists. The gap is likely related to the lack of senior women available to merit these roles. There are barriers to academic advancement of women, specifically the constraints of traditional sex roles, manifestations of sexism, and lack of effective mentors [32]. Career choice differences between men and women may also play a role. In order to increase the number of women represented in surgical journals, it is crucial that effective mentorship programs begin early in surgical training.

There are many factors to be weighed in determining authorship. Overall it is essential to provide framework and guidelines for authorship at the onset of each individual research project. Furthermore, effective mentor-mentee programs can assist to lay down fundamentals for determining authorship and establish clear roles for all individuals involved in the project.

Ethical Concerns for the "Basic Science" Surgeon-Investigator

Certain ethical concerns apply specifically to the basic science surgeon-investigator. It is essential that reproducible, unbiased scientific knowledge is produced. Open publication of methods and data, collaboration between labs, and peer review allow others to confirm or raise questions about results [33]. In recent years, publication of raw data from basic science studies in supplemental figures has become increasingly common [34]. We believe this practice represents a large step forward in the ethical conduct of research.

Replication and repetition must be taken into consideration in all experiments. Replication refers to multiple experimental runs independent of one another, probing variability between separate runs. In contrast, repeat measurements are taken during the same experimental run. It is advantageous to triplicate (or more) experimental runs for statistical reasons. This increases the sample size and thus precision and accuracy of the measurements. It is necessary to recognize that replicates do not necessarily allow interpretations to be made or allow us to draw conclusions about the hypothesis being tested. This is due to the idea that the samples are independent and therefore inferences can only be made about the population from which they are drawn. However, replicating data can act as an internal quality check on how the experiment was performed. Although replicability is important, it can be expensive or impractical in certain situations. In any case, all methods should be detailed, and scientists should be transparent about potential difficulties in replicability [35]. Communication with corresponding authors for clarification of methods is encouraged if questions of replicability arise.

Reproducibility assumes changes to be present in a distinctive setting, while replicability attempts identical conditions [36]. Reproducibility differs from replicability in the amount of variability present and relates to the generalizability of a finding. Casadevall et al. point out that when it is stated that something is reproducible, it is actually meant that it was replicated. Best practice involves repeating experiments on separate occasions, with each experimental run in triplicate. Results and figure legends in publication should be specific on how rigorously reproducibility was tested.

Finally, honest and accurate reporting of data is a fundamental ethical practice. Statistical outliers are frequently removed for analysis of data and are assumed to be nonsignificant. Other practices such as "massaging" data to make it fit, expected, or hoped-for outcomes should be discouraged. Outliers should still be reported even if not included in the final analysis, and the reason for exclusion should be explicatively stated. All authors and contributors to the final paper are accountable for the data that is reported and analyzed [33].

Concluding Remarks

- The ethical conduct of any human subject research requires that investigators be familiar with the best available guidelines as discussed in this chapter. These guidelines, if applied with an appreciation for the humanity of their subjects and their position as contributors to the common good, help to prevent unethical practices.
- Research into surgical therapy is intrinsically different from research into medical therapy; thus research practices necessarily differ. However, the guiding principles of ethical research conduct apply to surgical research.
- Quality improvement initiatives represent an ethical gray area between clinical practice and human subject research. Some forms of qual-

ity improvement are exempt from the rules that govern human subject research; however as a rule, investigation intended to produce generalizable knowledge, whether for broad publication or not, should be considered human subject research.

- Authorship should be discussed openly by all stakeholders at the onset of an investigation, rather than post hoc, and should be determined through objective evaluation of contributions cross-referenced with existing guidelines for authorship.

Glossary

Human subject research A systematic investigation designed to produce generalizable knowledge from observations of human subjects. This term applies broadly, and investigations classified as human subject research are generally subject to IRB review.

Protected health information (PHI) Personally identifiable health information (by which the identity of a study subject could be ascertained) maintained in a medical record that includes data on physical health, mental health, payment information, or genetic information.

Clinical equipoise A state in which two or more therapeutics exist that could treat a given condition; however a lack of strong evidence regarding superiority of either treatment exists. Equipoise is essential to the ethical conduct of clinical research.

Internal validity The relative truth of conclusions drawn through experimentation. Internal validity is directly related to the accuracy with which experimental conditions eliminate confounding and minimize bias. A study with high internal validity can make strong claims regarding causality, rather than simple associations.

External validity The extent to which the results of a study apply to the population being modeled. A study with high external validity is highly generalizable to large patient populations.

Quality improvement Any systematic, data-guided analysis of healthcare processes to improve the quality of care by measuring adherence to evidence-based guidelines of clinical best practice.

References

1. Meakins JL. Innovation in surgery: the rules of evidence. Am J Surg. 2002;183:399–405.
2. Axelrod DA, Goold SD. Maintaining trust in the surgeon-patient relationship: challenges for the new millennium. Arch Surg. 2000;135:55–61.
3. Brock C. Risk, responsibility and surgery in the 1890s and early 1900s. Med Hist. 2013;57:317–37.
4. Shuster E. Fifty years later: the significance of the Nuremberg code. N Engl J Med. 1997;337:1436–40.
5. World Medical Association (AMM). Helsinki declaration. Ethical principles for medical research involving human subjects. Assist Inferm Ric. 2001;20:104–7.
6. Reverby SM. Listening to narratives from the Tuskegee syphilis study. Lancet. 2011;377:1646–7.
7. Rothstein MA. Currents in contemporary ethics. Research privacy under HIPAA and the common rule. J Law Med Ethics. 2005;33:154–9.
8. Menikoff J, Kaneshiro J, Pritchard I. The common rule, updated. N Engl J Med. 2017;376:613–5.
9. Human D. Conflicts of interest in science and medicine: the physician's perspective. Sci Eng Ethics. 2002;8:273–6.
10. Kass NE, et al. The research-treatment distinction: a problematic approach for determining which activities should have ethical oversight. Hastings Cent Rep. 2013;Spec No:S4–S15.
11. Snyder J, Gauthier C. Evidence-based medical ethics: cases for practice-based learning. New York: Humana Press; 2008.
12. Emanuel EJ, Wendler D, Grady C. What makes clinical research ethical? JAMA. 2000;283:2701–11.
13. Horton R. Surgical research or comic opera: questions, but few answers. Lancet. 1996;347:984–5.
14. Weil RJ. The future of surgical research. PLoS Med. 2004;1:e13.
15. Bennett CL, Stryker SJ, Ferreira MR, Adams J, Beart RW Jr. The learning curve for laparoscopic colorectal surgery. Preliminary results from a prospective analysis of 1194 laparoscopic-assisted colectomies. Arch Surg. 1997;132:41–4.; discussion 45.
16. Paradis C. Bias in surgical research. Ann Surg. 2008;248:180–8.
17. McCulloch P, et al. No surgical innovation without evaluation: the IDEAL recommendations. Lancet. 2009;374:1105–12.
18. Strasberg SM, Brunt LM. Rationale and use of the critical view of safety in laparoscopic cholecystectomy. J Am Coll Surg. 2010;211:132–8.

19. Lynn J, et al. The ethics of using quality improvement methods in health care. Ann Intern Med. 2007;146:666–73.

20. Varkey P, Reller MK, Resar RK. Basics of quality improvement in health care. Mayo Clin Proc. 2007;82:735–9.

21. Paasche-Orlow MK, Taylor HA, Brancati FL. Readability standards for informed-consent forms as compared with actual readability. N Engl J Med. 2003;348:721–6.

22. Horng S, Grady C. Misunderstanding in clinical research: distinguishing therapeutic misconception, therapeutic misestimation, and therapeutic optimism. IRB. 2003;25:11–6.

23. Sihvonen R, et al. Arthroscopic partial meniscectomy versus sham surgery for a degenerative meniscal tear. N Engl J Med. 2013;369:2515–24.

24. Ruffin JM, et al. A co-operative double-blind evaluation of gastric "freezing" in the treatment of duodenal ulcer. N Engl J Med. 1969;281:16–9.

25. Cobb LA, Thomas GI, Dillard DH, Merendino KA, Bruce RA. An evaluation of internal-mammary-artery ligation by a double-blind technic. N Engl J Med. 1959;260:1115–8.

26. Horng S, Miller FG. Is placebo surgery unethical? N Engl J Med. 2002;347:137–9.

27. Surgeons, A.C.o. Revised statement on health care industry representatives in the operating room. (2016).

28. Djulbegovic B, et al. The uncertainty principle and industry-sponsored research. Lancet. 2000;356:635–8.

29. Editors, I.C.o.M.J. Recommendations for the conduct, reporting, editing and publication of scholarly work in medical journals. (2017).

30. Agel J, DeCoster TA, Swiontkowski MF, Roberts CS. How many orthopaedic surgeons does it take to write a manuscript? A vignette-based discussion of authorship in orthopaedic surgery. J Bone Joint Surg Am. 2016;98:e96.

31. Marusic A, Bosnjak L, Jeroncic A. A systematic review of research on the meaning, ethics and practices of authorship across scholarly disciplines. PLoS One. 2011;6:e23477.

32. Jagsi R, et al. The "gender gap" in authorship of academic medical literature--a 35-year perspective. N Engl J Med. 2006;355:281–7.

33. Capri A. E.A. Scientific Ethics. Visionlearning. 2009;2

34. Piwowar HA, et al. Towards a data sharing culture: recommendations for leadership from academic health centers. PLoS Med. 2008;5:e183.

35. Peng GC. Moving towards model reproducibility and reusability. IEEE Trans Biomed Eng. 2016;63(10):1997–8.

36. Casadevall A, Fang FC. Reproducible science. Infect Immun. 2010;78:4972–5.

Suggested Literature

Emanuel EJ, Wendler D, Grady C. What makes clinical research ethical? JAMA. 2000;283(20):2701–11.

Horng S, Miller FG. Is placebo surgery unethical. NEJM. 2002;347:137–9.

Lynn J, Baily MA, Bottrell M, Jennings B, Levine RJ, Davidoff F, et al. The ethics of using quality improvement methods in health care. Ann Intern Med. 2007;146(9):666–73.

McCulloch P, Altman DG, Campbell WB, Flum DR, Glasziou P, Marshall JC, et al. No surgical innovation without evaluation: the IDEAL recommendations. Lancet. 2009;374(9695):1105–12.

Surgical Ethics and Diversity

Judith C. French and R. Matthew Walsh

Key Points

- Surgeons have an ethical obligation to ensure all patients, regardless of their personal characteristics, receive the same quality of care.
- Healthcare disparities persist despite decades of acknowledgment of their existence, and they are pernicious and prevalent in surgery.
- Implicit biases influence our personal interactions with patients and colleagues, and personal biases should be explored.
- Cultural competence training for surgeons may help alleviate some of the issues with healthcare disparities and surgical workforce diversity.
- Regardless of their role in surgery, everyone can make a contribution to alleviate bias in patient care and in the surgical workforce.

J. C. French · R. M. Walsh (✉)
Department of General Surgery, Cleveland Clinic, Cleveland, OH, USA
e-mail: frenchj2@ccf.org; walshm@ccf.org

Introduction

The connection between ethics and diversity may not be readily apparent. To understand their relatedness, we must first grasp the meaning of ethics. Ethics are standards of behavior, standards of right and wrong. Ethical or moral codes are shaped by fairness and obligations to society. This is where the concepts of ethics and diversity intersect. Surgeons have an ethical obligation to ensure all patients, regardless of their personal characteristics, receive the same quality of care. Established surgeons also have an obligation to their peers or those who would like to join the field. The commitment to ethical hiring and working standards entails making certain all individuals have the same opportunities that are free from discriminatory practices.

Ethics also encompasses examining your own beliefs and conduct, which is the purpose of this chapter. We need to understand where we are with regard to diversity and how we can modify our practices to be more ethically sound. The importance of diversity as it relates to surgery will be discussed. The current status of surgical healthcare disparities and the surgical workforce will be highlighted along with the role bias plays in these areas. We will then focus on current responses from within the medical community and ways we can help to modify our ethical conduct. But before we turn a discerning lens upon surgery, let's first look to another occupational field that has been attempting to address diversity issues for years.

Lessons Learned from Business

The world of business has long realized the positive implications of having a diverse and inclusive workforce. An overwhelming number of global executives agree that diversity in a company is a necessity to drive innovation [1]. Studies have shown a positive correlation between increased racial and gender diversity and company performance [2–4]. These findings in business have helped shape a commitment by numerous companies to grow and maintain diversity and inclusion among their employees and expand products and services to a diverse population.

While global businesses have recognized and, in many cases, taken steps to increase diversity in their workforce, they are still struggling with gender diversity in leadership positions. In 2014 [5] a global sample of 21,980 companies reported that nearly 60% of them had no women on their corporate boards and over 50% had no women senior executives ("C-suite") and less than 5% had a female CEO. This same study revealed a positive correlation between the proportion of women with senior executive positions and company profitability. Correlational research cannot determine cause and effect, but it would seem prudent to recognize the value of diversity at all levels of an institution. Strategic pathways (or pipelines) need to be developed to help women move from manager positions to executive roles, and surgery needs to take the same approach to develop more women leaders in the field.

The Social Contract of Diversity

The globalization and international connectivity in the world has led to a diaspora of cultures. People interacting across different cultures throughout their lives. *Culture* is a social construct created through identification of similar characteristics in individuals. These characteristics can be more easily recognized or visible (e.g., race or gender), or they can be more difficult to discern visually (e.g., socioeconomic status or sexual orientation). People may openly belong to and embrace a particular culture, or they may be assigned to that culture unknowingly by other individuals. Shared experiences within these cross-cultural worlds allow members to have interactions which can lead to an expanding and collective knowledge. But what if this knowledge created by a particular group is never shared outside of that group? What if the knowledge gained was kept only to members of that culture? It is stifling for a population as a whole to not share ideas across cultures, to not learn equally from one another.

This is what diversity can do for everyone: diversity can lead to an increase in knowledge by sharing ideas with individuals who come from different life experiences and breakdown cultural barriers. In order to benefit from the shared knowledge of other cultures, those in the majority have to make room for them to exist within the mainstream. The unwritten social contract requires everyone to be open to input from other cultures and embrace the diversity of individuals. Embracing diversity means acceptance, not mere tolerance. To tolerate by definition means to "put up with." People tolerate a squeaky wheel on their grocery cart; it's an annoyance or hindrance that we put up with to accomplish a shared goal to get our groceries to the checkout. Cultural diversity deserves more than to be tolerated. Acceptance is more challenging and demanding, typically requiring an individual to confront cultural bias and norms. Without true acceptance of diversity, we will never achieve our full potential as a global society. Surgeons therefore have an ethical obligation to embrace diversity to improve practice. We see a diversity of patients; we work side by side with a diversity of peers; and we help train the next generation of diverse global surgeons. Failure to recognize and accept diversity will harm our field and ultimately harm those we are supposed to serve (see Box 1).

Box 1 Ethical Scenario
In the Patient's Shoes
Consider the following scenario. Are the ethical obligations to the patient met?

A transgender male patient, Sam Jones, arrives at the breast surgery clinic and finds himself to be the only male patient in the waiting room. There is some confusion at the front desk as the staff can find no records for a male patient with that name. Sam indicates that he has received treatment before at another clinic at that hospital under the name Samantha Jones. Once his records are located, the receptionist's tone changes from friendly to curt as Sam is told to take a seat.

The nurse who eventually leads Sam back to the exam room expresses a very cold demeanor toward him, and the less than warm reception Sam is receiving is making him even more anxious about being there. Even though Sam is experiencing pain from a lump he found in his right breast, he is considering walking away from the clinic. Sam convinces himself to stay until he talks to the surgeon.

Upon entering the room, the surgeon begins questioning Sam about his gender identity. Sam indicates he has not received any transformation surgeries, and he still possesses all of his original breast mass. The surgeon indicates Sam will need a breast biopsy and begins urging Sam to seek psychiatric care to "figure out the problem with your gender status." Sam walks away at the end of the visit and never returns to that clinic.

Issues with Diversity in Surgery

Healthcare Disparities

We have already begun to see the impact of ignoring the importance of diversity within healthcare, including surgery. *Healthcare disparities* are defined by the NIH as "differences in the incidence, prevalence, mortality, and burden of diseases and other adverse health conditions that exist among specific population groups" [6].

These population groups can be based on race, ethnicity, gender, sexual orientation, age, geographic location, socioeconomic status, and disability. A search of PubMed (search terms "healthcare disparities" and "healthcare disparities surgery") can reveal the pervasiveness of this issue and how healthcare's acknowledgment of it has grown over time (Fig. 1). Healthcare disparities still persist though despite decades of acknowledgment of their existence [7, 8], and they are pernicious and prevalent in surgery.

A study conducted in 2010 [9] revealed that black patients with comorbidities did not receive surgery for early-stage lung cancer as frequently when compared to white patients with similar comorbidities. The authors suggest communication between those black patients and their cancer physicians was not as effective as communication with white patients. Treatment for lung cancer is obviously very complex but even when basic conditions present themselves disparities still exist. Black patients with appendicitis and equal access to healthcare as other groups received laparoscopic appendectomy less often than whites [10]. Interestingly, in this same study, Hispanic patients received laparoscopic appendectomies more often than whites which is explained by the fact that in this particular patient population, Hispanics make up the majority and subsume the advantages of being such. When treated in hospitals with higher patient diversity, African-American patients have improved outcomes for cirrhosis and alcoholic hepatitis, gastrointestinal hemorrhage, gastrointestinal obstruction, inflammatory bowel diseases, and laparoscopic cholecystectomy [11].

Unfortunately, healthcare disparities based on race have been found in many surgical areas for breast cancer surgical approach [12, 13] colorectal cancer screening and outcomes [8, 12], diverticulitis [14], and cholecystectomy [15]. No surgical specialty is immune. The patients we are ethically bound to treat to our fullest capabilities are suffering through no fault of their own. Surgeons need to provide the same, high-quality care for all patients, but we must do more than acknowledge the existence of healthcare disparities. We have to look at our own practice of medi-

Fig. 1 Timeline of
healthcare disparities
publications in PubMed
(search conducted
September 2017)

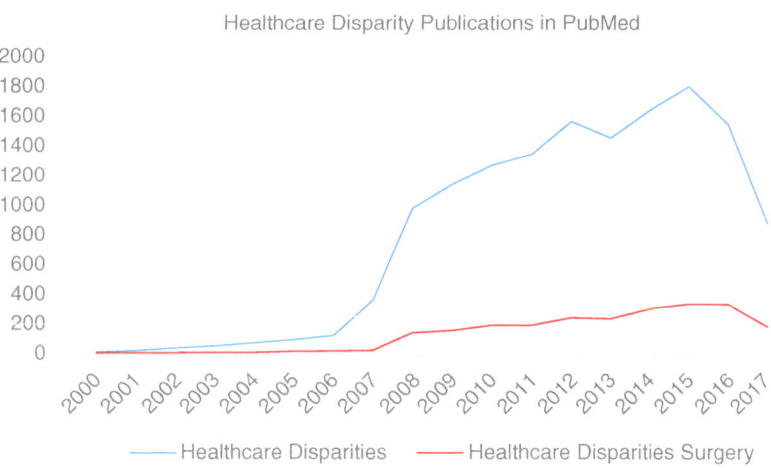

Healthcare Disparity Publications in PubMed

——— Healthcare Disparities   ——— Healthcare Disparities Surgery

*Search conducted September 2017

cine to determine if we are unknowingly making treatment decisions differently for a group of people, and ultimately we must objectively make changes when warranted.

Surgical Workforce

One proposed method to help reduce and eventually eliminate healthcare disparities is to increase the diversity of the workforce that treats patients, which would in turn reduce potential bias patients face in the healthcare setting [16, 17]. In order to appropriately care for a diverse population of patients, the surgical workforce needs to mirror that diversity. The US Census [18] indicates that blacks and African-Americans make up just over 13% of the population; however, the AAMC workforce survey [19] reveals that blacks and African-Americans represent only 4% of the physician workforce. Hispanic or Latinos comprise over 17% of the US population [18] but represent only 4% of physicians [19]. When looking specifically at race and representation in surgery (general, colorectal, etc.), as of 2013 black or African-Americans and Hispanic or Latinos combined make up less than 10% of the entire active surgical workforce. This is obviously not representative of the population which should be a cause for concern. Overall, nonwhite physicians care for a large portion of minority and non-

English speaking patients and increasing the diversity of the physician workforce may help reduce some of the healthcare disparities these patient populations face [20].

The metaphor of a leaky pipeline is often used to describe the loss of diversity along the training path to becoming a physician and a surgeon. Throughout our training, we face barriers that make moving forward exceedingly difficult and sometimes impossible. People are "leaked" out of the pipeline along the way with various groups being effected more at certain points than others with some lost from the pipeline before they even leave primary school [21]. Researchers have begun taking a closer look at these barriers to determine not only their causes but to determine solutions on how they can be alleviated.

A focus group study published in 2016 [22] revealed several perceived (and very real) barriers to medical and dental careers. The participants in the focus groups were undergraduate, underrepresented minority college students. The students highlighted four key barriers: "inadequate institutional support and resources; limited personal resources and social/family conflict; lack of access to information, mentoring, and advising; and societal barriers." Even if someone can figure out the path to becoming a surgeon, which can be an endeavor in itself without guidance, the steps necessary require many physical, mental, and financial resources. In 1978 there were 542 black male medical school matriculants, but in 2014 that number dropped to 515 [23]. While medical schools have experienced increases in the number of admissions, this increase is not reducing the leaky pipeline for everyone.

When examining gender, the US Census indicates that women make up 50.8% of the population [18]. The AAMC workforce data [19] reveals women represent 34% of all physicians; however women only represent 19% of general surgeons which is the highest percentage of women in pure surgical specialties (orthopedic surgery has the lowest percentage of women at 5%). Data over the years from the AAMC shows an upward trend of women residents in general surgery with an estimated equivalency to male residents in the year 2028 [24]. The trend for women with full professorships is not as steep and not expected to reach equivalency with men who hold full professorships until the year 2096 [24]. The Association of Women Surgeons reports that in 2017 there are 18 women surgical department heads across all of the USA and Canada which is reminiscent of the lack of women in leadership positions in business [25]. Women are experiencing barriers to becoming surgeons and to becoming leaders in the field.

A survey study conducted in 2014 [26] examined various diversity characteristics including certain invisible diversity traits; researchers found that surgical participants perceived their field to be less diverse in regard to gender/sexual identity than nonsurgical participants. This same study also found that surgical participants did not deem workforce diversity related to gender/sexual identity as important as participants in nonsurgical specialties. Surgeons acknowledged their field is not diverse in gender/sexual identity but do not value the importance of this diversity among their peers. As stated earlier, surgery has an ethical obligation to grow with diversity in mind.

The state of the surgical workforce right now is not very diverse specifically when compared to the overall population of patients we treat. Previous research involving medical students has shown that not only is the perception of diversity influenced by their own demographics, but these students were more comfortable with and valued it more when their medical class had increased diversity [27]. In other words, we need to be around diverse people to comprehend the benefit, but a conundrum arises when we realize the pipeline for some groups of people is leaking like a sieve. Surgical workforce diversity needs to increase regardless through an acceptance of cultural diversity (see Box 2).

Box 2 Discrimination from the Other Side
What to Do when Patients Discriminate
Much is written about patients receiving suboptimal treatment from physicians based on the patient's race, ethnicity, gender identity, religion, or sexual orientation. What happens when it is the patient or their family who is discriminating against the doctor? Whitgob et al. (2016) suggest the following:

- Determine the medical needs of the patient.
  Is there time to find another care provider?
- Attempt to build rapport by discussing the patient/family's underlying fear(s).
  What are they really concerned about?
- Don't take it personal.
  In the end the issues lie with the individual refusing care or making disparaging remarks.
- If working with trainees, protect them as much as possible in these situations.
  Support the trainee in front of the patient/family.

Response from Healthcare

In 2013 The American College of Surgeons released a statement calling for "optimal access to quality care," but studies conducted since then indicate disparities still exist with equal access. Access to care is definitely a necessary first step because without access, nothing we do in regard to our practice of surgery will matter for patients. Other surgical professional organizations have recognized the value of access, diversity and inclusion, and the needs of their diverse patient populations and have formed subcommittees to help address these key concerns within their specialty [28–30]. Progress of these various groups impacting patient or surgical workforce outcomes has yet to be shown, but forming the groups, or recognition of the need, is the first step.

The NIH established the Office of Minority Programs in 1990 and in 2010 that office transitioned to the National Institute on Minority Health and Health Disparities (NIMHD). The NIMHD is responsible for "all minority health and health disparities research activities conducted and supported by the NIH institutes and centers" [31]. In 2016 the institute began a research program geared toward analyzing and ameliorating surgical care and outcomes adversely affecting different racial, ethnic, and socioeconomic status populations [32]. Numerous grants have been awarded and continue to be awarded through this office in hopes of eliminating healthcare disparities for everyone.

Cultural Barriers

Bias

Implicit bias refers to unconscious associations everyone possesses toward characteristics such as race, gender, and age. Implicit bias is different from explicit bias in that explicit biases can be purposefully masked. Implicit biases are involuntary, mental shortcuts that develop over time through both direct and indirect (i.e., media representations) means. For example, a medical student is on her last month of her clinical courses. She has been exposed to several elderly patients both in the clinic and on rounds. Several of these patients were hard of hearing, and now when this student enters the room of an elderly patient, she automatically, and unknowingly, raises her voice. She has made an unconscious association between age and hearing ability. Implicit biases

may seem innocuous like the previous example, and they may not always be framed in a negative light (e.g., associating people of Asian descent with a high aptitude in mathematics). Nevertheless, these are biases that influence our personal interactions, and healthcare professionals are not immune to these biases [33].

In 1998 a nonprofit organization, Project Implicit, began with the goals of educating people on unconscious biases and collecting data from individuals that take one of the many implicit association tests (IAT) they offer online [34]. These various tests all have the same function which is to test the strength of associations we hold between two concepts. The tests rely on speed of recognition meaning that we connect two concepts faster when we feel those concepts are more related to each other (we unconsciously associate those concepts). So what are the concepts? The Gender-Career IAT requires test takers to association female/male with career/family. The Weapons IAT draws out associations between skin color (black/white) with harmless objects or weapons. The results given to test takers vary from no association between the two concepts to slight, moderate, or strong association between the concepts. There are over a dozen IATs available through the organization, and while not perfect, they offer an opportunity to begin to understand things about yourself you may have never realized existed.

The impact of implicit biases on patients can vary, but the trouble is we don't realize we are being biased. Do you associate age with being hard of hearing? Do you associate black skin and poor pain tolerance? Do you associate HIV/AIDS with gay men? Self-assessment and self-reflection is needed to determine the answers to questions like these. Possessing implicit biases does not make someone a "bad" person, but we need to understand how our implicit biases are impacting the care we provide to patients and limiting our ability for acceptance of diversity. Turning a discerning lens on yourself is not easy and can make many people feel uncomfortable. The only way you can begin to change though is to recognize that you may have biases that are getting in the way of providing the same quality care to all patients regardless of their personal characteristics.

Are these biases affecting the surgical workforce as well? From the data presented earlier, we can see that the makeup of surgery does not represent the patient population we treat, specifically with regard to race and gender. Those percentages decline even farther when we look to leadership positions or even faculty positions at academic medical centers. As we progress along our career path, we rely on mentors and advisors to help us along the way and part of that relationship comes in the form of letters of recommendation. Previous research [35] in academia but outside of surgery has shown differences in words used to describe male and female applicants. Women more often received words like "warm" and "kind" to describe them, while men received words like "ambitious" or "self-confident" as descriptors of their personality. No study to this date has looked at gender differences in letters of recommendation in the field of surgery.

Once the trainees are in the field, are we treating them in the same manner regardless of gender? Research has now shown that women residents receive less autonomy in the operating room when compared to men [36]. Women are not being given the same opportunities to build surgical skills and techniques, which could be due to implicit bias or other factors: female trainees tend to underestimate their abilities, while male trainees tend to overestimate their abilities [37]. When looking at gender differences in narrative direct observation feedback, male emergency medicine trainees who were struggling received consistent feedback from different attending physicians, whereas female trainees received conflicting feedback from various attending physicians. Overall the personality traits that were rewarded in trainees are those most often associated with being male (e.g., decisive, confident leader, etc.) [38]. While this research focused on emergency medicine trainees, surgery is more than likely not immune to these same issues. Fairness in an assessment system encompasses both equity (comparability of opportunities to learn and demonstrate abilities) and equality (equal practice or treatment) [39].

We have to ensure that all trainees regardless of their personal characteristics are given fair opportunities.

Path Forward

Cultural Competence

Cultural competence provides surgeons the ability to effectively communicate with not only a diverse population of patients but also a diverse surgical workforce. Cultural competence is the awareness of one's own cultural viewpoint, attitude toward differences in other cultures, knowledge of different cultural practices, and the skills to interact with other cultures. Cultural competence is different from *cultural sensitivity* in that the latter consists of an awareness of differences in cultures alone while treating others with respect and dignity. Being culturally competent is an indication that people can effectively interact with those from other cultures. Numerous studies exist which show cultural competence training improves patient satisfaction and increases treatment compliance [40], which could be important factors in reducing healthcare disparities for many patient populations. Governing bodies in medical education have embraced the concepts of cultural competence. The Liaison Committee for Medical Education (LCME) guidelines for medical schools require some form of cultural competence training to occur during undergraduate medical education (Standard 7.6 for 2018–2019). Within graduate medical education, the Accreditation Council for Graduate Medical Education (ACGME) requires cultural competence to be taught to residents (Common Program Requirement IV.A.5.d for 2017–2018).

How do we see the impact of different cultures in our practice? In most North American and European countries, avoiding eye contact with another person can be seen as a sign of someone trying to be deceptive or feeling shameful. In many African, Asian, and Latin countries, avoiding eye contact can be seen as conveying respect. When engaging in conversation, some cultures tend to prefer closeness and even touching (less personal space), while others prefer talking to each other at a distance and avoid physical contact. When people from different cultures come together, they bring their cultural uniqueness with them as well. If someone is not looking at you while they speak, it does not necessarily mean they are trying to be deceitful. If a colleague continues to step away from you during a conversation, it does not mean they are not engaged in the discussion. The message we communicate to someone ultimately may not be what we intend. All interpretations of discourse are left up to the receiving parties. We need to not jump to conclusions about people just because they are acting in a way that is different from what we are used to.

Cultural competence is a goal to strive for, and a good place to start is by adopting cultural sensitivity and treating others with respect and dignity. Once cultural sensitivity is reached, we can begin to look at other cultures and recognize the similarities and differences from our own, we can be open to exploring the biases we hold, and we can start making changes to our practice of surgery to better help our patients and communicate with our peers. Through practice with diversity concepts and patience though, we must try to obtain cultural competence.

A Personal Commitment to Diversity

Surgeons are leaders, but effective leaders lead by example across all aspects of healthcare. What you can personally do depends on your role in the surgical workforce (Fig. 2). As stated earlier, everyone needs to see they have an ethical obligation to embrace diversity, but how can you act beyond that? First of all, be aware of implicit biases you hold. A good initial step is to take an implicit bias test (https://implicit.harvard.edu/implicit/takeatest.html). Engage in a period of self-reflection and self-assessment to determine how your biases could be impacting your interactions. Second, monitor your practice habits with regard to patients. Determine if you are making similar recommendations and treatment decisions for patients from different genders or cultural backgrounds. Don't be afraid to ask others to help you monitor your practice. Third, be open to working with a diversity of peers by trying to help people feel welcome. Fourth, speak up when you see or experience biases in the system. If you feel that treatment pathways established for your area are biased against certain types of patients, let your concerns be known. If you feel hiring decisions of other surgeons in your department are being made that are blind to the diversity of

Fig. 2 Guidelines to improving patient care and surgical workforce based on diversity

Guidelines to Improving Patient Care and Surgical Workforce Based on Diversity

For Everyone:
- Be aware of implicit bias with patients
- Monitor practice habits periodically
- Be open to working with a diversity of peers
- Speak up when you see biases in othersor in the "system"

For Educational Leaders:
- Be mindful of bias in trainee selection processes
- Ensure your evaluation system is monitored for biases
- Educate trainees on healthcare disparities and bias
- Establish mentorship program for trainees

For Department Chairs:
- Take an honest look at the diversity of your department
- Monitor your hiring practices of surgeons
- Ensure employee engagement for everyone
- Support faculty development in cultural competence, implicit bias, and healthcare disparities

your patient population, speak up to those making those decisions. Again, we are all not always aware of our implicit biases.

If you are an education leader (medical school admissions, clerkship director, residency program director, etc.), you have additional commitments to diversity. Be mindful of biases in your trainee selection processes. Take a hard look at the trainees that you are offering positions to or ranking high for your program. If your classes are lacking in diversity, a deeper dive into why will be needed. Ensure your evaluation system is monitored for biases. Are all of your trainees being given equal opportunities to develop and showcase their skills? If racial minorities or women are scoring lower on evaluations overall, as the educational leader for the program, it is your responsibility to determine the cause (e.g., implicit bias from other faculty evaluators). Integrate curriculum (see Box 3 for suggestions) in your program to educate trainees on healthcare disparities and bias. Not only is cultural competence training required by the LCME and the ACGME, the sooner we learn to incorporate diversity concepts into our patient interactions and treatment decisions, the sooner we can make cultural competence a habit and not just "something else I have to think about." Establish a mentorship program for your trainees. We all need people to look up to, confide in, and get help from to reach the next level.

Box 3 Ideas for Training and Faculty Development
Ideas for Training and Faculty Development
The most logical place to start before beginning any curriculum development is to conduct a needs assessment to determine what has or is currently being taught in the curriculum. Early conversations with key stakeholders can ensure buy-in, which is needed for successful implementation of any topic. Goals and objectives can then be created to target your curriculum needs and will help you determine an educational

---

strategy. Several resources have been identified below to help you move your needs forward:

Implicit Bias
- *Project Implicit Self-Assessment Tests*: Several implicit association tests or (IAT) that can be integrated into workshops or taken as stand-alone assessments.
- *Source*: https://implicit.harvard.edu/implicit/takeatest.html
- *Exploring Unconscious Bias in Academic Medicine*: A 30-minute video from the Association of American Medical Colleges discussing the role implicit bias plays in the plateau of diversity at institutions.
- *Source*: https://www.aamc.org/initiatives/diversity/learningseries/346528/howardrossinterview.html

Cultural Competence
- *Cultural self-awareness workshop*: A 2-hour, in-house, workshop exploring culture, identities, and experiences. Best suited for groups of 20–24 learners.
- *Source*: Elliott D. Cultural self awareness workshop. MedEdPORTAL. 2009;5:1128. https://doi.org/10.15766/mep_2374-8265.1128
- *Think Cultural Health Cultural Competence Toolkits*: The toolkits provide a wide variety of topics (from a curriculum self-assessment tool to information on caring for a religiously diverse population) that can be integrated into a training curriculum depending on the needs.
- *Source*: https://www.thinkculturalhealth.hhs.gov/resources/library

Leadership
- *American College of Surgeons "Surgeons as Leaders: From Operating Room to Boardroom Course:"* A 3-day, large group course highlighting five content areas of leadership. This course is offered periodically by the American College of Surgeons.

- *Source*: https://www.facs.org/education/division-of-education/courses/surgeons-as-leaders
- *Association of American Medical Colleges Leadership Development Program*: Numerous topics from change leadership to conflict management are offered for people with various levels of leadership experience.
- *Source*: https://www.aamc.org/members/leadership/

If your role is that of a department chair or equivalent, begin by taking an honest look at the diversity of those working in your department. If the diversity of your department is not representative of the patient population you care for at your institution, take the necessary steps to address the issue (e.g., recruit with diversity in mind). Specifically, you need to make certain you hire and retain a diverse surgical workforce. With such low numbers of diversity in surgery, this may take time to address. Ensure employee engagement for everyone through workplace interventions and career mentoring. Employee engagement entails guaranteeing a work environment safe from sexual harassment and discrimination, and many organizations engage in institute wide educational interventions on these topics. Engagement through career mentoring can be achieved by establishing a longitudinal mentorship program for all faculty. Fully support faculty development in cultural competence, implicit bias training, and healthcare disparities (to include those specific to their field of surgery). Be the leader in the change you want to see.

Concluding Remarks

- Surgeons have an ethical obligation to ensure all patients, regardless of their personal characteristics, receive the same quality of care.
- No surgical specialty is immune from healthcare disparities.
- Implicit biases influence our personal interactions with patients and colleagues.

- Cultural competence provides surgeons the ability to effectively communicate with not only a diverse population of patients but also a diverse surgical workforce.
- Depending on your role in the surgical workforce, surgeons can make a positive impact on diversity in different ways.

Glossary

Fairness Equity, comparability of opportunities to learn and demonstrate abilities, and equality, equal practice or treatment.

Cultural competence The awareness of one's own cultural viewpoint, attitude toward differences in other cultures, knowledge of different cultural practices, and the skills to interact with other cultures.

Cultural sensitivity An awareness of differences in cultures alone but still treating others with respect and dignity.

Culture A social construct created through identification of similar characteristics in individuals.

Healthcare disparities Differences in the incidence, prevalence, mortality, and burden of diseases and other adverse health conditions that exist among specific population groups based on race, ethnicity, gender, sexual orientation, age, geographic location, socioeconomic status, and disability.

Implicit bias Unconscious associations everyone possesses toward people's characteristics such as race, gender, and age.

References

1. Forbes Insights. Global diversity and inclusion: fostering innovation through a diverse workforce. New York: Forbes; 2011.
2. Carter NM, Wagner HM. The bottom line: corporate performance and women's representation on boards (2004-2008): New York: Catalyst Inc; 2011.
3. Desvaux G, Devillard S, Sancier-Sultan S. Women at the top of corporations: making it happen. New York: McKinsey and Company; 2011.
4. Herring C. Does diversity pay?: race, gender, and the business case for diversity. Am Sociol Rev. 2009;74:208–24.

5. Noland M, Moran T, Kotschwar B. Is gender diversity profitable? Evidence from a global survey. Washington, DC: Peterson Institute for International Economics; 2016. p. WP16–3.
6. National Institute on Minority Health and Health Disparities. History. https://www.nimhd.nih.gov/about/overview/history/. Retrieved 28 Aug 2017.
7. Greenwald HP, Polissar NL, Borgatta EF, McCorkle R, Goodman G. Social factors, treatment, and survival in early-stage non-small cell lung cancer. Am J Public Health. 1998;88:1681–4.
8. Govindarajan R, Shah RV, Erkman LG, Hutchins LF. Racial differences in the outcome of patients with colorectal carcinoma. Cancer. 2003;97:493–8.
9. Cykert S, Dilworth-Anderson P, Monroe MH, Walker P, McGuire FR, Corbie-Smith G, Edwards LJ, Bunton AJ. Factors associated with decisions to undergo surgery among patients with newly diagnosed early-stage lung cancer. JAMA. 2010;303:2368–76.
10. Lee SL, Yaghoubian A, Stark R, Shekherdimian S. Equal access to healthcare does not eliminate disparities in the management of adults with appendicitis. J Surg Res. 2011;170:209–13.
11. Okafor PN, Stobaugh DJ, van Ryn M, Talwalkar JA. African Americans have better outcomes for five common gastrointestinal diagnoses in hospitals with more racially diverse patients. Am J Gastroenterol. 2016;111:649–57.
12. Ojinnaka CO, Luo W, Ory MG, McMaughan D, Bolin JN. Disparities in surgical treatment of early-stage breast cancer among female residents of Texas: the role of racial residential segregation. Clin Breast Cancer. 2016;17:e43–52.
13. Fedewa SA, Flanders WD, Ward KC, Lin CC, Jemal A, Goding Sauer A, Doubeni CA, Goodman M. Racial and ethnic disparities in interval colorectal cancer incidence: a population-based cohort study. Ann Intern Med. 2017;166:857–66.
14. Lassiter RL, Talukder A, Abrams MM, Adam BL, Albo D, White CQ. Racial disparities in the use of laparoscopic surgery to treat colonic diverticulitis are not fully explained by socioeconomics or disease complexity. Am J Surg. 2017;213:673–7.
15. Gahagan JV, Hanna MH, Whealon MD, Maximus S, Phelan MJ, Lekawa M, Barrios C, Bernal NP. Racial disparities in access and outcomes of cholecystectomy in the United States. Am Surg. 2016;82:921–5.
16. US Surgeon General. Elimination of health disparities. https://www.surgeongeneral.gov/priorities/prevention/strategy/elimination-of-health-disparities.html. Retrieved 23 Oct 2017.
17. American Medical Association. Reducing disparities in health care. https://www.ama-assn.org/delivering-care/reducing-disparities-health-care. Retrieved 23 Oct 2017.
18. US Census Bureau. QuickFacts 2016. https://www.census.gov/quickfacts/fact/table/US/PST045216. Retrieved 25 Oct 2017.
19. Association of American Medical Colleges. Diversity in the physician workforce: Facts & Figures 2014. AAMC 2014.
20. Marrast LM, Zallman L, Woolhandler S, Bor DH, McCormick D. Minority physicians' role in the care of underserved patients: diversifying the physician workforce may be key in addressing health disparities. JAMA Int Med. 2013. Retrieved October 25, 2017 from http://org.salsalabs.com/o/307/images/JAMA_Int_Med_Marrast.pdf.
21. Lupkin S. Deep roots for lack of minorities in American medical schools. MedPage Today 2016. https://www.medpagetoday.com/publichealthpolicy/medicaleducation/56730. Retrieved 25 Oct 2017.
22. Freeman BK, Landry A, Trevino R, Grande D, Shea JA. Understanding the leaky pipeline: perceived barriers to pursuing a career in medicine or dentistry among underrepresented-in-medicine undergraduate students. Acad Med. 2016;91:987–93.
23. Association of American Medical Colleges. Altering the course: Black males in medicine. AAMC 2015. https://members.aamc.org/eweb/upload/Altering%20the%20Course%20-%20Black%20Males%20in%20Medicine%20AAMC.pdf. Retrieved 25 Oct 2017.
24. Sexton KW, Hocking KM, Wise E, Osgood MJ, Cheung-Flynn J, Komalavilas P, Campbell KE, Dattilo JB, Brophy CM. Women in academic surgery: the pipeline is busted. J Surg Educ. 2012;69:84–90.
25. Association of Women Surgeons. Female chairs of departments of surgery. https://www.womensurgeons.org/in-practice/leaders-in-surgery/. Retrieved 23 Oct 2017.
26. French JC, O'Rourke C, Walsh RM. A current assessment of diversity characteristics and perceptions of their importance in the surgical workforce. J Gastrointest Surg. 2014;18:1936–43.
27. Elam CL, Johnson MMS, Wiggs JS, Messmer JM, Brown PI, Hinkley R. Diversity in medical school: perceptions of first–year students at four southeastern U.S. medical schools. Acad Med. 2001;76:60–5.
28. Society for Surgery of the Alimentary Tract. Diversity & inclusion liaison committee. http://ssat.com/about/committees.cgi?s=1&cc=DTF. Retrieved 25 Oct 2017.
29. American Society for Metabolic and Bariatric Surgery. Committees. https://asmbs.org/about/committees. Retrieved 25 Oct 2017.
30. American Society of Transplantation. Women's health community of practice. https://www.myast.org/communities-practice/women%E2%80%99s-health-community-practice-whcop. Retrieved 25 Oct 2017.
31. National Institute on Minority Health and Health Disparities. NIH announces Institute on Minority Health and Health Disparities. https://www.nih.gov/news-events/news-releases/nih-announces-institute-minority-health-health-disparities. Retrieved 25 Oct 2017.
32. National Institute on Minority Health and Health Disparities. NIH launches research program to reduce health disparities in surgical outcomes. https://www.nih.gov/news-events/news-releases/nih-launches-research-program-reduce-health-disparities-surgical-outcomes. Retrieved 25 Oct 2017.

33. FitzGerald C, Hurst S. Implicit bias in healthcare professionals: a systematic review. BMC Med Ethics. 2017;18:19.

34. Project Implicit. About us. https://implicit.harvard. edu/implicit/aboutus.html. Retrieved 24 Oct 2017.

35. Madera JM, Hebl MR, Martin RC. Gender and letters of recommendation for academia: agentic and communal differences. J Appl Psychol. 2009;94:1591–9.

36. Meyerson SL, Sternbach JM, Zwischenberger JB, Bender EM. The effect of gender on resident autonomy in the operating room. J Surg Educ. 2017. In press; https://doi.org/10.1016/j.jsurg.2017.06.014.

37. Flyckt RL, White EE, Goodman LR, Mohr C, Dutta S, Zanotti KM. The use of laparoscopy simulation to explore gender differences in resident surgical confidence. Obstet Gynecol Int. 2017; https://doi. org/10.1155/2017/1945801.

38. Mueller AS, Jenkins TM, Osbourne M, Dayal A, O'Connor DM, Arora VM. Gender differences in attending physicians' feedback to residents: a qualitative analysis. JGME. 2017;9:577–85.

39. Colbert CY, French JC, Herring ME, Dannefer EF. Fairness: the hidden challenge for competency-based postgraduate medical education programs. Perspect Med Educ. 2017;6:347–55.

40. Betancourt JR, Green AR, Carrillo JE, Ananeh-Firempong O. Defining cultural competence: a practical framework for addressing racial/ethnic disparities in health and health care. Public Health Rep. 2003;118:293–302.

Suggested Literature

Association of American Medical Colleges. Diversity learning series. https://www.aamc.org/initiatives/diversity/learningseries/.

Banaji MR, Greenwald AG. Blindspot: hidden biases of good people. New York: Bantam; 2016.

Dreachslin JL, Malone B. Diversity and cultural competence in health care: a systems approach. San Francisco: Jossey-Bass; 2013.

National Institute on Minority Health and Health Disparities. Overview. https://www.nimhd.nih.gov/about/overview/.

Project Implicit. Take a test. https://implicit.harvard.edu/implicit/takeatest.html.

Think Cultural Health. Culturally and linguistically appropriate services. https://www.thinkculturalhealth.hhs.gov/clas/standards.

Why and How to Teach Surgical Ethics?

Jonathan K. Chica and Jason D. Keune

Introduction

When it comes to patient care in surgery, there is often evidence as to how to manage patients one encounters that is organized according to organ system or disease process. However, when it comes to managing ethical issues that might arise in surgical practice, things are not always as rooted in data, and therefore the approach may not be as straightforward. Although ethics may seem superfluous when compared to the gravity of a patient with an acute gunshot wound to the inferior vena cava or a new diagnosis of hepatocellular carcinoma in an already cirrhotic patient, it plays a fundamental role in the global care of patients. As technology continues to advance in medicine, physicians will continue to be challenged and will continue to have moments where what is "right" may not be obvious.

The importance of ethics has been recognized by the major accrediting body for residency training programs in the United States. In 1999, the Accreditation Council of Graduate Medical Education (ACGME) selected and endorsed six "core competencies," which describe the foundational skills that every physician should possess [1]. The ACGME Core Competencies are meant to inform resident education. Ethics is within the scope of the competency "Professionalism." One subcomponent of this competency describes how surgeons should adhere to ethical principles, including the following:

- Compassion, integrity, respect
- Responsiveness to patient needs that supersede self-interest
- Respect for patient privacy and autonomy
- Accountability to patients, society, profession
- Sensitivity and responsiveness to diverse patient population

Most medical school curricula include the four principles that Beauchamp and Childress propose as starting points for clinical medical ethics which are respect for autonomy, nonmaleficence, beneficence, and justice [2]. Despite these concepts' teachability in the classroom setting, they can fall short when applied to the complex ethical terrain of day to day clinical practice. The history of ethical thought traces back to ancient Greek times, with the well-known Hippocratic oath that is still held sacred by physicians. Without direction, these principles and the meaning that they have can be easily forgotten.

In this chapter, the problems of WHY and HOW surgical ethics should be taught will be addressed.

J. K. Chica (✉) · J. D. Keune
St. Louis University, St. Louis, MO, USA
e-mail: jonathan.chica@health.slu.edu;
jason.keune@health.slu.edu

Why

Although ethics is often part of the medical school education and is also a subcomponent of the ACGME competencies that guide surgical residency training programs, ethics in surgical education is not often discussed explicitly. Downing et al. [3] reported in 1997, in a survey sent to the program directors of all of the accredited general surgery residency programs in the United States, that 28% of programs that responded offered no formal ethics education, 48% held only one teaching event in ethics, and only 24% conducted two or more such activities. In a survey that was done of the surgical house staff at Washington University in 2002 regarding their experiences with ethical issues, it was discovered that house staff encountered ethical dilemmas often and had a desire to have more opportunities to discuss them, feeling insufficiently prepared with situations: thus, a case-base format for teaching surgical ethics was started [4]. In this section, we present three arguments about why surgical ethics should be taught in surgical residency programs: the need to attend to the "hidden curriculum" of medicine, the problem of moral disengagement, and the need to handle moral emergencies.

Hafferty and Franks [5] make the point that formal instruction in ethics makes only a small contribution in the education of physicians, stating that the critical determinants of physicians' identities, including their moral formation, lie not within the formal curriculum but in a more subtle "hidden curriculum." In their paper, *The hidden curriculum, ethics teaching, and the structure of medical education*, published in *Academic Medicine* in 1996, they bring out three observed themes or beliefs about medicine that are evident when it comes to matters of ethics. The first belief is that errors committed in the past can be corrected, and avoided in the future, only if there is a greater presence of a formal curriculum throughout medical education. The second belief argues that one's moral character is developed prior to medical school, and even formal education in ethics will not decisively reshape a student's ethical conduct in the future. The third belief is that although one's character and foundations are

molded by personal and family values during upbringing, "the most influential vehicle involves informal processes such as "general clinical experience," peer interactions, "ward rounds," and "role models" rather than formal coursework in ethics or related topics" [6]. The authors consider that any formal curriculum should attend to this "hidden curriculum" in order for it to have any impact on the training of physicians. We suggest also that an unexamined hidden curriculum may be tinted with racial, socioeconomic, gender, and other biases; thus a formal curriculum is needed to reveal and correct such bias.

Moral disengagement is a concept from social psychology that can play a big role in surgical ethics education. Moral disengagement is the idea that in certain contexts, like membership in a group, individuals can reason convincingly that certain moral standards do not apply to themselves [7]. This concept allows a way of looking at the reasoning that an individual can act in a certain way, which augments the basic moral cognition approach that is common. Several authors have argued that this sort of reasoning is pervasive throughout humankind [7].

Bandura has described some of the mechanisms by which people can become disengaged: moral justification, euphemistic labelling, advantageous comparison, and displacement of responsibility. Moral justification implies that people must first justify their action morally prior to acting – even if this is harmful conduct. When actions are morally justified in this way, "pernicious conduct is made personally and socially acceptable by portraying it as serving socially worthy or moral purposes" [8]. Euphemistic labelling is a concept in which one makes "harmful conduct respectable and reduce personal responsibility for it" [9, 10] by utilizing "sanitizing language." Advantageous comparison is another morally disengaging behavior in which one minimizes one's actions by comparing it against something that is portrayed as worse. Displacement of responsibility is yet another mechanism in which there is a legitimate authority that accepts the responsibility for the effects of their conducts. As an example of this, Bandura cites a memoir of Colonel Burton C. Andrus,

American Commandant of the Nuremburg Prison, who explains that Nazi prison commandants and their staffs divested themselves of personal responsibility for their unprecedented inhumanities. Although this may seem like an exaggeration compared to the morally bounded life of the surgeon, if there is no accountability to which surgeons are held, as humans, we are bound to repeat history. One of the ways that surgeons are held accountable is the tradition of having morbidity and mortality conference, but even with these conferences, the role of surgical ethics may easily be "swept under the rug" if it is not highlighted or incorporated in some way.

Although there may exist a "hidden curriculum" as described before, the teaching of ethics should parallel ethical issues as they arise during the training experience, beginning in the basic science years, and continuing into clinical training, including surgical residency [5]. With that in mind, the teaching of ethics should evolve and be deliberate to avoid wrong morality. In her titled *Moral Understandings*, Walker defines morality as "a socially embodied medium of understanding and adjustment in which people account to each other for the identities, the relationships, and the values that define their responsibilities" [11]. Avoiding wrong morality can only occur if there is a form of accountability and education regarding ethical dilemmas that are frequently encountered during residency and in practice.

As surgeons, we are constantly being faced with challenges that present themselves quickly and have to be addressed expeditiously. As such, surgeons are constantly faced with moral emergencies, since a subset of these challenges can have morally ambiguous components. Kwame Anthony Appiah describes this in his book, *Experiments in Ethics*, in which moral emergencies have the following features [12, 13]:

- A decision of what to do needs to be made in a very short period of time.
- There is a clear and simple set of options.
- Something of great moral significance is at stake.
- No one else is as well placed as you are to intervene.

Such features should be familiar to the practicing surgeon, especially the necessity of quick decision-making and the situation of the surgeon in a context when no one else around is prepared to intervene. In the case of surgical emergencies endowed with ethical ambiguity, the surgeon is usually the most well placed to intervene, and with that comes great responsibility. Ethical issues that are time-sensitive may arise, such as those associated with the end of life, surrogate decision-making, futility-related issues, and do-not-resuscitate orders in the operating room.

New ethical challenges are sure to arise as surgery progresses into the future. As technology continues to advance, ethical dilemmas that may have not been present in the past start to be revealed. This march of technology in contemporary surgery is prominent and strong, and with the development of new technologies, concomitant new ethical dilemmas will need to be met with sophistication and experience. We should not wait until ethical issues arise to start preparing to address them.

Ethics training is a self-perpetuating issue. If there isn't a more formal integration in surgical training, the lack of faculty expertise in clinical ethics in the future will only be perpetuated. The lack of faculty expertise in ethics is the most commonly cited reason that it is not taught [14]. However, if surgical ethics is included with more presence now, future surgeons will be more inclined to incorporate it into their clinical practice and pass this along to future generation of surgeons. Several physician-ethicists have argued that teaching clinical ethics improves the quality of patient care, by acknowledging that a serious medical decision involves two essential and necessary components: a technical decision that requires application of basic scientific and clinical knowledge, and a moral decision that takes into account what ought to be done for individual patients [3, 15].

How

Since surgical ethics is something that should be taught formally, the greater question is how should it be taught. One may argue that there isn't one right way to teach surgical ethics, but many ways in which it can be taught. It is important to understand that people have different learning styles and accommodating many learning styles may pose a challenge when it comes to teaching surgical ethics. In one study, a survey was done from general surgery residents, concluding that even though 88% of surgical residents had previous exposure to formal medical ethics within medical school, 93% continue to believe that an ethics curriculum is an important part of their education during residency [16]. When program directors were asked how a surgeon should acquire knowledge in ethics, education during residency (82%) and experience during residency (88%), in addition to personal life experience (84%), were the major determinants [3]. Teaching ethics throughout general surgery training may be challenging, given the limitation of time that residents in surgery generally have due to their clinical duties. There have been different studies regarding how to teach surgical ethics. Below, we describe several ways that may be easily incorporated into surgical training.

Case-Based Format

In the realm of ethics instruction, the pedagogical value of focusing on individual cases or a case-based comparison has been well known [17]. This pedagogical mode relies on the concept of casuistry, "revived" and developed by Stephen Toulmin and Albert Jonsen [18]. Casuistry is "that part of Ethics which resolves cases of conscience, applying the general rules of religion and morality to particular instances in which 'circumstances alter cases', or in which there appears to be a conflict of duties" [19]. About casuistry, John Arras writes, "According to this rehabilitated form of casuistry, the greatest confidence in our moral judgments resides not at the level of theory, where we endlessly disagree, but rather at the level of the case, where our intuitions often converge without the benefit of theory" [20].

A successful case-based approach in a general surgery training program has been described in the literature. At Washington University School of Medicine, a monthly case-based session lasting 1 hour was held for 5 years in which all residents in the general surgery training program were required to attend. Faculty and fellows in surgery were invited to attend, as well as other residents, fellows, and faculty from other specialties. They met in a classroom setting to maintain hierarchy to a minimum, with the residency program director serving as the moderator and a surgeon-ethicist and a PhD ethicist helping facilitate the discussion. Other hospital members including hospital Ethics Committee as well as individuals from nursing, chaplaincy, and palliative care were welcome to attend. During the session, a resident was asked to present the case in brief oral format, with less emphasis on the medical aspects and more emphasis on the social and ethically challenging aspects of the case. After the 5 years were completed, a survey to all house staff was repeated, with the conclusion that they felt more prepared to make decisions compared with the group surveyed 5 years before [4].

Formal Curriculum

Another way that surgical ethics may be taught is by incorporating a formal curriculum into surgical training. A study was conducted among surgical residents of the University of Pittsburgh in which the program used the American College of Surgeons published educational resource in surgical ethics, *Ethical Issues in Clinical Surgery* [21], as the curricular text. This text introduces core issues central to the ethical practice of surgery, providing learning objectives, case scenarios, questions for discussion, a glossary, and resources for further reading. In the study, they incorporated four 60-minute, faculty-facilitated, didactic seminars organized to cover specific ethics content and integrated with case-based discussions, with each session spaced out every 3–4 weeks. This educational intervention "increased both knowledge about surgical ethics ($P = 0.013$) and confidence in dealing with competition of interests ($P = 0.001$), professional obligations ($P = 0.011$), truth telling ($P = 0.013$), confidentiality ($P = 0.011$), end-of-life issues ($P = 0.007$), and surrogate decision making ($P = 0.052$)." A questionnaire was also done both before and after the educational intervention. Prior to the intervention, only 57% felt that ethics training should be a "standard" aspect of surgical residency, compared with 70% after the intervention [22].

Incorporating Ethics into M&M

The concept of morbidity and mortality (M&M) conference traces back to the beginning of the twentieth century, through the work of Ernest Amory Codman, a surgeon at Massachusetts General Hospital. In 1983, it became required by the Accreditation Council for Graduate Medical Education for training program certification in General Surgery [23]. Traditionally, the goal of M&M conference is to provide a forum for faculty and trainees to explore specific management details of particular cases wherein morbidity and mortality occurred, with emphasis in revisiting errors to gain insight without blame or derision [24]. The deliberate and explicit incorporation of ethical issues into M&M conference is a recent development that shows promise.

Anji Wall and colleagues at Vanderbilt University [25] performed an observational study of surgical M&M at their institution to identify ethical issues. Of the 123 cases they captured, 79 (64%) discussed at least one ethical issue. The most common issues involved the role of palliative care, withholding and withdrawing life-sustaining treatment, risk-benefit analysis, and reporting medical errors. They concluded that M&M conferences raise a spectrum of complex ethical issues that provide a roadmap for focusing surgical ethics education on common scenarios. Karen Devon [26] performed a qualitative study in which ethical issues were introduced into regularly scheduled M&Ms at five sites in the Division of General Surgery at the University of Toronto in Canada. The Ethics M&M occurred for one 30-minute period per month during which ethics issues surrounding care of the complication were discussed. A resident would prepare a case with the help of a surgeon mentor and another faculty member would moderate the confidential discussion. This created increased awareness of ethics regarding patient care where there is a safe physical and intellectual space to debrief difficult cases.

This form of teaching may be appealing to many surgical programs given that it is something that can be easily applied and incorporated into what is commonly considered "protected time." There wouldn't be a need for changes with scheduling where clinical duties or responsibilities would be affected.

Conclusion

- Ethics is pervasive throughout surgical training and the surgical career. Even the most fundamental notion that one should strive to achieve good outcomes has an ethical ring to it.
- Ethics should not be taught by simple enculturation, but rather deliberately to avoid biases found in the "hidden curriculum" or with an aim to avoid moral disengagement.
- There is a wide range of pedagogical methods to teach ethics.
- Surgeons are, if anything, "doers" in the world, and if ethics asks, "What should one do?," then ethics seems like an apt field of study for surgeons.

References

1. Edwards FD, Frey KA. The future of residency education: implementing a competency-based educational model. Fam Med. 2007;39(2):116.
2. Beauchamp TL, Childress JF. Principles of biomedical ethics. Oxford: Oxford University Press; 2001.
3. Downing MT, Way DP, Caniano DA. Results of a national survey on ethics education in general surgery residency programs. Am J Surg. 1997;174(3):364–8.
4. Klingensmith ME. Teaching ethics in surgical training programs using a case-based format. J Surg Educ. 2008;65(2):126–8.
5. Hafferty FW, Franks R. The hidden curriculum, ethics teaching, and the structure of medical education. Acad Med. 1994;69(11):861–71.
6. Pellegrino ED, Hart RJ, Henderson SR, Loeb SE, Edwards G. Relevance and utility of courses in

medical ethics: a survey of physicians' perceptions. JAMA. 1985;253(1):49–53.

7. Gabor T. Everybody does it!: crime by the public. Toronto: University of Toronto Press; 1994.

8. Bandura A. Selective moral disengagement in the exercise of moral agency. J Moral Educ. 2002;31(2):101–19.

9. Lutz WD. Language, appearance, and reality: double-speak in 1984. ETC: A Review of General Semantics. 1987;44:382–91.

10. Laird CG. The legacy of language: a tribute to Charlton Laird. Reno: University of Nevada Press; 1987. p. 103–19.

11. Walker MU. Moral understandings: a feminist study in ethics. Oxford: Oxford University Press; 2007.

12. Appiah A. Experiments in ethics. Cambridge, MA: Harvard University Press; 2008.

13. Liao SM. Appiah's experiments in ethics: chapter 3 2008 [Ethics Etc, A forum for discussing contemporary philosophical issues in ethics and related areas]. Available from: http://ethics-etc.com/2008/03/31/appiahs-experiments-in-ethics-chapter-3/.

14. Ledbetter E. Ethics education in medicine. Adv Pediatr. 1991;38:365–87.

15. Pellegrino ED, Siegler M, Singer PA. Teaching clinical ethics. J Clin Ethics. 1990;1:175–80.

16. Angelos P, DaRosa DA, Derossis AM, Kim B. Medical ethics curriculum for surgical residents: results of a pilot project. Surgery. 1999;126(4):701–7.

17. Keefer M, Ashley KD. Case-based approaches to professional ethics: a systematic comparison of students' and ethicists' moral reasoning. J Moral Educ. 2001;30(4):377–98.

18. Jonsen AR, Toulmin SE. The abuse of casuistry: a history of moral reasoning. Berkeley: Univ of California Press; 1988.

19. Dictionary OE. Oxford English dictionary online: JSTOR; 2007. https://en.oxforddictionaries.com/. Accessed november 17, 2017.

20. Arras J. Theory and bioethics: metaphysics research lab, Stanford University; 2016 [cited 2017 Oct. 24]. Winter 2016: Available from: https://plato.stanford.edu/archives/win2016/entries/theory-bioethics/.

21. McGrath MH, Risucci DA, Schwab A. Ethical issues in clinical surgery: for residents. Chicago: American College of Surgeons; 2007.

22. Thirunavukarasu P, Brewster LP, Pecora SM, Hall DE. Educational intervention is effective in improving knowledge and confidence in surgical ethics—a prospective study. Am J Surg. 2010;200(5):665–9.

23. Liu V. Error in medicine: the role of the morbidity and mortality conference. Virtual Mentor. 2005;7(4)

24. Kravet SJ, Howell E, Wright SM. Morbidity and mortality conference, grand rounds, and the ACGME's core competencies. J Gen Intern Med. 2006;21(11):1192–4.

25. Wall A, Tarpley M, Heitman E. Quantification and categorization of ethical issues discussed at surgical morbidity and mortality conferences. J Am Coll Surg. 2015;221(4):S63–S4.

26. Snelgrove R, Ng S, Devon K. Ethics M&Ms: toward a recognition of ethics in everyday practice. J Grad Med Educ. 2016;8(3):462–4.

The Surgeon, the Patient, and the Healthcare System: Access, Equity, and Fairness

Alexis G. Antunez and Andrew G. Shuman

Introduction

The goal of this section is to summarize and provide the necessary background and vocabulary regarding how academics have approached these issues before delving into the details that apply to surgical practice.

The diverse nature of surgical practice implies a litany of pathways by which a patient might arrive in an operating room. Access to the appropriate surgeon and supporting team can be a complex ordeal in either elective or emergency circumstances and reflects myriad and inevitable system-level barriers. At every point along this journey, there is an obligation to optimize equitable and efficient service that mitigates inherently unfair or default stratification due to socioeconomic circumstances and other institutionalized disparities. This increasing global complexity forces us to ensure that the ethics of access are addressed directly.

A. G. Antunez
University of Michigan Medical School,
Ann Arbor, MI, USA
e-mail: lexia@umich.edu

A. G. Shuman (✉)
Department of Otolaryngology-Head and Neck Surgery, University of Michigan Medical School,
Ann Arbor, MI, USA

Center for Bioethics and Social Sciences in Medicine,
University of Michigan Medical School,
Ann Arbor, MI, USA
e-mail: shumana@med.umich.edu

Health service researchers have recently begun investigating how surgical care is accessed and delivered on a population level. Work in recent years has focused on barriers to access and the patient-level factors contributing to disparities in surgical care including race [1], gender, insurance status [2], income, education, geography [3], and nationality [4]. This corpus is also useful in demonstrating that disparities transcend surgical specialties and individual diseases. The modern surgeon must understand how patients interact with social, financial, and political systems whose activities determine individual health outcomes. The ethics of access in surgery is therefore a timely and necessary area of study for those surgeons in both clinical and academic practice settings.

While the four ethical principles are intertwined with surgical practice at every level, justice is particularly concerned with the *how* of many of these decisions. For this reason, justice often intersects with practice, policy, law, and administration. This exploration of the ethics of surgical access considers varying perspectives on justice, from countries, to institutions, to individual surgeons.

Introduction to Terms

Equality implies that all people are given the same amount of a limited resource, whereas equity means that people are provided for according to their needs [5]. Rationing schemas are invoked when there are not enough resources for all who need them, and thus consistent ways of expressing values and setting priorities are necessary to standardize these decisions. It is helpful to initially organize these value frameworks into two categories: those that use the outcomes of allocation to determine whether their criteria were fulfilled (consequentialist) and those that are focused on regulating the process whereby resources are allocated (non-consequentialist). Eventually there emerge more intricate theories that use a combination of these criteria to assess fairness.

The consequentialist viewpoint is one framework that determines fairness based on the ultimate results of an allocation system [6]. Oversimplified, it asserts that justice is maximized when the "most good" is achieved for the most people, and so this framework assesses justice at the population level. Consequentialist philosophies are diverse and impossible to summarize as a cohesive theory, but they are not widely or exclusively employed to measure justice in modern health systems. The relative "goodness" of an outcome must be assessed separately from its moral weight, since it would be tautological to assign ethical value to an event and then use it to argue that a policy is ethical using a consequentialist framework. Bentham's classical utilitarianism delineates good along the lines of pleasure and pain. Later, Mill and Moore refined pleasure into higher and lower classifications and endeavor to differentiate it from intrinsic value, respectively. Hume and Brandt exclude fanatical and irrational pleasures from their frameworks [7]. Utilitarianism in healthcare justice would therefore assess an intervention or organizational system based on the relative amount of reasonably preferred outcomes it promotes while minimizing suffering for those it disadvantages. Of course, a complex description and philosophical analysis of these theories exceeds the scope of this chapter.

Distributive justice is an approach that seeks to allocate resources so that inequities are minimized. In this framework, people who are disadvantaged by circumstances of birth, tragic accidents, or similarly random factors should be provided with resources commensurate with their increased needs. In some instances, their needs should be prioritized over the more advantaged in certain aspects of allocation. Rawls' libertarianism is an example of distributive justice theory [8]. Rawls first establishes that all people have the same claim to basic rights and liberties. This can be described as an egalitarian principle, wherein the *process* of allocation occurs in a way that allows all people to have the same opportunities to acquire goods and resources [9]. Put in another way, a healthcare system would be egalitarian if it facilitates access regardless of patients' individual characteristics. Rawls finally asserts that if access to basic rights and liberties, as well as to positions of power in society, is the same for everyone, then any systems that promote unequal treatment should benefit those people in society with access to the fewest resources.

Turning briefly to the egalitarian ideas presented above, processes that facilitate equal treatment can be interpreted in different ways. While there is variation between countries, most nationalized healthcare systems treat citizens equally in theory by making government-sponsored healthcare available to all, while multi-payer systems give all people the same opportunity to buy services either directly or via insurance markets. Both of these approaches will have different effects on access, because in actuality, their schemas still necessitate unequal treatment in some form. The former model necessarily treats people differently when deciding how to rationalize limited resources. Even if all people have the same access to a purported universal healthcare, when dealing with expensive, novel, or exceptionally scarce health resources, some people will be prioritized over others. And within the latter system, while the opportunities to purchase services are not restricted, people without the financial means to do so will be unable to access care creating an essentially libertarian model. Of course, many complex systems merge aspects of both structures, further confounding our ability to make oversimplified categorizations thereof.

To simplify Rawls' theory, it is reasonable to say that there are immutable circumstances that cause disparities, but to redistribute resources based on outcomes will necessarily take something from those "better off." All people can claim some right to their resources. By defining how we allocate equitably, a framework must define the basic liberties that should be afforded to everyone, and which privileges may be relegated or reserved to those with more privileges, however that may be defined. This idea overlaps with Beauchamp and Childress' discussion of justice, specifically their conception of a "fair basic minimum" [10].

Theories of communitarianism ethics contrast the libertarians. While the libertarians highlight individualism, communitarians view groups of people as the fundamental moral unit of society. Modern pragmatic ethics is a branch of communitarianism concerned with people as "socially situated selves in communities" and asserts that moral meaning comes from relationships between people [11]. This requests that questions of justice critically examine at situational circumstances with a keen consciousness of biases that accompany judgments. This is illustrated in the ways that communitarians argue against commodification of organs. Since solid organ transplants come from diverse members of a given community, a system for selling organs would need to ensure that all members of the community have access to this scarce, lifesaving resource [12]. It follows that in communitarian ethics frameworks, transplants should not be rationed or sold in the same way as other medical services, because they are a resource shared by and for the group. Furthermore, this ideology would also favor implementing a system that encourages gradual shifting of the moral culture to increase donations and thereby decrease scarcity [13]. By reducing the need for rationing in this way, the inequities associated with the current state of transplants are lessened. By analyzing the biases in how transplants are obtained and allocated, communitarians illustrate ways to ethically improve the process. The UNOS system is a paradigmatic example of how this plays out in practice.

The goal of this section was to summarize and provide the necessary background and vocabulary regarding how academics have approached this problem before delving into the details that apply to surgical practice.

Just Access Within Nations

Governmental policies about surgical access demonstrate how ethical frameworks are applied. Different countries have developed various strategies to deal with the challenges of fair access to healthcare. These organizational systems clearly impact access and can do so at multiple levels of decision-making. Governments regulate which technologies and new medications may be licensed and available, establish the networks in which health systems operate, and organize payment systems for the healthcare marketplace.

Countries with systems of private insurance tend to have a more libertarian perspective on healthcare provision. Within these systems, insurance is a major arbiter of access. Being uninsured restricts the care that a patient can receive, and even the insured face access limitations when doctors or hospitals do not accept a certain type of insurance or when service requires significant deductibles or co-pays. Within nationalized healthcare systems, insurance status does not determine basic access. However, care will necessarily be limited in other ways. Cost-effectiveness, cost-utility, and cost-benefit studies ideally weigh relative benefits, harms, and thus relative value of a given medical intervention. These analyses then allow lawmakers to assess the utility and necessity of that intervention and ultimately decide whether it should be covered under their national health plan. This is one explicit way to determine what constitutes a fair basic minimum [9]. For example, in the UK, the National Institute for Health and Care Excellence (NICE) was established to conduct cost-effectiveness research and implement decision analyses into practice guidelines [14]. This sort of value-based care is one way to translate abstract priorities into real-time decisions. Rigorous research methodology is a cornerstone of NICE's successful implementation of data into decisions that influence care allocation and access.

Even within cost-effectiveness studies, the judgments about how much money a given unit of suffering is worth can be just as subjective or biased as those made in other systems. Some care is rationed in subtle, nearly undetectable ways, like social structure and interpersonal interactions. This is even more nebulous, but qualitative research methods are one way in which these patterns are being characterized and assessed [15]. Ultimately, no country's system of organization avoids the recurring problem that some people will not receive care they need and would be rightfully entitled to if there were a surplus of medical resources. Simply put, providing medical care for 7.6 billion people inevitably involves trade-offs and recognition of inherently limited resources. So, rationing necessarily occurs whether by default or in a more deliberate manner [16]. Whether a country uses waiting lists, insurance plans, or economic analyses, it has to choose one value or service (or individual) over another. Some decisions are made easier by situational urgency or randomness, but often the entire situation is reduced to a subjective decision.

It is the explicit nature of rationing in governmental healthcare systems that can cause moral distress. The implicit rationing that takes place in egalitarian systems is not more just, and the growing body of health disparities research is rightly pulling attention to those marginalized populations. They people are not only prevented from accessing care, but their experience is also more difficult to capture in research studies. This kind of system functions to effectively discriminate based on race and income in many countries. A society's values are reflected in whether and how citizens are systematically prevented from accessing healthcare.

Once patterns of disparities become apparent, states may take action to increase access for the marginalized. Fundamental to this process are the logistical considerations. For example, patients' first contact point with health system can be hugely influential in the timeliness, quality, and type of care that they receive. This first contact point varies by patient demographic factors. The Emergency Medical Treatment and Labor Act (EMTALA), passed in 1986 in the United States,

requires that all patients presenting to an emergency department be treated and stabilized. But while it guarantees emergent care, it does not ensure access otherwise [17]. Disadvantaged patients can obtain care when their health is in crisis, but their long-term management needs continue to go unmet. These patients' health outcomes are different as a result of the type of care they can access, and this legal intervention has changed the populations that are seen in outpatient versus emergency settings. EMTALA is intended to improve access in a fair way but fails to fully realize this goal.

In comparison, countries with socialized medicine have been more successful in improving access to care, be it emergency, outpatient, or surgical in nature. However in resource-poor settings, it would not matter whether EMTALA existed if there simply is not enough medical equipment or personnel to meet demand [18]. With regard to surgical care, mandating emergency care is also appropriate for surgical diseases usually amenable to elective surgery that may suddenly become acute, like cholecystitis. If these patients were not guaranteed emergency surgery and on-call surgeons were unavailable, there would be serious gaps in access to care. National awareness of repeated patterns of disparities in access can lead to improvements in care allocation, but the exact implementation and impact of such programs should be carefully monitored.

Institutions

Medical systems are organizational structures that can also influence how care is distributed, and the relationships between hospitals are an additional mechanism that determines how patients' needs are prioritized. Further exploring the impact of EMTALA, its history is rooted in inappropriate transfers that were based on patients' lack of insurance and socioeconomic resources [19]. Currently, while it is legitimate to initiate transfer of stable surgical patients due to surgical subspecialist unavailability at referring institutions, there are still a sizable proportion of surgical patients transferred inappropriately [20]. This may unduly burden referral centers and academic institutions [21]. Importantly, unregulated transfer patterns between hospitals can create and exacerbate disparities in access to specialized care [22]. Minimizing inappropriate surgical transfers can more effectively utilize the collective resources of hospital networks and also reduce the burden of travel and follow-up for patients transferred unnecessarily.

Diversity of faculty, staff, and trainees within institutions may not, at first, seem like an issue of access. However, increasing diversity in the medical profession improves access in two ways – there is a direct effect on access from the actions of individual doctors, and there is an indirect effect on the inclusiveness of an institution. By increasing diversity in the medical workforce, it is more likely that patients will share an aspect of their identity with their physician, be it race, sex, gender, sexual orientation, age, disability, class, or the like. This can enhance communication and lead to better therapeutic relationships [23]. But perhaps even more importantly, diversity within healthcare institutions improves the quality of care, and not only for the marginalized [24]. The preceding discussion has demonstrated that organizations, whether they are international, governmental, or local, are frequently the actors making decisions about how care is allocated. In administrative, clinical, or research teams, and especially in challenging surgical cases, cognitive diversity improves the group's ability to solve problems [25]. Recruiting and promoting diverse learners throughout advanced clinical training will eventually make healthcare institutions more effective in solving access problems for patients. Sociological study reveals the clear threat of groupthink and hierarchies, which can be counteracted by increasing the diversity of physicians, researchers, and healthcare workers [26].

Interpreter services are an excellent example of intra-institutional diversity as a means of ensuring just access to surgical care. The ability to comprehend treatment options, ask questions of a surgeon, and express preferences is taken for granted when patient and physician speak the same language. Language barriers limit surgical access because patients who cannot understand their provider are more likely to experience errors in care [27]. When patients are provided with the tools to communicate with their physician, people who otherwise would have been unable to understand and be understood by their surgeon now can have a similar therapeutic relationship. As populations around the world become more diverse, institutions will increasingly be called upon to lead efforts to increase diversity in the workplace and thus uphold higher standards for just access to surgical care.

Finally, priorities within institutions can impact access. In setting goals to guide daily decisions, hospitals impact which patients are seen and when, as well as which surgical specialties have precedence to use resources. Hospitals may have different ends in mind when setting priorities, but all need to prioritize financial stability at some level in order to function. Surgery requires a remarkable amount of coordinated care in order to occur: specialized equipment, personnel, and inpatient/outpatient services. Certain types of surgery may bring in more revenue for the hospital or require less expensive resources. But if procedures or surgeons that are more lucrative are able to monopolize limited operating room time, other surgeries may be delayed at a detriment to the population. Healthcare organizations must find a way to reconcile their priorities with available resources, as well as with their bottom line. One approach that marries abstract goals to practical decisions is the "describe, evaluate, and improve" strategy, which can guide administrators in making their choices concordant with overall goals for fairness [28]. The people who have the power to make and enforce these choices should pay attention to the effect that logistical decisions have on the actual experiences of patients within the system, as well as how they impact the stakeholders on the ground.

Justice in Global Health

Just access on a global scale primarily deals with issues of epidemiology, economics, and humanitarianism. Inequity in global surgery can trace its roots to the unequal distribution of resources across countries. While this chapter will not provide whole historical context of international disparities in health, it is a necessary fact to recognize at the outset. Given that populations have varied needs and governments and nongovernmental organizations have mixed abilities to meet them, the question of how to act to mitigate inequity in global surgery concerns itself less with the past and more with current workflows and strategies.

Worldwide, 5 billion people lack access to safe surgical care, and this unmet need is often absent from discussions of global medical aid [29]. Including surgical care in campaigns to improve health in low- and middle-income countries is feasible, necessary, and increasingly common. Obstetric, oncologic, cardiovascular, and trauma procedures are large contributors to the burden of disease and so require particular attention. Dr. Mark Shrime has led academic inquiry in global surgical ethics, most recently with assessments of the inadequate financial contributions to build surgical infrastructure [30]. His team has also found that transportation is a major barrier to accessing surgical care in low-resource settings, and eliminating this obstacle can double the number of appointments that patients attend [31]. Geographic disparities in accessing surgical care continue to be significant, in both developed and developing nations [32, 33]. However, these research teams continue to investigate setting-specific solutions for improving access to surgery in low- and middle-income countries (LMICs).

In 2017, the director general of the World Health Organization (WHO) spoke about fairness in health provision across borders when addressing the Human Rights Council, further highlighting the growing public and academic interest in equitable delivery of global surgery [34]. The WHO functions to oversee nations and acts to provide healthcare to citizens when political and social unrest prevent governments from doing so. They also serve to compile data from around the globe and advocate for equity in healthcare provision, such as differential medication pricing for LMICs. Certain groups of people, such as migrant populations that move for agricultural work, may not fall under the purview of any single government, and so humanitarian organizations that provide healthcare without regard for national borders are often their only source of help. This classifies as a form of virtue-based care, since this is a situation in which the duty to provide just care is not fulfilled, but access for marginalized populations is accomplished via an appeal to charity and global unity. So, humanitarian groups' obligation to provide care is fundamentally different in nature than that of elected governments. Some of these nongovernmental organizations (NGOs) are faith-based efforts, which have additional priorities and motivations. Some humanitarian groups differ in their tendency to express public political stances, and there have been instances where raising awareness about atrocities in armed conflict has subsequently prevented groups from accessing populations in need of care; in other words, political climate and unintended consequences require strategic action [35].

There are some international surgical endeavors that may be well intentioned but must guard against unintended consequences. Medical tourism, or "voluntourism," broadly construed, involves self-limited mission trips into resource-poor settings. This practice often stems from well-intentioned physicians who may be unfamiliar with basic principles of global health ethics and care delivery in different settings. Such efforts, however laudatory, risk failing to develop relationships with a community and potentially leaving patients without adequate follow-up care or recourse in the event of complications if not well integrated with existing resources or otherwise planned appropriately [36]. While this topic is not our primary focus, its relationship to just access is important to highlight. Surgeons owe patients in resource-poor settings the same consideration and standards of care as the patients in their home countries, of course bounded within local resource constraints [37]. Interventions thus need to be culturally sensitive and integrated with community stakeholders in order to ensure both safety and continuity. Visiting surgeons' greatest potential value likely involves training local providers in strategic efforts to build or grow self-sustained systems of care.

The Individual Surgeon's Obligation

The individual surgeon has a duty to promote the welfare of each individual patient, which may conflict with the practical realities of complex medical and social institutions. The increasing diversity of colleagues and patients is a phenomenon that requires the modern surgeon's attention, understanding, and empathy [38]. Cultural barriers to care can break down the therapeutic relationship, but just as physicians train to deliver bad news in a compassionate and professional manner, so they must also learn and practice culturally suitable ways of interacting with patients that come from different backgrounds. And while the cultural competence of physicians is far from the only required criterion to ensure access, its absence can be a serious and detrimental barrier to care.

Generally, individual physicians can decide whether and how they will provide care for a patient based upon scope of practice, professional judgment, and a myriad of other factors. Exceptions include when failing to do so causes the patient to be abandoned or in emergent settings [39]. This is relevant to access in situations where patients' health conditions put the healthcare providers caring for them at risk. When taking care of patients jeopardizes their own wellbeing, physicians, nurses, and other healthcare personnel face a more ethically complex dilemma. These situations become morally fraught when the patients with limited access are also those with debilitating and deadly diseases that are difficult to treat. This especially applies to surgeons simply due to the increased personal risk inherent to performing invasive procedures. Vulnerable patients are more likely to be marginalized, which may be due to infectious diseases, poverty, cultural barriers, and other reasons [40]. The classic example of this has been people caring for trauma victims at the beginning of the HIV/AIDS crisis [41]. Providers were not entirely sure how a new, morbid, and stigmatized virus was transmitted and were understandably concerned about how their obligation to take care of patients with communicable disease conflicted with the risk to other patients, as well as their own safety. The practical and ethical solutions involving universal precautions have informed how

we deal with these issues broadly, but only once we can truly understand both the nature of the posed risk and its impact upon those involved [42].

Recently, ophthalmologic surgery on Ebola survivors demonstrates similar corollaries. Some Ebola survivors develop delayed uveitis, but at the same time that this phenomenon was being recognized, researchers also found that viable virus was detectable in the aqueous humor [43]. Ophthalmologists were understandably hesitant to operate, given the limited information and risk for serious illness [44]. Surgeon self-preservation and safety is thereby a delicate balance against our professional duty to treat vulnerable patients and one which will result in continual struggles in varying circumstances.

The answer in both cases of HIV and Ebola ultimately came from gathering more knowledge about transmission and educating providers about how to protect themselves while carrying out their work. There is an institutional responsibility to protect employees, which can be viewed as a sort of social contract between providers and their employers [16]. If reasonably safe working conditions cannot be provided, then the professional obligation to provide medical services is weakened. Not all surgeons feel compelled to aid the sickest and most vulnerable patients in settings of great personal risk, but they nonetheless have a virtue-based duty to do so. Dr. Christine Grady writes that the duty to care for these patients can still be derived from a professional obligation to practice courage and impartiality [45]. Especially in situations where there may not be anyone better equipped than the provider at hand, that person has a stronger obligation to face the risks of treating that patient. Framing this as an issue of just access may make the case for helping more compelling, but the balance is also influenced by professional ethics. Dr. Christian Vercler's argument that surgical care is supererogatory runs parallel to Grady's concept of courage [46]. He describes it in the context of the surgeon's obligation to continue to care for "the 'hateful patient'" but it can easily be extended to caring for people with diseases that put providers at risk. In these situations, a surgeon is the only person physically able to provide the kind of care that a given patient needs, so this unique skillset confers a greater obligation to act.

Medical professionalism naturally invokes virtue ethics in matters of individual conduct. The standards of professional behavior for physicians are stringent, proportional to their relatively amplified power, privilege, and personal knowledge of their patients [47]. Virtuous conduct habitually practiced over a career can bring doctors closer to the exalted ideals of the profession, including integrity, humility, and self-improvement [48]. Physicians encounter structural injustice in various ways through myriad roles in a health system – for example, in caring for their patients, in advocacy roles, and via research to create generalizable knowledge. It is at this intersection of doctors' power as individuals and structural injustice where virtues become relevant to just access, because surgeons will often find themselves in situations where they are the only person determining whether a disadvantaged patient is able to receive surgical care. A physician who cultivates a sensitivity for fair access will be far better equipped to take advantage of these opportunities to ensure that the sick and vulnerable have the advocates they so desperately need.

Putting this awareness into action is a way to apply abstract virtues to concrete practice. Surgeons with an appreciation for existing injustices in access may be more inclined to deliberately expand their referral base to include underserved areas. In some systems such as the USA, decisions regarding which forms of insurance will be accepted are another example thereof. Further, when clinically evaluating disadvantaged patients, a surgeon conscious of social determinants of health will have a more complete view of the person they are treating and may be better equipped to tailor treatment to their particular needs. Finally, these individual virtues are critical in cases where an institutional policy may drive inequities in access to care and may even directly harm vulnerable populations [49].

Conclusion

Small differences in access to healthcare, and surgery in particular, can magnify social inequalities. In this way, injustice can seep into the medical field, which has an opportunity and an obligation to mitigate unfair practices. At every level, there is a duty to understand how disparities in access occur, to educate key stakeholders about the realities of these situations, and to address injustice in ways that map values onto actions. Humanitarian groups and global surgery programs should continue to further academic inquiry in the field of access. They should continue their laudable work with a focus on local sustainability. When governments recognize how they ration care, they can ensure that organizational systems work to provide a fair basic minimum for all people. When resources above that benchmark are allocated, they can be shared within communities in a way that promotes social justice. Institutions should make a deliberate effort to prioritize programs and business structures that enhance fair access for patients. Investigating the details of patients' experiences and seeking out ways to eliminate barriers to care, especially for vulnerable populations, are an excellent way to initiate this process. Surgeons who cultivate an appreciation for their patient's varied abilities to access care will be best equipped to reduce inequities on the level of individuals. Surgery is unique in healthcare for its resource intensiveness, the immediacy of its impact, and the variety of its morbidity, and all of these attributes are brought to bear on situations where only certain people have access to this kind of care. Reducing inequities in surgical access can vastly improve patients' lives, but moreover, by fulfilling a challenging ethical imperative this high-stakes field, the profession can set a standard for promoting justice in healthcare overall.

Works Cited

1. Morris AM, Rhoads KF, Stain SC, Birkmeyer JD. Understanding racial disparities in cancer treatment and outcomes. J Am Coll Surg. 2010;211(1):105–13.
2. Scott JW, Havens JM, Wolf LL, et al. Insurance status is associated with complex presentation among emergency general surgery patients. Surgery. 2017;161(2):320–8.

3. Dimick J, Ruhter J, Sarrazin MV, Birkmeyer JD. Black patients more likely than whites to undergo surgery at low-quality hospitals in segregated regions. Health Aff (Project Hope). 2013;32(6):1046–53.

4. Fitzmaurice C, Allen C, Barber RM, et al. Global, regional, and national cancer incidence, mortality, years of life lost, years lived with disability, and disability-adjusted life-years for 32 Cancer groups, 1990 to 2015: a systematic analysis for the global burden of disease study. JAMA Oncol. 2017;3(4):524–48.

5. Braveman P, Gruskin S. Defining equity in health. J Epidemiol Community Health. 2003;57(4):254–8. https://doi.org/10.1136/jech.57.4.254.

6. Ruger JP. Health and social justice. Lancet. 2004;364(9439):1075–80. https://doi.org/10.1016/S0140-6736(04)17064-5.

7. Häyry M. Utilitarianism and bioethics. In: Ashcroft RE, Dawson A, Draper H, McMillan J, editors. Principles of health care ethics. Chichester: John Wiley & Sons, Ltd.; 2006. p. 57–64. Retrieved from: http://onlinelibrary.wiley.com/doi/10.1002/9780470510544.ch8/summary

8. Rawls J, Kelly E. Justice as fairness: a restatement. Cambridge, MA: Harvard University Press; 2001. Print.

9. McCullough LB, Jones JW, Brody BA. Surgical ethics. Oxford: Oxford University Press; 1998.

10. Beauchamp TL, Childress JF, Oxford University Press. Principles of biomedical ethics. New York/Oxford: Oxford University Press; 2013.

11. Hester DM. Community as healing. Lanham: Rowman and Littlefield Publishers, INC; 2001.

12. Sharp LA. Denying culture in the transplant arena: technocratic medicine's myth of democratization. Camb Q Healthc Ethics. 2002;11(2):142–50. Cambridge Core, Cambridge University Press. https://doi.org/10.1017/S0963180102112060.

13. Etzioni A. Organ donation: a communitarian approach. Kennedy Inst Ethics J. 2003;13(1):1–18. Project MUSE. https://doi.org/10.1353/ken.2003.0004.

14. NICE | The National Institute for Health and Care Excellence. (n.d.). [CorporatePage]. Retrieved from https://www.nice.org.uk/.

15. Janz NK, Mujahid MS, Hawley ST, Griggs JJ, Hamilton AS, Katz SJ. Racial/ethnic differences in adequacy of information and support for women with breast cancer. Cancer. 2008;113(5):1058–67. https://doi.org/10.1002/cncr.23660.

16. Scheunemann LP, White DB. The ethics and reality of rationing in medicine. Chest. 2011;140(6):1625–32. https://doi.org/10.1378/chest.11-0622.

17. Centers for Medicare & Medicaid Services (CMS), HHS. (2003). "Medicare program; clarifying policies related to the responsibilities of Medicare-participating hospitals in treating individuals with emergency medical conditions. Final rule." Fed Regist 68(174):53222-53264

18. Wolf LA, Perhats C, Delao AM, Moon MD, Clark PR, Zavotsky KE. It's a burden you carry: describing moral distress in emergency nursing. J Emerg Nurs. 2016;42(1):37–46. https://doi.org/10.1016/j.jen.2015.08.008.

19. Zibulewsky J. The Emergency Medical Treatment and Active Labor Act (EMTALA): what it is and what it means for physicians. Proc (Bayl Univ Med Cent). 2001;14(4):339–46.

20. Crichlow RJ, Zeni A, Reveal G, Kuhl M, Heisler J, Kaehr D, et al. Appropriateness of patient transfer with associated orthopaedic injuries to a level I trauma center. J Orthop Trauma. 2010;24(6):331–5. https://doi.org/10.1097/BOT.0b013e3181ddfde9.

21. Esposito TJ, Crandall M, Reed RL, et al. Socioeconomic factors, medicolegal issues, and trauma patient transfer trends: is there a connection? J Trauma. 2006;61:1380–6; discussion 1386–1388.

22. Tyler PD, et al. Racial and geographic disparities in Interhospital ICU Transfers. Crit Care Med. 2018;46(1):e76–80. PubMed. https://doi.org/10.1097/CCM.0000000000002776.

23. Martin KD, Roter DL, Beach MC, Carson KA, Cooper LA. Physician communication behaviors and trust among black and white patients with hypertension. Med Care. 2013;51(2):151–7.

24. Kirch DG, Nivet M. Increasing diversity and inclusion in medical school to improve the health of all. J Healthc Manag. 2013;58(5):311–3.

25. Page S. The diversity bonus. NJ: Princeton University Press; 2017.

26. Beran TN, Kaba A, Caird J, McLaughlin K. The good and bad of group conformity: a call for a new programme of research in medical education. Med Educ. 2014;48(9):851–9. https://doi.org/10.1111/medu.12510.

27. Meuter RF, Gallois C, Segalowitz NS, Ryder AG, Hocking J. Overcoming language barriers in healthcare: a protocol for investigating safe and effective communication when patients or clinicians use a second language. BMC Health Serv Res. 2015;15:371.

28. Martin D, Singer P. A strategy to improve priority setting in health care institutions. Health Care Anal. 2003;11(1):59–68.

29. Meara JG, Leather AJ, Hagander L, et al. Global surgery 2030: evidence and solutions for achieving health, welfare, and economic development. Lancet. 2015;386(9993):569–624.

30. Gutnik LA, Dielman J, Dare AJ, Ramos MS, Riviello R, Meara JG, et al. Funding flows to global surgery: an analysis of contributions from the USA. Lancet (London, England). 2015;385(Suppl 2):S51. https://doi.org/10.1016/S0140-6736(15)60846-7.

31. Shrime MG, et al. Effect of removing the barrier of transportation costs on surgical utilisation in Guinea, Madagascar and the Republic of Congo. - PubMed - NCBI. https://www-ncbi-nlm-nih-gov.proxy.lib.umich.edu/pubmed/29225959. Accessed 15 Dec. 2017.

32. Massenburg BB, et al. Assessing the Brazilian surgical system with six surgical indicators: a descriptive and modelling study. - PubMed - NCBI. https://www-ncbi-nlm-nih-gov.proxy.lib.umich.edu/pubmed/28589025. Accessed 15 Dec. 2017.

33. Forte T, et al. Geographic disparities in surgery for breast and rectal Cancer in Canada. https://www-ncbi-nlm-nih-gov.proxy.lib.umich.edu/pmc/articles/PMC3997449/. Accessed 15 Dec 2017.

34. Chan M. Keynote address. Paper presented at: Human Rights Council panel on promoting the right to health through enhancing capacity-building in public health 2017; Geneva, Switzerland.

35. Brauman R. Médecins Sans Frontières and the ICRC: matters of principle. Int Rev Red Cross. 2012;94.888:1523–35. Web.

36. Snyder J, Dharamsi S, Crooks VA. Fly-by medical care: conceptualizing the global and local social responsibilities of medical tourists and physician voluntourists. Glob Health. 2011;7:6.

37. Murthy SS, Eyal N, Norheim OF, Ruan DT, Ntakiyiruta G, Robert R. Standard of care versus second-best: ethical dilemmas in surgery for high risk papillary thyroid cancer in low and middle-income countries. J Cancer Policy. 2015;6(Supplement C):8–10. https://doi.org/10.1016/j.jcpo.2015.08.006.

38. Changoor NR, Udyavar NR, Morris MA, et al. Surgeons' perceptions toward providing care for diverse patients: the need for cultural dexterity training. Ann Surg. 2017;269(2):275-82.

39. American Medical Association Code of Medical Ethics, Opinion 1.1.2[a].

40. Stevens P. Diseases of Poverty and the 10/90 Gap. International Policy Network, November 2004.

41. Grady C. Acquired immunodeficiency syndrome. the impact on professional nursing practice. Cancer Nurs. 1989;12(1):1–9.

42. Wolf LE. Universal precautions. Virtual Mentor. 2005;7(10). *journalofethics.ama-assn.org*, https://doi.org/10.1001/virtualmentor.2005.7.10.ccas3-0510.

43. Varkey JB, Shantha JG, Crozier I, Kraft CS, Lyon GM, Mehta AK, et al. Persistence of Ebola virus in ocular fluid during convalescence. N Engl J Med. 2015;372(25):2423–7. https://doi.org/10.1056/NEJMoa1500306.

44. Grady D. Ebola's legacy: children with cataracts. The New York Times. October 19, 2017.

45. Wolf L, Ulrich CM, Grady C. Emergency nursing, Ebola, and public policy: the contributions of nursing to the public policy conversation. Hastings Cent Rep. 2016;46:S35–8. https://doi.org/10.1002/hast.630.

46. Vercler CJ. Surgical ethics: surgical virtue and more. Narrat Inq Bioeth. n.d.;5(1):45–51.

47. Pellegrino ED, Thomasma DC. The virtues in medical practice. Oxford: Oxford University Press; 1993.

48. McCammon SD, Brody H. How virtue ethics informs medical professionalism. HEC Forum. 2012;24(4):257–72. https://doi.org/10.1007/s10730-012-9202-0.

49. Lenzer J. Whistleblower charges medical oversight bureau with corruption. BMJ. 2004;329(7457):69. *www.bmj.com.proxy.lib.umich.edu*. https://doi.org/10.1136/bmj.329.7457.69.

Ethics in Global Surgery

Anji E. Wall

Introduction

There is no standard definition for global surgery, but it generally refers to the provision of surgical care in low-resource settings, mainly low- and middle-income countries (LMICs). While there are many different paradigms, the majority of global surgical interventions are comprised of short-term medical volunteer missions by providers travelling developed countries. These missions are through non-governmental organizations (NGOs), religious groups, and academic institutional partnerships. While the characteristics of these missions can vary considerably, they share a common context of placing medical volunteers in unfamiliar places, with unfamiliar languages and cultural beliefs and severely limited resources. It is no surprise that surgical volunteers in this type of setting are faced with ethical challenges. The purpose of this chapter is to develop an argument for the importance of global surgical missions, discuss recent advancements in the field of global surgery, outline the benchmarks that organizations and individuals should strive for in the development of global surgical interventions, and provide a methodology for addressing clinical ethical issues that arise in global surgical missions.

A. E. Wall (✉)
Annette C. and Harold C. Simmons Transplant Institute, Baylor University Medical Center, Dallas, TX, USA

The Burden of Surgical Disease

In order to argue that global surgery is an important component of global health, it is essential to understand how rampant and disabling surgical disease is throughout the world. The Lancet Commission estimates that about 30% of the global burden of disease is surgical. Moreover, they found that 5 billion people worldwide are unable to access surgical services, with the majority of these individuals residing in LMICs [1]. Of the 234 million operations performed each year, 73.6% are on the richest third of the world, while 3.5% are on the poorest third [2, 3]. The leading causes of death worldwide include cardiovascular disease, trauma, and cancer, all of which are disease classes in which surgery is an essential element of disease management. These numbers clearly show that there is a huge unmet need for the care of surgical diseases in the LMICs.

Beyond the obvious mortality associated with surgical disease, these conditions also confer a huge burden of morbidity. Providing basic surgical care in LMICs has been estimated to be able to prevent the loss of 77.2 million disability-adjusted life years annually [4]. The bulk of surgical disease is made up of injuries, obstetric complications, perinatal conditions, congenital anomalies, malignancies, cataracts, and glaucoma [5]. In fact, it is estimated that children lose six times more productive years from burns and subsequent contractures than from war.

© Springer Nature Switzerland AG 2019
A. R. Ferreres (ed.), *Surgical Ethics*, https://doi.org/10.1007/978-3-030-05964-4_14

Taking into account injuries alone, an estimated 20 million children are injured each year, and 875,000 die from their injuries [2]. Those who do not die are often left with lifelong disability. Looking at current trends, the WHO estimates that by 2030, traffic accidents will rise from the ninth to the fifth leading cause of death globally, and the majority of these injuries will continue to be in LMICs, which currently account for greater than 90% of road traffic morbidity and mortality [6]. Much of the morbidity and mortality associated with road traffic accidents requires surgical management. Surgical care plays a major role in preventing and minimizing morbidity and mortality in younger patients who are the most vulnerable to traumatic injuries.

Another common condition in developing countries is cleft lip and palate. Approximately 190,000 children are born each year with this condition [7]. SmileTrain estimates the global backlog of cleft disease is 1.1 million people [8]. Muntz and colleagues explored the cost of clef lip and palate to individuals in the Philippines based on the fact that these conditions affect communication thereby affecting which jobs are available to them [9]. Income from jobs requiring communication skills was roughly double that of jobs not requiring communication skills. Individuals with cleft deformities are therefore less likely to hold high-paying jobs and are more likely to live in poverty. As Curci puts it: "Inadequate access to timely surgical care not only leads to unnecessary death, but inhibits the ability of survivors to lead productive lives" [10]. Surgical care for cleft lip and palate can improve quality of life and the economic productivity of patients.

Untreated surgical disease is prevalent in LMICs. It preferentially causes morbidity and mortality in young patients, who have the most productive life years ahead of them. It is expensive when disability takes away individuals' ability to work, be educated, and contribute to society. Global surgical interventions have the opportunity to improve the quality and quantity of life of the patients affected by untreated surgical disease, most of whom are in LMICs.

The Making of Essential Global Surgery

Historically, surgery has not been a major component of global health initiatives or basic health packages because of perceptions that surgical disease is only a small component of the global burden of disease and that surgical care is too expensive and specialized to implement in LMICs [2]. In 2008, Paul Farmer and Jim Kim described surgery as the "neglected stepchild of global health" [11]. As detailed above, there is a plethora of data that shows the significant burden of surgical disease. Moreover, there are successful examples of sustainable, cost-effective surgical enterprises in LMICs [12].

Thankfully, negative and dismissive attitudes toward global surgical interventions have changed. Mock labeled 2015 a banner year for global surgery because of three major events [13]. First, the *Disease Control Priorities, 3rd Edition* (DCP3), was published [14]. It describes the cost-effectiveness of large-scale health interventions. It has nine volumes, the first of which is dedicated to essential surgery. It estimates that if the 44 essential surgical procedures identified by authors were available worldwide, they would avert 1.5 million deaths. Moreover, the cost of these procedures is only $10–$100 per disability-adjusted life year, which is on par with interventions such as immunizations for infectious diseases [13]. The second event identified by Mock is the publication of findings by the Lancet Commission of Global Surgery. This paper ultimately argues that surgery should be viewed as an "indivisible, indispensable part of health care" [1]. Moreover, it recommends template national surgical plans for LMICs as well as surgical indicators to measure outcomes of implementing these plans. The final event cited by Mock is the World Health Assembly resolution on surgical care. This resolution is entitled "strengthening emergency and essential surgical care and anesthesia as a component of universal health coverage" [15]. It demonstrates a commitment by the World Health Organization to focus on surgical care. These three exciting events have ignited a dedication within the world community to focus on making global surgery an essential component of global health.

Quality of Surgical Care in LMICs

While access to surgical care in LMICs is limited, it is not altogether absent. About 3.5% of surgical procedures performed are in resource-restricted settings. However, the surgical care available in LMICs is not the same as that available in developed countries. Surgery is relatively safe in developed countries, with overall mortality related to surgical interventions estimated between 0.4% and 0.8%. Mortality rates are at least 10 times higher in developing countries [16]. There is also a two- to fourfold higher risk of anesthesia-related mortality in LMICs when compared to high-income countries [17]. Overall, surgery is less available and less safe in LMICs.

There are many barriers to safe and high-quality surgical care in LMICs. First, the focus of global health has historically been on infectious diseases, so the infrastructure in health systems of LMICs is primarily directed at the prevention and treatment of conditions such as HIV/AIDS, tuberculosis, malaria, and other infections. Obviously, the infrastructure needed for surgical care is very different from that needed for infectious disease prevention and treatment. Infrastructural barriers to care range from absence of operating rooms to nonfunctional operating rooms due to inadequate or broken equipment to functional operating rooms without a consistent source of clean water or electricity [18, 19]. Often, the supplies needed for surgical interventions including suture, instruments, and mesh are in short supply.

Personnel are also limited, and lack of skilled surgical providers has been cited in many studies as a primary barrier to the provision of surgical care in LMICs [2, 20]. Limited access to care, inability to pay, and the use of traditional medicine as the first intervention all contribute to late presentations of surgical disease, which can complicate operative management [19]. In addition,

when surgical missions are involved in providing care, there are often language and cultural differences between providers and their patients.

Political instability and civil war can also create barriers to care. Gulland (2013) describes the toll the war has put on the Syrian health-care system as "systematic destruction" [21]. The war has destroyed both health-care infrastructure and the health-care workforce. Without these elements, it is impossible to provide access to surgical care. And last, but not the least, there is a component of indifference toward the health care of the poor by many governments in these countries.

Organizational Ethics and Global Surgery

There is an obvious need for high-quality, safe, surgical care in LMICs, and there are many avenues that can be taken to address this need. These options range from short-term, service-based trips to long-term academic partnerships. This section discusses the common benchmarks that all surgical initiatives in LMICs should strive for, which are based in the common goal of global health in general and global surgery in particular, which is to provide maximal benefit to the population being served. These benchmarks are preparedness, competence, collaboration, sustainability, continuity of care, and outcomes monitoring [22].

Preparedness in the setting of global surgery requires acquiring knowledge of the location, culture, and language. In addition, volunteers must also get a sense of the capacity for surgical care in the area where they are planning to go. It is essential that volunteers know how many operating rooms they will have, what resources are available in operating rooms, the ward and ICU capacity, availability of common perioperative medications (e.g., antibiotics, analgesics), and local support staff among other things. There are several tools available for capacity assessment for surgical interventions. Surgeons Overseas developed a 100+ questions capacity assessment tool which asks about personnel, infrastructure, procedures, equipment, and supplies (PIPES) [23]. This type of instrument should be used prior to starting a mission to determine if the mission is feasible, identify the likely barriers surgical care, and decide what supplies should be brought with volunteers.

Competence refers to the level and focus of volunteers' training. Surgical volunteers should participate in missions that are in line with their expertise both for the purposes of patient safety and for provider comfort. Neither patients nor providers benefit when surgical volunteers work outside of their skill set or comfort zone. Surgeons may be asked to perform procedures that are outside of their comfort zone. If they have prior knowledge of common procedures that they will be asked to do, they can prepare for this prior to starting the mission. If this situation comes up during a mission, they should be prepared to say no if they are not comfortable or confident in their ability to do the procedure.

Collaboration as well as teaching and training local human resources is essential for a productive global surgery mission. Specifically, volunteers must reach out to local medical providers and community leaders during the planning stages of their missions and continue communication during and after the mission. Local collaborators can help develop specific goals and expectations that are based in community needs. Collaboration helps global surgical groups to design missions that are aimed at addressing the needs of the area where they are going as defined by medical providers and community leaders in that area. This fosters a positive relationship from the start, encourages buy-in from the community, and establishes a common vision for the mission.

Sustainability is paramount to global surgery initiatives. Short-term, isolated, procedure-based missions provide immediate benefit to those lucky enough to have procedures performed and have good outcomes from those procedures. But, when these groups leave, there is often no one left to continue to work. Sustainability, in a nutshell, is the objective of aid organizations to bring communities to the point where the organizations are no longer needed. When a community has the capacity in infrastructure, resources, and personnel to address its own surgical needs, the solution is permanent. Sustainability can be achieved through many avenues including education of local providers, continued presence through rotating teams, continued presence through virtual tools, and monetary support for continuing surgical care in the area.

The next benchmark, which goes hand in hand with sustainability, is continuity of care. While an operation is an isolated event, the follow-up care and the management of long-term complications are not. Surgical teams must recognize their obligation to patients even after they have left and that they have a plan for providing care after the mission, either with local providers, through a continued organizational presence with rotating volunteers, or a virtual consultation presence. Without continuity of care, patients can have delayed complications with associated morbidity and mortality, and the surgical care proved will fail.

The final benchmark for surgical missions is outcomes monitoring. Historically, surgical missions have measured this benefit in numbers of procedures performed. They have not recorded nor reported outcomes, thereby making the assumption that all patients who undergo procedures have positive outcomes and do not succumb to morbidity or mortality. As we know from surgical practice in the developed world, even in near-ideal conditions, surgical patients do suffer morbidity and mortality, so it only makes sense that patients in LMICs would be in the same situation even if volunteer surgeons are not in an area long enough to see these complications. The metric for measuring and reporting the benefit of medical missions has to change. Most US surgical programs collect data through the National Surgical Quality Improvement Program (NSQIP), which helps to monitor trends within hospitals and identify areas where hospitals are below average. This allows institutions to identify areas that are in need of improvement. While there are barriers to monitoring outcomes in LMICs, such as losing patients to follow-up and limited resources and personnel to do the outcomes monitoring, these are not insurmountable. If the goal of global surgery is to maximize benefit to the populations being served, then it is paramount that outcomes are monitored so that benefit can be quantified and surgical care can be improved.

These benchmarks for global surgery should be used by organizations to assess their missions and by individual surgical volunteers to decide how they can best contribute to global surgery. While they cannot prevent ethical issues from occurring, they can help minimize some of the root causes of ethical issues through planning, early communication, continued support, and a dedication to quality improvement.

Assessing Ethical Issues in Global Surgery

A surgical resident from the United States travels to Africa for an away rotation. During the rotation, she encounters a patient with an ulcerated breast cancer invading the abdominal wall. Chest X-ray does not show lung metastasis. The attending surgeon offers a radical mastectomy but not chemotherapy or radiation. In discussing the situation with the attending surgeon, he explains that chemotherapy is not available and radiation is only done at one center in the country. It is prohibitively expensive for patients.

Radical mastectomy without chemotherapy or radiation is not the standard practice in the United states and the resident is troubled by the situation of offering suboptimal treatment. She wonders if it is appropriate to offer mastectomy alone given the high risk of recurrence.

This case demonstrates how limitations dictate the care that can be offered for patients in LMICs. When faced with limited options and unsure about what the best course of action is, case analysis methods are helpful for thoroughly assessing the situation.

Even with the best laid plans for interventions, ethical issues are still bound to arise at the clinical level. Therefore, surgeons should be prepared to identify, analyze, and resolve ethical issues that they encounter just as they should be prepared for the surgical interventions they are likely to perform. The context of surgical volunteer work in LMICs is characterized by limitations and differences [24]. Limitations come in the form of instruments, operating rooms, anesthesia support, medications, electricity, running water, and supplies such as mesh, suture material, and sterile gloves. Differences between patients and providers include different languages, cultures, understandings of medicine, goals and values, as well as different approaches to the surgical informed consent process. While these characteristics may exist in high-income countries, they are more pronounced and universally present in LMICs.

There are many great techniques for addressing clinical ethical issues including the four-box method by Jonsen, Siegler, and Winslade, the seven-question method by Bernard Lo, and the root cause analysis by Jim Dubois [25–27]. Each of these methods directs the physician to ask a set of questions or determine a set of facts about the situation and use this information to guide the decision. These methods are designed for Western providers in their home setting and assume a shared culture, language, and basic understanding of medicine among stakeholders.

The methodology that I have proposed for addressing ethical issues in global health uses the methods described above for guidance but adds in the contextual features of differences and limitations. It is specifically designed for medical volunteer workers in LMICs. For example, it uses a mini-ethnography approach to ask patients explicitly about their understandings of the medical problems, what they think can be done about it and what they think will happen with and without intervention [28]. It seeks to identify the norms and values of all of the stakeholders involved and the limitations of the situation.

The methodology is outlined in Tables 1 and 2. These tables provide the list of questions that should be asked to various stakeholders when analyzing an ethical issue that arises during a

Table 1 Questions for medical aid workers and patients [29]

Category	Medical aid worker questions	Patient questions
Medical facts	What is the patient's diagnosis? What are the most prominent symptoms? What is the cause of the patient's health problem? What can be done to treat this problem? What is the prognosis for this patient? What do you expect the outcome of treatment to be?	What do you call your medical problem? What effect has this problem had on your life? What is the cause of your medical problem? What have you done to treat this problem? Has this intervention been successful? Do you know what else can be done by a doctor to treat this problem? What do you think will happen to you because of this problem? What do you fear about this medical problem? What do you fear about the treatment of this problem?
Values	What is your goal for medical intervention with this patient? What values are important to you in this case?	What is your goal for medical intervention in your condition? What values are important to you in this case?
Norms	What ethical norms are important in this case? What professional norms are important in this case? What legal norms are important in this case?	What ethical norms are important in this case? What professional norms are important in this case? What legal norms are important in this case?
Limitations	What constraints does time put on the treatment options? What constraints do limited medical resources put on the treatment options? Are there any other limitations to the treatment options?	What are the constraints on your ability to adhere to treatment options? Are there any treatment options that you would not be able to adhere to? Why? Are there any other limitations to the treatment options?
Stakeholders	Is there local medical staff to consult about this case? Are there other important stakeholders who should be consulted?	Does anyone help you make medical decisions? Should anyone be told about your medical care?

Table 2 Questions for local medical providers and additional stakeholders [29]

Category	Medical personnel questions	Other stakeholder questions
Medical facts	What is the patient's medical diagnosis? What are the most prominent symptoms? What is the cause of the patient's health problem? What can be done to treat this problem? What is the prognosis for this patient? What do you expect the outcome of treatment to be?	What do you call the patient's medical problem? What effect has this problem had on your life? What is the cause of the patient's medical problem? Do you know what else can be done by a doctor to treat this problem? What do you think will happen to the patient because of this problem? What do you fear about this medical problem? What do you fear about the treatment of this problem?
Values	What is your goal for medical intervention with this patient? What values are important to you in this case?	What is your goal for medical intervention in the patient's condition? What values are important to you in this case?
Norms	What ethical norms are important in this case? What professional norms are important in this case? What legal norms are important in this case?	What ethical norms are important in this case? What professional norms are important in this case? What legal norms are important in this case?

Table 2 (continued)

Category	Medical personnel questions	Other stakeholder questions
Limitations	What constraints does time put on the treatment options? What constraints do limited medical resources put on the treatment options? Are there any other limitations to the treatment options?	Are there any treatment options that the patient would not be able to adhere to? Why? What are the constraints on the patient's ability to adhere to treatment options? Are there any other limitations to the treatment options?
Stakeholders	Are there additional stakeholders who should be consulted?	Are there additional stakeholders who should be consulted?

medical mission to an LMIC. They basically outline how to perform an "ethics history and physical." Just as with a template history and physical, these tables are meant to provide guidance for approaching an ethical issue. There will be cases where some of the questions are not applicable or when a particular question or category needs to be explored in greater depth. At the end of the analysis, the provider should have a deeper understanding of the root cause or causes of the ethical issue (e.g., miscommunication, different understandings of the situation, variation among stakeholder values or goals). In addition, they will have identified the limitations and barriers to various options that might be considered in solving the ethical problem. For example, in the case described, the surgical resident is focused on recurrence as an outcome measure for the surgical procedure being considered. However, the attending surgeon may be more focused on preventing morbidity and improving quality of life by removing the ulcerated mass so that the patient does not have to perform wound care or be burdened with a malodorous lesion that could lead to marginalization from her community.

Determining a Course of Action

The methodology presented above helps the stakeholders analyze the root cause of ethical issues and the limitations to the options. After this exercise, the goal is to identify options that fit within the limitations that have been identified. If there are multiple options, they can be assessed using a decision aid. The decision aid presented in this section was initially designed for making decisions about public health ethics but can be modified to assist with clinical ethics decision-making [30].

The decision aid provides five considerations to use in analyzing each feasible option.

1. The first consideration is necessity. Stakeholders must determine if it is necessary to infringe on identified values or norms to achieve the goals.
2. Next, they should decide if the action is likely to be effective in achieving the desired goal.
3. Third is the concept of proportionality. Assessing proportionality requires stakeholders to decide if the desired outcome is important enough to infringe on the identified values and norms.
4. The forth consideration is that of least infringement. Stakeholders must determine if the option has been designed to minimize infringement on the norm or value that it is in conflict with.
5. Finally, the stakeholders should evaluate their decision-making for proper process. In clinical ethics decision-making, it is important to make sure that the right person (either patient or surrogate) is making the decision and that they have done so with adequate information, time, and without undue manipulation (Table 3).

Table 3 Process for deciding among available options [30, 31]

Effectiveness: Is the option likely to achieve the desired goal?

Proportionality: Do the expected benefits outweigh the infringement of the option of the norms and values?

Necessity: Is infringement on the norms and values necessary?

Least Infringement: Has infringement on the norms and values been minimized?

Proper Process: Do stakeholders agree that the process for making a decision was fair and transparent? Would they be comfortable sharing this with others?

References

1. Meara JG, Leather AJM, Hagander L, et al. Global surgery 2030: evidence and solutions for achieving health, welfare, and economic development. Lancet. 2015;386(9993):569–624. https://doi.org/10.1016/S0140-6736(15)60160-X.
2. Bowman KG, Jovic G, Rangel S, Berry WR, Gawande AA. Pediatric emergency and essential surgical care in Zambian hospitals: a nationwide study. J Pediatr Surg. 2013;48(6):1363–70. https://doi.org/10.1016/j.jpedsurg.2013.03.045.
3. Wall AE. Ethics in global surgery. World J Surg. 2014;38(7):1574–80. https://doi.org/10.1007/s00268-014-2600-5.
4. Debas HT, Donkor P, Gawande A, Jamison DT, Kruk ME, Mock CN. In: Debas HT, Donkor P, Gawande A, Jamison DT, Kruk ME, Mock CN, editors. Essential surgery: disease control priorities, Third Edition (Volume 1); Washington, DC: The World Bank; 2015. https://doi.org/10.1596/978-1-4648-0346-8.
5. Cometto G, Belgrano E, De Bonis U, et al. Primary surgery in rural areas of southern Sudan. World J Surg. 2012;36(3):556–64. https://doi.org/10.1007/s00268-011-1403-1.
6. Mathers CD, Loncar D. Projections of global mortality and burden of disease from 2002 to 2030. Samet J, editors. PLoS Med. 2006;3(11):e442. https://doi.org/10.1371/journal.pmed.0030442.
7. Corlew DS. Estimation of impact of surgical disease through economic modeling of cleft lip and palate care. World J Surg. 2009;34(3):391–6. https://doi.org/10.1007/s00268-009-0198-9.
8. Ozgediz D, Kijjambu S, Galukande M, et al. Africa's neglected surgical workforce crisis. Lancet. 2008;371(9613):627–8. https://doi.org/10.1016/S0140-6736(08)60279-2.
9. Muntz HR, Meier JD. The financial impact of unrepaired cleft lip and palate in the Philippines. Int J Pediatr Otorhinolaryngol. 2013;77(12):1925–8. https://doi.org/10.1016/j.ijporl.2013.08.023.
10. Curci M. Task shifting overcomes the limitations of volunteerism in developing nations. Bull Am Coll Surg. 2012;97(10):9–14.
11. Farmer PE, Kim JY. Surgery and global health: a view from beyond the OR. World J Surg. 2008;32(4):533–6. https://doi.org/10.1007/s00268-008-9525-9.
12. Riviello R, Ozgediz D, Hsia RY, Azzie G, Newton M, Tarpley J. Role of collaborative academic partnerships in surgical training, education, and provision. World J Surg. 2010;34(3):459–65. https://doi.org/10.1007/s00268-009-0360-4.
13. Mock C. A banner year for global surgery: now how to make it make a difference on the ground. World J Surg. 2015;39(9):2111–4. https://doi.org/10.1007/s00268-015-3154-x.
14. Mock CN, Donkor P, Gawande A, et al. Essential surgery: key messages from disease control priorities, 3rd edition. Lancet. 2015;385(9983):2209–19. https://doi.org/10.1016/S0140-6736(15)60091-5.
15. Price R, Makasa E, Hollands M. World Health Assembly Resolution WHA68.15: "strengthening emergency and essential surgical care and anesthesia as a component of universal health coverage"—addressing the public health gaps arising from lack of safe, affordable and accessible surgical and anesthetic services. World J Surg. 2015;39(9):2115–25. https://doi.org/10.1007/s00268-015-3153-y.
16. Kwok AC, Funk LM, Baltaga R, et al. Implementation of the World Health Organization surgical safety checklist, including introduction of pulse oximetry, in a resource-limited setting. Ann Surg. 2013;257(4):633–9. https://doi.org/10.1097/SLA.0b013e3182777fa4.
17. Bainbridge D, Martin J, Arango M, Cheng D, Group FTE-BP-OCORE. Perioperative and anaesthetic-related mortality in developed and developing countries: a systematic review and meta-analysis. Lancet. 2012;380(9847):1075–81. https://doi.org/10.1016/S0140-6736(12)60990-8.
18. Hoover EL, Cole-Hoover G, Berry PK, et al. Medical care on the brink: the need for re-engineering health-care services in sub-Saharan Africa. J Natl Med Assoc. 2005;97(3):397–404.
19. Sharma K, Costas A, Shulman LN, Meara JG. A systematic review of barriers to breast cancer care in developing countries resulting in delayed patient presentation. J Oncol. 2012;2012(8):1–8. https://doi.org/10.1155/2012/121873.
20. Choo S, Papandria D, Goldstein SD, et al. Quality improvement activities for surgical services at district hospitals in developing countries and perceived barriers to quality improvement: findings from Ghana and the scientific literature. World J Surg. 2013;37(11):2512–9. https://doi.org/10.1007/s00268-013-2169-4.
21. Gulland A. Medical students perform operations in Syria's depleted health system. BMJ. 2013;346(may14 2):f3107. https://doi.org/10.1136/bmj.f3107.
22. Wall AE. Benchmarks for international surgery. Arch Surg. 2012;147(9):796–7. https://doi.org/10.1001/archsurg.2012.696.

23. Groen RS, Kamara TB, Dixon-Cole R, Kwon S, Kingham TP, Kushner AL. A tool and index to assess surgical capacity in low income countries: an initial implementation in Sierra Leone. World J Surg. 2012;36(8):1970–7. https://doi.org/10.1007/s00268-012-1591-3.

24. Wall A. The context of ethical problems in medical volunteer work. HEC Forum. 2011;23(2):79–90. https://doi.org/10.1007/s10730-011-9155-8.

25. Jonsen AR, Siegler M, Winslade WJ. Clinical ethics. 8th ed: New York, NY: McGraw Hill Professional; 2015.

26. Lo B. Resolving ethical dilemmas. Philadelphia: Lippincott Williams & Wilkins; 2013.

27. DuBois JM. A framework for analyzing ethics cases. Ethics in Mental Health Research; 2008.

28. Kleinman A, Benson P. Anthropology in the clinic: the problem of cultural competency and how to fix it. PLoS Med. 2006;3(10):e294. https://doi.org/10.1371/journal.pmed.0030294.

29. Wall AE. Ethics for International Medicine. UPNE; 2012.

30. Childress JF, Faden RR, Gaare RD, et al. Public health ethics: mapping the terrain. J Law Med Ethics. 2002;30(2):170–8.

31. Wall A, Angelos P, Brown D, Kodner IJ, Keune JD. Ethics in surgery. Curr Probl Surg. 2013;50(3):99–134. https://doi.org/10.1067/j.cpsurg.2012.11.004.

The Anesthesiologist and the Surgeon: Two Professionals Sharing the Command of the Patient in the Operating Room

Anthony M. Roche and Gerald Dubowitz

Introduction

The operating room is a high-stress environment, where there are multiple competing interests in daily practice. These competing interests can stretch the will and ethical standard of any clinician in difficult situations. No matter what the situation is, the clinician's first ethical standard remains "primum non nocere" (first do no harm), probably the most well known of all physician ethical principles. It is arguably the first description of medical ethics and one which still holds true to this day. Whether or not a physician in practice today has taken some form of Hippocratic oath, it is this standard by which physicians are expected to hold themselves accountable in their day-to-day practice.

Clinical duty and providing care of quality may often be in conflict with institutionally mandated measures or production pressures to perform more surgeries. These pressures are commonly based on the requirements of meeting minimum health system targets or to increase revenues of for-profit or not-for-profit entities. Despite these pressures, it is the duty of the anesthesiologist and surgeon to take professional leadership in the operating room environment. However, this is not always guaranteed to be a straightforward decision-making process.

Arguably the single most important role of physicians in an operating room environment is that of fiduciary duty toward the patient. A fiduciary duty involves a trust relationship between a patient and the physician whereby the patient trusts his or her physician to make the best decisions on their behalf. The obligation of the physician is always to act in the patient's best interest and benefit [1]. Surgeons and anesthesiologists are the patient's fiduciary agents in the operating room, a role which carries a profound responsibility.

Not all operating room responsibilities are fixed. There are models of shared responsibility, with anesthesiologists working alongside certified registered nurse anesthetists (CRNAs), who in many states in the USA are independent practitioners. Indeed, in many countries, nurse anesthetists may be the only anesthesia providers, and the previously pervasive hierarchical model of doctors in charge is often still the case. Regardless of the medical hierarchy, it is crucial that surgeons and anesthesia providers of all backgrounds remain steadfast in their commitment to patients, always acting in their best interest. The same can be said for partnership with operating room (OR) nurses and other staff, ensuring that all OR members collectively care for their patients in an ethical and committed manner, respecting a patient's right to choose.

A. M. Roche (✉)
University of Washington, Seattle, WA, USA
e-mail: aroche@uw.edu

G. Dubowitz
University of California, San Francisco,
San Francisco, CA, USA

Although consent and the consent process is an important component of surgical care, it falls outside the scope of this chapter. This chapter rather focuses on principles of ethics, human rights, professionalism, and leadership.

Human Rights

According to the Oxford Dictionary, the definition of human rights is "a right that is believed to belong justifiably to every person" [2]. Furthermore, the United Nations Universal Declaration of Human Rights states in Articles 1 and 2 that "All human beings are born free and equal in dignity and rights. They are endowed with reason and conscience and should act towards one another in a spirit of brotherhood" and "Everyone is entitled to all the rights and freedoms set forth in This Declaration, without distinction of any kind, such as race, colour, sex, language, religion, political or other opinion, national or social origin, property, birth or other status. Furthermore, no distinction shall be made on the basis of the political, jurisdictional or international status of the country or territory to which a person belongs, whether it be independent, trust, non-self-governing or under any other limitation of sovereignty" [3]. These basic principles of human rights are entrusted to the medical professional to uphold in the practice of medicine. It is also therefore the medical professionals' responsibility to assure that at no point a patient's individual human rights are violated.

> **The Four Cornerstones of Healthcare Ethics**
> - Autonomy
> - Beneficence
> - Nonmaleficence
> - Justice

Ethical Principles

There are four cornerstones of healthcare ethics. These are the ethical standards to which both healthcare providers and institutions must adhere [4].

Autonomy

The principle of autonomy refers to self-determination, especially when it refers to decisions regarding medical care. It is a competent adult patient's right to make informed decisions about their medical care and forms the basis of informed consent and agreement to care. It is the personal rule of the self that is free from both controlling interferences by others and from personal limitations that prevent meaningful choice. Although autonomy usually refers to patient autonomy, it cannot be forgotten that all professionals are also entitled to their own autonomy. This is relevant not only when patients agree to treatment plans but is so too for individual professionals (e.g., surgeons and anesthesiologists) independently being able to make healthcare decisions.

Beneficence

Beneficence is an act that is performed for the benefit of others. In healthcare, physicians have both moral and professional obligations to help their patients. Their ethical duty is to balance risk versus benefit and strive to always remove risk of harm. They are expected to act in the patient's best interest, with the goal of promoting patient welfare or well-being. Beneficence and patient autonomy do not always go hand-in-hand and could lead to ethical dilemmas, e.g., a patient requiring coronary bypass surgery who smokes and refuses to stop smoking. The physician beneficence dilemma is to perform the surgery and to counsel the patient about smoking cessation. On the other hand, the patient's autonomous decision to continue smoking despite counseling and honest discussion with his or her physician should be respected.

Nonmaleficence

This principle relates to the necessity to (a) do no harm and (b) also not to cause any risk of harm. Clearly, this is not always easy to do, as there will always be a risk associated with any procedure. The challenge therefore is to balance what would be considered *acceptable* versus *unacceptable* risk. Only an honest, fair, balanced, and respectful discussion with the patient would be able to place all in context, and thereby patient and physician reach a collaborative acceptable risk-benefit decision. The knock-on legal aspects relating to risk are focused around liability and negligence, where liability would be deemed to be a risk or adverse event caused due to medical care provided. This does not imply negligence, as liability does not automatically indicate lack of practicing at a minimum standard. Negligence is care provided (or not provided) which does not meet a minimum professional standard, be it either by advertent negligence or recklessness (due to intentionally causing risk) or inadvertent negligence (by not providing adequate care). Checks and balances in the legal system help protect patients by holding physicians accountable for their acts.

Justice

Ferreres refers to justice in medicine as "… the fair allocation of resources" [5], whereas the Oxford English Dictionary refers to justice as "the quality of being fair and reasonable" [6]. Specifically in this area of healthcare, it is the role of social justice which clinicians knowingly or unknowingly practice. Physicians, who are held accountable by the Physician Charter, are reminded that they "… work actively to eliminate discrimination in healthcare, whether based on race, gender, socioeconomic status, ethnicity, religion, or any other social category" [7, 8].

Based on these principles, it is up to the physician, be they anesthesiologist or surgeon, to uphold their fiduciary duty to their patients of justice in the operating room. This is an especially unique setting, one where the patient, who may often no longer be able to speak for themselves due to the effects of anesthesia, has placed his or her trust in the surgeon or anesthesiologist to stand up and act on their behalf. It cannot be overstated that this is likely

the most important ethical role of a surgeon or anesthesiologist. The physician's role is therefore to steadfastly represent the patient's best interest, regardless of race, economic standing, gender identity, ethnicity, finances, or social setting.

Health Equity

Although arguably an issue closer related to humanitarianism, health equity is a crucial component of healthcare. It remains an ethical issue for leaders, especially in countries or system or even hospitals without universal access to healthcare. Health equity can be defined as "Attainment of the highest level of health for all people. Health Equity means efforts to ensure that all people have full and equal access to opportunities that enable them to lead healthy lives." Conversely, health inequities are "differences in health that are avoidable, unfair and unjust. Health inequities are affected by social, economic, and environmental conditions," and health disparities are "Differences in health outcomes among groups of people" [9].

The traditional physicians' Hippocratic oath obligates them to uphold ethical standards of care. The Hippocratic oath dating back to Hippocrates in BC460 was the first expression of medical ethics in medicine and set a number still highly pertinent ethical standards for the practice of medicine. Physicians' commitment is not only to the patient directly in their care but also to the society as a whole, and caring for populations at large remains a core principle of medical ethics. Public health is another cornerstone of healthcare, yet it is a skill set largely lacking among physicians. Due to physicians' ethical commitment to health for all, it is crucial for them to also be leaders in public health, specifically addressing health inequities and disparities. As a result it follows that surgeons and anesthesiologists play a vital role in leading efforts to address the inequities and disparities in surgical care at large. This remains valid for care both in the operating room, as well as creating systems for care of the poor and destitute.

Professionalism

A profession is a vocation or a calling to which the individual involved has trained to a higher level, reaching a minimum standard. Medicine is an example of a profession, and physicians remain among society's highest-trusted professionals. It is therefore crucial that physicians maintain a minimum standard of professionalism. The Royal College of Physicians in London, UK (RCP), described medical professionalism as "A set of values, behaviours, and relationships that underpins the trust the public has in doctors." Furthermore the 2005 RCP report proceeds:

> Medicine is a vocation in which a doctor's knowledge, clinical skills, and judgment are put in the service of protecting and restoring human well-being. This purpose is realised through a partnership between patient and doctor, one based on mutual respect, individual responsibility, and appropriate accountability.
> In their day-to-day practice, doctors are committed to: integrity, compassion, altruism, continuous improvement, excellence, working in partnership with members of the wider healthcare team.
> These values, which underpin the science and practice of medicine, form the basis for a moral contract between the medical profession and society. Each party has a duty to work to strengthen the system of healthcare on which our collective human dignity depends. [10]

The implied trust in the medical profession has been eroded in recent times, and the challenge is for the profession overall to regain society's trust. This can only be achieved by the profession as a whole, as well as by individual physicians acting with the highest ethical and professional standards.

It is this earned trust of patients to their anesthesiologists and especially their surgeons which creates a responsibility of these professionals to ethically fulfill their fiduciary duty as leaders of care in an operating room environment, both when the patient is aware as well as when they are under the effects of anesthesia. These professionals are expected to continually act in a manner which justifies this trust, which at times could put them in conflict with hospital management, other professions, or other individuals in the operating room environment. Regardless, it is their duty to act in their patient's best interest and lead the team to solving disagreements or conflicts.

Leadership

Surgeons and anesthesiologists are expected to be leaders in the operating room and perioperative environment. Despite excellent teaching of healthcare and clinical practice, unfortunately medical schools and residencies have not historically served their students or trainees effectively in leadership training. Although this is changing, it is still not sufficient in equipping these healthcare professionals for the enormity of their medical leadership role.

Leadership skills are a continuum. On one end of the spectrum, some individuals have an innate ability, possessing a natural skill set that enables them to effectively and calmly lead. On the other end of the spectrum are those who have few natural leadership skills. To them, leadership is based more on managing or following a set of rules, and they often (knowingly or unknowingly) find leadership positions difficult. No matter where the individual falls on this continuum, all need formal leadership training. There are skills that will assist in any number of scenarios, by equipping the individual to lead.

Examples of Leadership Skills
- Motivational and inspirational
- Patient-focused
- Lead by example
- Lead with skills and knowledge
- Effective organizational skills
- Interconnectivity
- Innovation
- Communication
- Independent thinker
- Visionary
- Change agent
- Serve staff and patients regardless of leadership style

Much has been written about leadership styles. The eight most effective leadership styles are charisma, innovation, command and control, pacesetter, laissez-faire, servant, situational, and transformational. There may be significant overlap in styles and leaders may possess more than one style. Different situations may call for different approaches or style, and the most effective leaders do so by innate ability, training, and experience. Surgeons and anesthesiologists require these skills as leaders in the operating room setting [1, 10].

Formal leadership training is required in physician training, along with training in management skills, finance, and business, along with an expanded public health focus. This is essential to better equip physicians for the leadership roles in surgery, anesthesia, perioperative medicine, and the operating room environment.

Surgeon-Anesthesiologist Relationship and Aligned Goals

There is a critical relationship between the surgeon and anesthesiologist. This extends beyond just care of the patient on the operating table but so too in the running of and leadership in the operating room itself. Due to the different backgrounds and focus of these specialties, mutual respect, partnership, and collaboration are key in the safe care of patients and leading the operating department. Not only are these two specialties codependent on each other, but they can also set the tenor in working with nursing leadership, technicians, and ancillary staff and administration.

Surgeons and anesthesiologists caring for the same patient can be described as an ideal synergy of skills, knowledge, expertise, and commitment. It follows therefore that as professionals, they function extremely well as a team in the vast majority of circumstances.

One could consider the unique surgeon-anesthesiologist relationship as aligned goals falling into one or more of the following categories: (1) patient safety, well-being, and satisfaction; (2) highest-quality, evidence-based care;

(3) efficient operating room utilization; (4) clinical governance; and (5) leading the entire perioperative team in morale and teamwork by creating a communicative, safe, and supportive environment.

Specifically, clinical governance is a concept widely described in the UK, Canada, Australia, and New Zealand. The UK's National Health Service (NHS) describes clinical governance as "Clinical governance is the way the NHS works to improve the quality of care patients receive and to maintain that high quality of care. It is about ensuring that patients get the right care at the right time from the right person and that it happens right first time" [11]. It comprises the following key clinical components: (1) risk management; (2) clinical audit; (3) education, training, and continued professional development; (4) evidence-based care and effectiveness; (5) patient and caregiver experience and involvement; and (6) staffing and staff management [12–15].

By incorporating the principles of clinical governance into surgical care, the OR team are equipped to provide best-practice evidence-based care. The team's goals are aligned, with ongoing quality improvement processes in place, thereby providing the ideal circumstances for best possible outcomes [14].

Different Perspectives and Viewpoints

It would be naive to assume that this is always the case. As highly trained individuals with independent skills and knowledge, conflicts will occur. It is the personal and professional duty of surgeons and anesthesiologists to respectfully work toward solving conflicts. The core of this conflict resolution needs to focus at all times on the patient's well-being.

There are multiple sources of potential conflict in or around the operating room, for example, clinical issues, changes in patient condition, professional conflicts with management, and nursing versus medical conflicts. In addition, interpersonal conflicts can occur between individuals and different specialties, as well as between surgeon and anesthesiologist.

As professionals, whatever the cause or situational details regarding the conflict, it is the surgeon and anesthesiologist's duty to lead resolution. This resolution may be particularly difficult when the conflict is between these two individuals.

Historically, the anesthesiologist worked *for* the surgeon, and this still remains the case in many countries at the present time, especially where anesthesia provision is not always provided by a physician anesthesiologist. The surgeon was, and often still is, considered in charge of the operating room. This traditional "captain of the ship" model implied that the surgeon is undoubtedly in charge of all activities in the operating room and that all responsibility rests on this individual's shoulders. This model remained until the 1980s, when it became apparent that anesthesiology departments were becoming entities in their own right, rather than being under the overall umbrella of surgery departments. Considering the complexity of hospitals and healthcare systems, it can no longer be presumed that a surgeon is responsible for everything that happens in an operating room, one where there are institutionally mandated process measures, nursing staff, scrub technicians, anesthesiologists, nurse anesthetists, anesthesia technicians, ancillary staff, environmental services, as well as laboratory, engineering, and management interactions. This represents a significant change in the operating room environment, one that has also significantly affected the relationship between surgeons and anesthesiologists, as well as the dynamics and responsibilities in the operating room. Teamwork and collaboration are key in providing the best for each patient.

As times and roles have changed, it is essential for surgeons and anesthesiologists to have a clear understanding of each other's role in the perioperative process. Each one of these physicians has a unique and expert skill set, which as a whole makes a formidable team to care for each patient. However, trust and mutual respect are crucial for the effective performance of this team. Additionally, despite the leadership roles of all of the professionals, in the operating room environment, they are completely dependent on the larger team involved in patient care, from clinic medical and paramedical staff, through assessment clinics, preadmission, operating room, recovery, and disposition locations. There are many "moving parts," each crucial for safe and effective care of each patient. It is upon the surgeon and anesthesiologist to recognize their role in the team and be prepared to lead it.

What if conflicts initially appear unresolvable? How should these professions work toward resolving their differences?

If the conflict is of a clinical nature, the urgency of the case in question should dictate how the team proceeds. In elective or otherwise nonemergent situations, the surgeon and anesthesiologist have the benefit of time to work on an amicable or sensible solution. This may require further investigation or consultation within or outside the specialties in question, suggesting it may require a full multidisciplinary discussion. In circumstances where differences in opinion still exist, open and honest discussion with patient and/or family may be required, providing them with all relevant benefit and risk information. Despite differences in opinion, the surgeon and the anesthesiologist have to be objective and unbiased in managing this family conference, allowing the patient and/or family to reach an informed decision.

Emergent situations require a greater degree of urgency in reaching consensus. It is equally important that both specialties pursue this within the context of doing what is best for the patient. Further consultation may not be an option, hence the responsibility for the surgeon and anesthesiologist to pursue a solution. In situations where no consensus can be reached, systems need to be created which can help reach resolution. Institutions have created the role of an operating room lead, usually an in-hospital physician (24 hours a day) who is tasked with making a best-judgment decision in resolving conflict.

High-Risk Patients

Based on their already described role of managing conflict or differences in opinion, surgeons and anesthesiologists are uniquely positioned to lead multidisciplinary high-risk clinics and evaluation panels. This is due to their unique perspectives on surgical and medical disease, as well as their knowledge of projected disease course. These professionals are well equipped to create and guide these teams in workup, preparation, and perioperative optimization of high-risk patients but also to advise if surgery is not the ideal option, if risk profile outweighs potential benefits in specific patients.

International Health Ethics

Compared to its traditional roots in medicine and infectious disease, global health is an emerging field in surgery. Surgeons and anesthesiologists often travel to extremely low-resourced areas of the world to be involved in global surgery endeavors. The Lancet Commission on Global Surgery described the lack of access to surgery globally as a public health crisis, especially in middle- and low-income countries. The commission recommended a dramatic and urgent need for expansion in surgical capacity [16]. The most common type of global surgery program has been the specialist surgery camp or mission, where a number of procedures are performed on a population who usually do not have access to such services through their own healthcare systems. Common examples are surgeries for cleft lip and palate, orthopedic trauma, club foot, pediatric cardiac surgery, ophthalmology, burns, vesicovaginal fistulae repair, and so many more. Often these camps travel with an entire team of surgeons, operating room nurses, anesthesiologists, scrub technicians, and biomedical engineers, often all flown in from a high-income country. Much time and planning is involved in facilitating a successful camp. Funding is provided in a number of different ways, some from foundations, charities, and governments to private donors, personal expenses, or gifts. This has a significant potential to cause production pressures on the camps, especially to perform a large number of procedures in a short timeframe. There are multiple potential problems that arise, especially when considering much of the host sites infrastructure is usually worse than what these teams are used to having in their own institutions. Equipment is

either often not present at the proposed site or broken or partially broken, and unfamiliar tools and devices may be all that is available. Additionally, medications are often in alarmingly short supply, as are electrical power and oxygen, neither of which is necessarily guaranteed to be available and when they are may only be intermittently available. The question is how does the team proceed and what are the ethical considerations of (a) being there and (b) proceeding in situations where their familiarity and skill levels are potentially tested beyond their boundaries of safety? Furthermore, although the surgeon and anesthesiologist may be the leaders in their own home institutions, who is in charge when they visit a low-resource setting? Seeing as they brought a team, usually at substantial expense, does that allow the visiting team to be in charge, or are they guests of a host institution and should therefore submit to the host institution leaders? Although this appears like an easy and logical answer, the details of implementation are often vague and challenging in the real-life settings.

Although service-based surgical camps remain the most common global surgery endeavors, emphasis is now moving toward more sustainable partnerships in teaching and training, program development, and capacity building [17]. These are arguably more sustainable global health models, as the service-based model can disenfranchise the local medical community, often leaving them at the periphery of the care delivery being provided by the highly resourced visitors from abroad. Regardless of the model, ethical concerns remain.

If one pauses and considers the pillars of human ethics, being autonomy, beneficence, nonmaleficence, and justice, it is easy to see where problems could arise. Visiting professionals struggle to adequately consent individuals due to language barriers, and it could be perceived as coercion, rather than true patient autonomy. Beneficence may not be as big of a problem, as the visitors are usually well-intended, but nonmaleficence could easily be a stumbling block. Although the visitors do not wish to harm patients, their unfamiliarity with the host system, equipment, medications, staffing, and process could lead to unwanted risk or harm. In addition, follow-up is an important component of any surgical care, and plans may not always be in place for managing not only the routine follow-up but even more importantly complications relating to care. That could lead to questions of justice and self-determination for the patients affected and could clearly question how effective a visitor-led full informed consent process is in such settings.

The specifics of who's in charge echo loudly as cases are being performed. Imagine a situation where a visiting surgeon from the USA is operating on a child alongside a host surgeon in sub-Saharan Africa, and they encounter technical problems during the surgery, where these problems could lead to significant unintended harm. If there is a difference of opinion on what the course of action should be, even if the visiting surgeon is reasonably experienced in handling this situation, who makes the final decision? Who's in charge? Whose patient is it? What if the host surgeon's plan isn't likely to rescue the clinical dilemma? What if the visiting surgeon has an entire team from his home institution and everybody felt strongly that their US-based surgeon should make the decision? What would the impact be to the host surgeon, the program, and, obviously, the patient? How would they handle a family conference after surgery if things did not proceed well and the patient suffered complications? What recourse does the family have? Who takes the blame?

Most people in healthcare are well intentioned; this is especially true of the majority of those who take time to travel on global health missions and camps. However, being poorly prepared for the reality of what these providers will encounter once they reach these low-resource settings is a setup for personal and professional ethical dilemmas. Indeed, little focus is placed on the ethics of international health, and there is a paucity of ethics publications specifically regarding global surgical programs. This leads to poor preparation of many well-intended healthcare workers, most who have to figure it out on-the-fly. Furthermore, are the good intentions of surgeons, anesthesiologists, and their visiting teams enough to justify their programs, or could the mere existence of these programs be causing more harm than could be intended?

It can be difficult for leaders like surgeons and anesthesiologists from highly resourced centers and countries to honestly reflect on the potential for causing harm in low-resource settings. The mantra is often "any help is good help," and "people are receiving treatment they may not otherwise be receiving." Visiting professionals of all types may often believe that seeing as they come from impressive highly resourced and educated backgrounds, they know what is best for the host institutions or programs. Often, visitors lack meaningful of insight into the immense cultural, ethnic, and anthropologic determinants of healthcare practice in the new settings and may make sweeping statements or dictate rules of engagement for the host institutions. These institutions and local providers often acquiesce to these demands in the hope of funding, donations, and help with managing their population's disease burden. This type of paternalistic arrangement is effectively healthcare "colonialism" or "neocolonialism" as described to the authors of this chapter by a Ugandan colleague in 2015.

Holm and Malete wrote a compelling commentary in The Chronicle of Higher Education in 2010, calling it "Nine Problems That Hinder Partnerships in Africa." In their paper, they describe nine problematic themes endemic in partnership programs.

Holm and Malete's "Nine Problems That Hinder Partnerships in Africa"

1. Developed countries take academic lead.
2. Outside scholars dictate university curricula.
3. Visiting academics top-down approaches.
4. Developing countries cannot afford project costs.
5. Multiple donor partners.
6. Developed countries' researchers have an obligation toward donors.
7. Top-quality universities looking for comparable quality in developing countries.
8. Overexaggerated risk to staff and students in Africa.
9. New skills often taught in quick workshops.

The authors finish their paper by stating the following and is a challenge to leaders of global health programs:

> The challenges can be overcome, but not over a 1- or 2-day visit. They require the development of a relationship that stems from friendship, trust, and mutual respect, a relationship that comes with shared experiences, disagreements, conversations, and solving problems together. All of that is demanding, but not impossible [18].

The global financial "aid" community is another sobering example of good intentions not always improving the lives of those intended to be helped. Aid to poor countries, for example, by debt forgiveness or monetary donations, has often worsened the problems of the countries they were intended to help. Although this is not the case of all aid programs, it is indeed a reality in the aid industry. Whatever the net effect is for different programs, good or bad, they require intense personal and programmatic scrutiny for potential negative effects [19–22]. This introspection and evaluation should also be another cornerstone of global health programs, one where physician leaders lead the drive to assess impact, quality, and effectiveness, contrasted with negative effects on patients, their families, the health system served, and the host site healthcare professionals.

It requires a diligent, patient, and reflective approach to listening to what the host leaders and providers have to say and then to gently probe with pointed questions of knock-on effects to the host site. This is true not just of surgery camps, but also teaching, program-building, or research-based programs. That type of leadership should be demanded from physician leaders of these endeavors.

At the outset these programs should be under the leadership and guidance of the host institution physicians and healthcare workers who then partner with appropriate visiting professionals, those familiar with the host environment, as well as being leaders in their own institutions. This may be a stumbling block for many of the visiting physicians, who are task oriented, aiming at achieving the goals they may have set themselves or coerced from host leaders. Once again, the host institutions should lead these processes, with astute, respectful visiting physicians being key partners in a collaborative process. As partners, with a strong focus on healthcare and international health ethics, visiting teams can build programs with respect and sustainability for those who require it most – the patients in need.

As medical ethicists have struggled with paternalistic behavior which remains common in global health, progress has been made on tools for ethical engagement in partnerships. The Canadian Coalition for Global Health Research has published a collaborative groundbreaking tool for assessing and improving global health partnerships. The "Partnership Assessment Tool" focuses on five key themes for bilateral consensus [23]:

- Sustainability
- Knowledge production
- Knowledge translation
- Capacity development
- Innovation

Surgeons and anesthesiologists who embark on global health programs are required to lead with humility respect, keeping their teams focused on valuing and embracing the local communities and healthcare providers, to listen more than they speak, and to also serve with respect.

Summary

The operating room has developed into a high-pressure and complex medical environment, one with multiple and often distracting processes occurring at any point in time. Yet, at the core remains the patient, who has placed their trust in the operating room team to provide the best possible care and always act in the patient's best interest. There are systems in place, with operating room leadership and management, yet on the basis of the physician-patient contract, the patient has provided their surgeon and anesthesiologist fiduciary duty for their care. This is both an ethical and legal contract, which ideally places these physicians in the position of leading the perioperative team. It is their expectation and duty to manage and coordinate the complex flow in the operating room environment, taking responsibility to lead the team. In times of patient condition change, conflict, or differences in opinion, it is the role of the fiduciary agent to create an environment of problem solving. The pillars of ethical behaviors remain the same whether they occur in a high-income setting or the realms of global health in low- and middle-income countries. Patients deserve no less.

References

1. Chervenak FA, McCullough LB. The moral foundation of medical leadership: the professional virtues of the physician as fiduciary of the patient. Am J Obstet Gynecol. 2001;184(5):875–9; discussion 9–80
2. Press OU. Human right. Oxford University Press; 2018 [Definition of Human Right]. Available from: https://en.oxforddictionaries.com/definition/human_right.
3. Nations U. Universal declaration of human rights. Geneva; 1948. Available from: http://www.un.org/en/universal-declaration-human-rights/.
4. Beauchamp TL, Childress JF. Principles of biomedical ethics. 6th ed. New York: Oxford University Press; 2009.
5. Ferreres AR. Professionalism in the operating room. In: Jericho BG, editor. Ethical issues in anesthesiology and surgery. Heidelberg: Springer; 2015. p. 127–38.
6. Press OU. Justice. Oxford University Press; [Definition of Justice]. Available from: https://en.oxforddictionaries.com/definition/justice.
7. Kelly RJ, Nisynboim C. Fatigue and the care of patients. In: Jericho BG, editor. Ethical issues in anesthesiology and surgery. Heidelberg: Springer; 2015. p. 79–92.
8. Medicine AFABoI, Medicine A-AFACoP-ASoI, European Federation of Internal M. Medical professionalism in the new millennium: a physician charter. Ann Intern Med. 2002;136(3):243–6.
9. Institute HE. Health equity. San Francisco: San Francisco State University; 2018. Available from: https://healthequity.sfsu.edu/content/defining-health-equity.
10. Party RCoPW. Doctors in society. Medical professionalism in a changing world. London; 2005.
11. Trust UHBN. Clinical Governance Edgbaston. Birmingham; 2018. Available from: www.uhb.nhs.uk/clinical-governance.htm.
12. Philippon DJ, Braithwaite J. Health system organization and governance in Canada and Australia: a comparison of historical developments, recent policy changes and future implications. Healthc Policy. 2008;4(1):e168–86.
13. Scholefield H. Clinical governance: what foundation doctors need to know. Br J Hosp Med (Lond). 2005;66(10):M42–4.
14. Spark JI, Rowe S. Clinical governance: its effect on surgery and the surgeon. ANZ J Surg. 2004;74(3):167–70.
15. Braithwaite J, Travaglia JF. An overview of clinical governance policies, practices and initiatives. Aust Health Rev. 2008;32(1):10–22.
16. Meara JG, Leather AJ, Hagander L, Alkire BC, Alonso N, Ameh EA, et al. Global surgery 2030: evidence and solutions for achieving health, welfare, and economic development. Lancet. 2015;386(9993):569–624.
17. Kotagal M, Horvath K. Surgical delivery in under-resourced settings: building systems and capacity around the corner and far away. JAMA Surg. 2015;150(2):100–2.
18. Holm JD, Malete L. Nine problems that hinder partnerships in Africa. Commentary. The Chronicle of Higher Education. 2010. 06/13/2010.
19. Deaton A. The great escape: health, wealth, and the origins of inequality. Princeton: Princeton University Press; 2013.
20. Easterly W. The white man's burden: why the west's efforts to aid the rest have done so much ill and so little good. New York: Penguin Publishing Group; 2006.
21. Moyo D, Ferguson N. Dead aid: why aid is not working and how there is a better way for Africa. New York: Farrar, Straus and Giroux; 2009.
22. Sachs J. The end of poverty: economic possibilities for our time. New York: Penguin Books; 2006.
23. Research CCfGH. Partnership assessment tool: Canadian Coalition for Global Health Research; 2017. Available from: www.ccghr.ca/resources/partnerships-and-networking/partnership-assessment-tool/.

The Surgeon-Patient Relationship: Built Upon Trust

H. Alejandro Rodriguez and Carlos A. Pellegrini

Introduction

A trusting relationship and enduring between physicians and their patients lies at the very heart of the practice of medicine. This bond has its base on communication [1]: a bidirectional flow of information through which the physician obtains information about his or her patient, counsels on the nature of disease and available treatments, and provides necessary support. In turn, patients communicate to their physician the series of events that brought them to seek medical care, inquire about the specific impact of a given disease process, and participate in decisions regarding available treatments.

Trust has been described as an essential component of all human relationships [2]. Indeed, most surgeons will instinctively recognize that trust is crucial to a successful surgeon-patient relationship. But what, exactly, is trust? What are the barriers to trust? What are the mechanisms by which we can we promote trusting relationships in our everyday practice? Unfortunately, scarce methodical attention has been given to these concepts within the field of surgery. In this chapter, we will review the general concept of trust, its

barriers, ways to develop and maintain trust, and its relevance to an ethical surgeon-patient relationship.

Trust

> Trust is the foundation of society. Where there is no truth, there can be no trust, and where there is no trust, there can be no society. Where there is society, there is trust, and where there is trust, there is something upon which it is supported.
> – Frederick Douglass (circa 1818–1895)

The Oxford English Dictionary defines trust as the "firm belief in the reliability, truth, or ability of someone or something" [3]. Similarly, Merriam-Webster's Dictionary states that trust is "assured reliance on the character, ability, strength or truth of someone or something" [4]. In these definitions, multiple synonyms (e.g., reliance, truth) are used to describe trust. The sociologist Bernard Barber observed this tendency and astutely identified a common thread among definitions of trust: expectations [5]. Thus, Barber defined trust as the following expectations:

1. The persistence and fulfillment of the moral and social orders.
2. A technically competent role performance from those involved in social relationships and systems.
3. The involved parties will carry out their fiduciary obligations and responsibilities.

H. Alejandro Rodriguez
Department of Surgery, University of Washington, Seattle, WA, USA

C. A. Pellegrini (✉)
UW Medicine, University of Washington, Seattle, WA, USA
e-mail: pellegri@uw.edu

© Springer Nature Switzerland AG 2019
A. R. Ferreres (ed.), *Surgical Ethics*, https://doi.org/10.1007/978-3-030-05964-4_16

Another sociologist, Niklas Luhmann, describes trust as a "a basic fact of life," without which a person would be "prey to a vague sense of dread, to paralyzing fears." Luhmann's view is that daily life is filled with innumerable possible scenarios, too complex for any person to calculate or anticipate. Trust then is to reduce this complexity, "to behave as though the future were certain" [6]. From the above definitions, it follows that trust is essential for all human relationships and a functioning society.

All trusting relationships, however, introduce a measure of vulnerability. To trust is to subject oneself to the goodwill of others, to the belief that the moral order will persist. If either party in a relationship does not fulfill his or her obligations, it is done so at the detriment of the opposite party. Thus, while trust can be described as a necessary (and desirable) component of everyday life, it can be hazardous as it opens the possibility of exploitation.

Trust in Medicine

If trust is necessary for a functional society, it is to be expected that it is indispensable for the successful practice of surgery. Patients must put aside fears and place faith in the fact that a surgeon will be competent and will carry out his fiduciary duty. Indeed, Barber's definition of trust can be well adapted to medicine, as society generally expects physicians to:

1. Diagnose, counsel, and treat the patient.
2. Do the above in a technically proficient manner.
3. In all decisions, place the patient's best interest first and foremost.

Luhmann's concepts can be likewise applied to medicine; a surgeon must reduce complexity by helping patients to navigate through a multitude of available treatment options and possible outcomes. Without this reduction in complexity, we could well expect patients to be paralyzed with fear and to refuse medical or surgical care.

While definitions of trust between any two parties can be applied to medicine, there are a couple of aspects that make trust within the physician-patient relationship unique. First, the asymmetry of power should be considered. On one hand, the patient is partially forced into the relationship: she seeks care after a series of symptoms pushed her beyond a sufficient threshold [2]. She could be expected to be concerned, anxious, or otherwise suffering. The very action of presenting and requesting medical care is the first act of trust: "I will go to the doctor, surely she can figure out what is going on and help me get better." Finally, the patient lacks what she so sorely needs – the knowledge and expertise to heal herself. These processes (the suffering of illness, "blind" trust, and the knowledge gap) contribute to a state of vulnerability. Furthermore, while both parties (surgeon and patient) seek a positive outcome, it is for the patient, whose very existence may be threatened, for whom the stakes are higher. On the other hand, the surgeon holds the knowledge and the wisdom, can foresee a number of expected outcomes, and needs not personally suffer the anxiety of illness. Finally, physicians hold power as "gatekeepers" of medical care – a starting point from which further testing, consults, and treatment may result [7].

System Trust

Just as trust must exist between individuals, it is necessary within relationships between individuals and larger systems. In today's modern society, we rely not only on the expectations that a certain professional will honor her or his fiduciary duty but also a system that supports and validates the role of the professional [6]. To place trust in a lawyer is to also place trust in the law school she attended, the bar association that certifies her licensure to practice, and the court system or legal firm that employs her [5]. Thus, while a client's trust may rest mainly in his lawyer, a fraction of that trust is shifted to a larger network.

The same is true for medicine. We expect surgeons to be properly trained and licensed by the relevant authorities. Further, surgeons do not practice medicine by themselves. From doorway to the operating room, a patient will encounter an array of medical and administrative staff, who must all fulfill a certain moral and social order, do so competently, while honoring the fiduciary duty.

That medicine is not practiced in a vacuum is no surprise for physicians. However, many will eschew responsibility for other elements in the system. Imagine the following scenario: a surgeon is late for a patient appointment, largely in part due to a clerical error. The patient is irritated by this situation and voices his displeasure to the surgeon. The surgeon could respond "yes, the staff here is very inefficient, I'm no fan of them myself." In an effort to preserve his own bond of trust with the patient, this surgeon has distanced himself from the system. However, in the eyes of the patient, the surgeon is part of the system, and the distrust he harbors for it is likely to extend to the surgeon himself. Consider now an alternative response: "I am very sorry for this mistake and take full responsibility. My staff and I will work together to ensure it does not happen again." This surgeon has accepted that some measure of trust between himself and the patient will be lost but assured the patient that he trusts the system he is part of. We contend that the latter option allows for trust building to continue and is a preferable response. In fact, we believe that when the physician "builds" the patient's trust in the entire system, the physician is building his or her own trust from the patient [1].

The above example illustrates the fact that while system trust may represent a replacement of trust in the individual, the latter cannot be fully absconded. As Pellegrino and Thomasma point out in their treatise on clinical practice, trust is an ineradicable component of all human relationships [2].

Physician Trust in the Patient

Most explorations of trust in medicine focus on the trust patients attribute to their physicians; however a handful of authors have described the characteristics and importance of physician trust in the patient. For many physicians, the notion of reciprocal trust may be alien. However, this bidirectional flow of trust has been described by patients as a foundation on which trust can continue to be generated [8].

This concept was explored in great detail by Thorne and Robinson, who performed an analysis of healthcare relationships among chronically ill patients. The authors found that in the course of a chronic illness, patients and their family members developed specific knowledge and competencies in managing their disease. If these competencies were not validated by their healthcare provider, a loss of trust ensued. Conversely, if the patient's acquired competencies were acknowledged, a feeling of empowerment and trust was fostered [9].

In a qualitative analysis of interviews of patients who were active drug users and their physicians, Merrill et al. explored "mutual mistrust" and its consequences in healthcare delivery. Their results showed that physicians feared being deceived by their patients, which in turn led to a lack of engagement with key patient complaints. For their part, patients in this study were extremely sensitive to subtle negative cues from the physicians, which they took to be signs of intentional mistreatment [10].

Thus, if trust is to thrive in a surgeon-patient relationship, it must be reciprocal. The patient who feels trusted bolsters her trust in the physician. Conversely, the physician who does not trust his patient is unlikely to be trusted by the patient. Trust in medicine thus depends on a virtuous circle (Fig. 1) between the patient, the physician, and the system. Lack of trust (or the emergence of distrust) in any one part of the circle will cause the entire cycle to collapse.

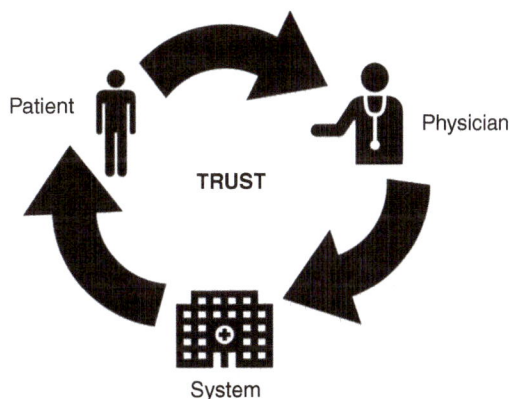

Fig. 1 The virtuous cycle of trust. A break in any part of the cycle weakens the whole system

Recent Trends in Trust and the Medical Profession

Many physicians hold the belief that public trust in physicians has steeply diminished over the past few decades. In truth, distrust for the medical profession has long been harbored by some sectors of society. Consider the following line from Anton Chekhov's play *Ivanov* (first performed in 1887): "Doctors are the same as lawyers; the only difference is that lawyers merely rob you, whereas doctors rob and kill you" [11]. Chekhov, who was a physician himself, based much of his early work on his own observations of society as well as his medical experience [12].

Satirists notwithstanding, the medical profession has historically enjoyed high rates of public trust. This continues to be true today; public polling from Gallup shows that in the United States, 65% of interviewees rate the honesty of physicians as "high" or "very high" [13].

This is not to say that trust in physicians has not changed in recent times. In the United States, the past 30 years has seen the rise and establishment of managed care as the main system for healthcare delivery. This system consists of delivering healthcare through certain organizations (insurance networks and health maintenance organizations, or HMOs), with the stated interest of reducing costs while improving the quality of care. In practice, managed care resulted in significant changes to medical practice, where access to certain diagnostic procedures, treatments, and referrals is constrained. This shift necessarily produced a new environment, where the burden of health maintenance and responsibility for outcomes is shared between physicians and systems [14].

Surveys performed in the late 1990s documented that most Americans believed that the rise of health maintenance organizations resulted in decreased quality of care. A 1997 survey reported that only 30% of Americans trusted their insurance plan to "do the right thing" [14]. In other words, patients did not believe that HMOs and other organizations would carry out the fiduciary duty. While focused on organizations, many authors expressed concerns that this decrease in systems trust translated to patient's trust in physi-

cians [15]. These concerns were validated by a 1997 survey that found that patients reported a higher level of trust in physicians who participated in the fee-for-service model than they did in those that participated in salaried models [16].

From a global perspective, increased access to the World Wide Web has changed how patients initiate access to healthcare. Recent polling has shown that 72% of Internet users have looked online for health information; the same number use online reviews as a first step for choosing a doctor [17, 18].

In a perfect world, online reviews of physician's performance are an excellent resource for patients, delivering transparency and accountability, which patients can use in their search for a physician. In reality, many physicians have expressed discomfort with online reviews. Specifically, concerns include lack of transparency (a doctor cannot know if someone posting a review is truly a patient or a malicious actor) and lack of utility (there are no defined metrics that tie high ratings to better outcomes). Furthermore, there is concern that only disgruntled patients will take the time to write a review [19]. However, quantitative analysis of online physician reviews shows the opposite. In a 2011 study of 4999 online physician reviews, Kadry and colleagues found that most patients assign their physicians a positive score [20].

Discomfort notwithstanding, patients have become accustomed to online reviews, and they are unlikely to go away. Most physician ratings exist in specialized websites (e.g., Yelp.com, healthgrades.com); however some healthcare organizations have gone a step further and adopted rating systems into their own websites [21]. This represents a concerted effort to increase transparency, one which can help patients address the knowledge gap, and empower them to feel more confident about their choice in physician and/or healthcare system. In this manner trust is fostered even before the initial patient encounter.

Another proposed benefit of online reviews is that they provide a system of immediate feedback [22], which may aid physicians in adjusting their own practice in response to perceived deficiencies. It must be noted however that physician rating scores and outcomes have not shown an association. A recent review compared 30-day risk-adjusted mortality rates following coronary artery bypass graft surgery (publicly reported in five US states) to online physician ratings. Here again most ratings were high (4.4 out of 5). No correlation between ratings and mortality rates was found, likely because ratings are based on of the quality of the surgeon-patient relationship, and not the patient's health or the quality of the operation [23].

Special Aspects of Trust in Surgery

While establishing a successful patient-physician relationship is a complex and delicate process in all areas of medicine, surgeons face unique challenges in this arena. When meeting a patient, a surgeon must assess the patient, counsel on the nature of disease, review therapeutic options, and discuss expected outcomes. If an operation is part of a proposed treatment, the surgeon must initiate a discussion on the risks and benefits related to a given procedure and engage in a thorough discussion known as "informed consent." While possibly "routine" to a surgeon, for a patient any operation represents a high-stakes intervention – one which requires a high level of trust in the surgeon and the medical system. Not uncommonly, the entire information gathering, counseling, and consent process occur during a single office visit or bedside consultation [24].

It is in these, often brief, visits that surgeons must achieve a high level of trust from their patients. Viewed objectively, there is perhaps no bigger act of trust than surrendering one's body to a surgical procedure.

The Role of Communication

If trust plays a central role in the surgeon-patient relationship, it follows that surgeons would do everything in their power to establish, maintain, and strengthen a trusting relationship with their patients at every opportunity. But how is this bond achieved? We contend that the prime manner to do so is through communication.

Communication, in its broadest sense, is the act of "imparting or exchanging of information by speaking, writing, or using some other medium" [25]. This process is indeed indispensable for the execution of the main functions of a medical interview: (1) determine and monitor the nature of the problem; (2) develop, maintain, and conclude the therapeutic relationship; and (3) carry out patient education and implementation of treatment plans [26]. In addition, the medical interview is the best place to provide moral support, discuss patient's preference, and empower the patient by providing information and resources that allow the patient to participate in his/her own care.

That patients care about meaningful communication has been well established in the literature. In a 1997 study, Levinson et al. studied the relationship between communication and malpractice claims among 65 surgeons and 59 primary care physicians. Three trained coders listened to recorded clips form patient encounters, and these clips were graded among three domains: content, process, and emotional affect. Here, process refers to statements of orientation (defining roles and expectations) as well as facilitation (encouraging patients to elaborate on comments and asking about their understanding). The authors found that primary care physicians who used more statements of orientation, used more humor, and encouraged patient communication were less likely to suffer litigation than those who did not. There was, however, no observed difference in malpractice claims and communication style among surgeons [27].

More than Words

Presumably, communication is mainly comprised of spoken (or sometimes written) words, and patients may judge the quality of the therapeutic relationship based on the content of these words. Yet a vast body of research shows that the quality of patient-physician communication is bound by much more than just spoken words. Subtle and unintentional cues such as tone of voice, eye contact, and body posture have been shown to convey meaning to patients [28].

A perhaps understudied aspect of communication in surgery is the physician's tone of voice. In one of the few studies to examine this, Ambady and colleagues examined 144 conversations recorded during routine medical visits between patients and community practice surgeons. Sixty-five general and orthopedic surgeons participated; two patient encounters were graded for each surgeon. Interactions were reviewed by 12 judges and rated on a 7-point scale on the following variables: warm/professional, concerned/anxious, hostile, and dominant. These variables were then compared to surgeon's history of malpractice claims. Their results showed that surgeons who were judged to be more dominant and less concerned/anxious were more likely to have been sued [29].

Another study by Levinson et al. evaluated racial disparities in surgeon-patient communication, comparing the latter to patient satisfaction among elderly white versus African American patients. Eighty-nine surgeons and 886 patients participated in the study. As in their previous research, three coders listened to brief audio clips, grading them on content for elements of informed decision-making and on process (as above). Finally, patients completed a patient satisfaction questionnaire. The authors found no significant differences in content of informed decision-making elements based on race. However, coders rated process elements (responsiveness, respect, and listening) higher in visits with white patients than in those with African American patients. Accordingly (and despite having received similar information as their white counterparts), African American patients reported significantly lower satisfaction scores [30].

Findings from the studies mentioned above reinforce the notion that it is not just what surgeons say but *how* we say it and what *emotions* we display when we talk that can foster good communication and enhanced trust.

There is an additional, albeit less studied, issue that can contribute to the generation of trust and strengthening of physician-patient relationship: the actions of the physician. Those actions may manifest during the medical interview, for

example, in terms of body language and expression – such as listening intently, asking follow-up questions, abstaining from entering data in the computer while the patient is speaking, and other such actions. But it goes beyond that delivering in a promise (e.g., "I will call this prescription to the pharmacy" or "I will let your referring physician know about this consultation" or "I will call you tomorrow to see how you are doing," etc.,) there are literally tremendous opportunities to "positively surprise" a patient, to deliver on a promise, to show respect, all elements that contribute to the generation of trust [1].

Teaching Communication Skills

In 2004, the United States Medical Licensing Examination introduced a new component, the Clinical Skills test. This exam aims to evaluate the presence of fundamental clinical skills necessary for safe patient care. One component of the test focuses on communication and interpersonal skills (CIS). As a result, medical schools in the United States adopted formal training for CIS into their curriculum [31]. However most of this training takes place during the 1st and 2nd year of a physician's training, *before* clinical exposure. Moreover, most postgraduate training programs do not include communication training in their curricula. Thus, while physicians receive education early in their career, residents (immersed in clinical practice) must improve their communication skills only by the occasional feedback or suggestion [32].

In a 2004 study, Yudkowsky et al. examined the communication and interpersonal skills of 22 general surgery residents. Standardized patients were used in order to examine communication in six challenging scenarios: giving bad news, informed consent, treatment refusal, domestic violence, patient education, and history and physical exam. Resident's capacity to maintain a patient-centered approach was judged on a 5-point scale. Overall scores were 3.7/5 and were lowest in the informed consent and domestic violence scenarios. There were no significant differences in communication skill scores between PGY2 and PGY3 residents. On a posttest survey, 92% of residents reported that feedback from the standardized patients was useful [33]. Results from this study show that communication skills do not necessarily improve organically (i.e., without a formal curriculum) and that residents value feedback that allows them to improve.

A final aspect to consider in medical communication is that, in a therapeutic relationship, both physician and patient share a common goal: the reestablishment or improvement of health. However, either party approaches the problem from a different perspective. The patient lacks medical expertise; yet he is an expert not only on his own symptoms but also on his goals and priorities. Conversely, the physician is an expert on the details of the nature of disease and treatment, but before meeting the patient, she is largely ignorant of the patient's concerns and preferences. The union of both agendas should thus be the basis of an effective and humanistic medical interview [34].

The Conclusion

Key concepts of trust and its centrality to the surgeon-patient relationship have been discussed. As the world adapts to new societal paradigms and technologies, surgeons must also react to respond to their patient's needs. In order to foster a strong bond with his or her patient, the surgeon must not only to be trustworthy but must foster trust in the patient and medical system. To accomplish this, we must deliver knowledgeable and more importantly, compassionate care. In the words of Francis Peabody [35], "the secret of the care of the patient, is in caring for the patient."

References

1. Pellegrini CA. Trust: the keystone of the patient-physician relationship. ACS. 2017;224:95–102. https://doi.org/10.1016/j.JAmCollSurg.2016.10.032.
2. Pellegrino ED, Thomasma DC. The virtues in medical practice. New York: Oxford University Press; 1993.
3. Trust. Available at: https://en.oxforddictionaries.com/definition/trust. Accessed 2 Dec 2017.
4. Trust. Available at: https://www.merriam-webster.com/dictionary/trust. Accessed 2 Dec 2017.
5. Barber B. The logic and limits of trust. New Brunswick: Rutgers University Press; 1983.
6. Luhmann N. Trust and power. Cambridge: Polity Press; 2017.
7. Mechanic D, Schlesinger M. The impact of managed care on patients' trust in medical care and their physicians. JAMA. 1996;275:1693–7. https://doi.org/10.1001/jama.275.21.1693.
8. Thom DH, Wong ST, Guzman D, et al. Physician Trust in the Patient: development and validation of a new measure. Ann Fam Med. 2011;9:148–54. https://doi.org/10.1370/afm.1224.
9. Thorne SE, Robinson CA. Reciprocal trust in health care relationships. J Adv Nurs. 1988;13:782–9.
10. Merrill JO, Rhodes LA, Deyo RA, et al. Mutual mistrust in the medical care of drug users. J Gen Intern Med. 2002;17:327–33.
11. Chekhov AP. Ivanov. In: The Oxford Chekov: Ivanov; Platonov; the Seagull. Oxford: Oxford University Press; 1967.
12. McLellan MF. Literature and medicine: physician-writers. Lancet. 1997;349:564–7.
13. Honesty/Ethics in Professions. Available at: http://news.gallup.com/poll/1654/honesty-ethics-professions.aspx. Accessed 28 Jan 2018.
14. Davies HT, Rundall TG. Managing patient trust in managed care. Milbank Q. 2000;78:609–24; iv–v.
15. Axelrod DA, Goold SD. Maintaining trust in the surgeon-patient relationship: challenges for the new millennium. Arch Surg. 2000;135:55–61.
16. Kao AC, Green DC, Zaslavsky AM, et al. The relationship between method of physician payment and patient trust. JAMA. 1998;280:1708–14.
17. Fox S, Duggan M. Health Online 2013. Pew internet & American Life Project. 2013. Available at: http://www.pewinternet.org/2013/01/15/health-online-2013/. Accessed Jan 26 2018.
18. Loria G. How patients use online reviews. Available at: https://www.softwareadvice.com/resources/how-patients-use-online-reviews/.
19. Segal J. The role of the internet in doctor performance rating. Pain Physician. 2009;12:659–64.
20. Kadry B, Chu LF, Kadry B, et al. Analysis of 4999 online physician ratings indicates that most patients give physicians a favorable rating. J Med Internet Res. 2011;13:e95–13. https://doi.org/10.2196/jmir.1960.
21. Lee V. Transparency and trust – online patient reviews of physicians. N Engl J Med. 2017;376:197–9. https://doi.org/10.1056/NEJMp1610136.
22. Hill D, Feldman SR. Online reviews of physicians: valuable feedback, valuable advertising. JAMA Dermatol. 2016;152:143–4. https://doi.org/10.1001/jamadermatol.2015.3951.
23. Okike K, Peter-Bibb TK, Xie KC, Okike ON. Association between physician online rating and quality of care. J Med Internet Res. 2016;18:e324–9. https://doi.org/10.2196/jmir.6612.
24. Clapp JT, Arriaga AF, Murthy S, et al. 2017 Surgical consultation as social process: implications for shared

decision making. Ann Surg. https://doi.org/10.1097/SLA.0000000000002610.

25. Communication. Available at: https://en.oxforddictionaries.com/definition/communication. Accessed 15 Dec 2017.

26. Goold SD, Lipkin M. The doctor-patient relationship. J Gen Intern Med. 1999;14:S26–33. https://doi.org/10.1046/j.1525-1497.1999.00267.x.

27. Levinson W, Roter DL, Mullooly JP, et al. Physician-patient communication. The relationship with malpractice claims among primary care physicians and surgeons. JAMA. 1997;277:553–9.

28. Ong LM, de Haes JC, Hoos AM, Lammes FB. Doctor-patient communication: a review of the literature. Soc Sci Med. 1995;40:903–18. https://doi.org/10.1016/0277-9536(94)00155-M.

29. Ambady N, LaPlante D, Nguyen T, et al. Surgeons' tone of voice: a clue to malpractice history. Surgery. 2002;132:5–9. https://doi.org/10.1067/msy.2002.124733.

30. Levinson W, Hudak PL, Feldman JJ, et al. It's not what you say …. Med Care. 2008;46:410–6. https://doi.org/10.1097/MLR.0b013e31815f5392.

31. Gilliland WR, Rochelle JL, Hawkins R, et al. Changes in clinical skills education resulting from the introduction of the USMLE™ step 2 clinical skills (CS) examination. Med Teach. 2008;30:325–7.

32. Levinson W, Lesser CS, Epstein RM. Developing physician communication skills for patient-centered care. Health Aff. 2010;29:1310–8. https://doi.org/10.1377/hlthaff.2009.0450.

33. Yudkowsky R, Alseidi A, Cintron J. Beyond fulfilling the core competencies: an objective structured clinical examination to assess communication and interpersonal skills in a surgical residency. Curr Surg. 2004;61:499–503. https://doi.org/10.1016/j.cursur.2004.05.009.

34. Smith RC. The Patient's story: integrating the patient- and physician-centered approaches to interviewing. Ann Intern Med. 1991;115:470–7. https://doi.org/10.7326/0003-4819-115-6-470.

35. Peabody FW. The Care of the Patient. JAMA. 1927;88:877–82. https://doi.org/10.1001/jama.1927.02680380001001.

The Transformation and Challenges of the Surgeon–Patient Relationship

Piroska K. Kopar

Key Points Summary
- The surgeon–patient relationship is unique in the field of medicine.
- The surgeon–patient relationship is informed by the history of the doctor–patient relationship and its legal and bioethical underpinnings.
- Patient autonomy and societal resource allocations weigh more heavily in the surgeon–patient relationship than before.
- Surgeons are under competing interests and professional obligations.
- Navigating conflicting obligations through honesty, transparency, and reflection can help preserve patients' trust in surgeons.
- Ethics curricula can provide the space and the language for meaningful discussion with our patients and each other.

Introduction

Bioethicists, like philosophers, are concerned with wisdom for its own sake. Their work is devoted to the pursuit of the truth not for any other reason than to know it. They question everything that may be an accepted fact of life to most others, for not-to-question is to take on faith or custom, a notion entirely contrary to the practice of philosophy. Physicians, on the other hand, especially surgeons, tend to be more practically minded. Surgeons perceive a problem that needs to be fixed, and they devise interventions based on a combination of their previous experience and on predicted results. Factors that do not affect the outcome are all but immaterial.

While we do not need to make bioethicists of all surgeons, the more familiar we are with our profession's implicit contract with society, and the more aware we are of the ethical dimensions of our practice, the better doctors we will be to our patients. In this chapter the words "surgeon" and "doctor" will be used. I will first address the uniqueness of our field within medicine, followed by a brief historical overview of the transformation and challenges of the surgeon–patient relationship. I will examine the legal and bioethical foundations of our practice and will explore two of our ethical principles, autonomy and justice, in greater detail. I will review the variety of conflicts of interest and obligations surgeons face and reflect on our emerging challenges. I will end the chapter with some practical considerations for the ethical conduct to best optimize the surgeon–patient relationship.

P. K. Kopar (✉)
Surgical Critical Care, Acute Care and Trauma
Surgery, Yale School of Medicine,
New Haven, CT, USA
e-mail: piroska.kopar@yale.edu

© Springer Nature Switzerland AG 2019
A. R. Ferreres (ed.), *Surgical Ethics*, https://doi.org/10.1007/978-3-030-05964-4_17

The Surgeon–Patient Relationship Is Unique

The surgeon–patient relationship is unique from other medical encounters for several reasons. When a trauma happens, when a cancer grows, and when an organ fails, a surgeon is called. Surgeons cut to heal and hurt to cure: surgeons, by nature, are invasive. While all physicians regularly participate in life or death decisions, the surgeon's involvement is the most immediate and the most tangible. Lives are saved and bodies are left behind. The surgeon must go onto another operation, give someone else a chance, and be someone else's hope. As the medical Charles Bosk has observed, surgeons respond to adverse events startlingly differently to the way medical doctors do. When the patient of an internist dies, his colleagues ask "What happened?," but when the patient of a surgeon dies, the colleagues ask "What did you do?" [1].

As surgeons, we are often not afforded the luxury of deliberation and so we must learn to think quickly, almost automatically, and act with scientific objectivity. The responsibility assumed by the surgeon, best exemplified by intraoperative dynamics, does not foster equality [2]. In the surgeon–patient relationship, a certain degree of paternalism is not only permissible, but in fact necessary, as patients place their trust in their surgeon while they are under anesthesia. It is because of this transferring of agency and power that trust plays such a crucial role in the surgeon–patient relationship [3].

As surgeons, we are trained under the tutelage of our masters for longer than most. The nature of our work mandates physical proximity to our teachers for hours on end, shared experiences of failure with those we operate with when outcomes fall behind what we had hoped for, and the burden of guilt when we make a mistake. It is hard not to become emotionally attached to "the way I have trained." To quote the stoic philosopher Marcus Aurelius: "Your way of thinking will be influenced by the nature of the objects you most often represent, for the color of the soul comes from representations" [4].

We are connected to our patients in a way that we feel grants us special "ownership." When speaking about how one acquires ownership of one's land and home, John Locke makes the argument that you may declare something as yours if you have invested physical work in it: if you have plowed and planted and farmed a piece of land, then you may call it and the fruits it produces yours [5]. Similarly, surgeons seem to have a unique sense of ownership of their patients related to a physical connection, a mixing of body parts, and an exercise of muscle memory in operating on them. We all routinely defer judgment in the patient's continued care to the operating surgeon. Patients, however, are moral agents themselves, and we must be mindful not to let our feeling of ownership cloud our judgment in their postoperative care. The concept of surgical buy-in relates to this idea. Some surgeons embrace the notion that once a patient has agreed to undergo an operation, he now "owes" a certain amount of effort at recovery to his surgeon [6]. Conversely, when we cause an iatrogenic injury, we are much more reluctant to allow our patient to "give up" [7].

Surgeons, we tell ourselves, are doctors who can fix things: physicians who can operate. Paradoxically, the very attributes that make one a good surgeon may also subtract from one's abilities as a physician. The efficiency paramount in a hands-on field might translate into impersonal haste, and the distance to preserve objectivity may limit compassion. The guidance offered to patients intended in the spirit of beneficence may deteriorate into loss of respect for patient autonomy. To be a good surgeon *and* a good physician is a constant balancing-act of virtues and behaviors that may, at times, conflict. Recognizing ethical and conflicts and knowing how to resolve them becomes especially challenging. To internalize the ethical conduct of a surgeon *and* a physician requires education and, most of all, practice.

The Doctor–Patient Relationship: A Historical Perspective

The doctor–patient relationship has evolved significantly over the centuries. Although there is plenty of evidence of medical and surgical practice predating Hellenic times, we trace the sanctity of the duties of the physician to Hippocrates.

Notably, the original Hippocratic oath was much different from its current version recited at white coat ceremonies across US medical schools today. At its inception, the Hippocratic oath offered practical guidelines for interactions with patients (e.g., it forbade sex with patients). Included in the Hippocratic corpus as interpreted by Greeks, Romans, and Western societies to follow were the concepts of nonmaleficence, beneficence, and confidentiality. The code of the physician–patient relationship was founded on the competence and knowledge of the physician, extending to and including decisions about what course of action might be best for the patient. Such a paternalistic model of medicine served as the paradigm of the healing relationship for about 2500 years and, in many societies around the globe, persists today [3]. The first record of physician–patient relationship and informed consent could be traced in Plato's *The Laws* (book XI), where he points out the difference between free citizens and slaves.

The notion of patients' rights did not enter the medical scene until after World War II. History texts from this period are filled with documentation of the atrocities committed in the concentration camps of Nazi Germany against those considered subhuman by the Aryan race, as well as against their very own nationals who were elderly, frail, or mentally retarded. Many of these transgressions were committed by those in the medical profession: physicians and nurses. Medical procedures and human experimentation without consent, including on vulnerable populations and children, were rampant. The Helsinki Declaration, calling for the protection of basic human rights, was born from the aftermath of World War II. While the Helsinki Declaration has no legal standing, it forms the cornerstone of the ethical underpinnings of human experimentation in civilized nations [8].

Supplemental Story: Doctors Doing the Devil's Work
The atrocious acts of many doctors and nurses during the 1930s and 1940s in Nazi Germany is a shameful part of our profession's collective history and easy to dismiss as a thing of the past. Looking back today, we find these practitioners' choices morally repulsive, inexcusable, and completely unrelatable. What could possibly explain, let alone justify, the systematic killing of nursing home residents, the sterilization of mentally retarded people, and the cruel medical experimentation on children; let alone the systematic murder of millions of people in concentration camps? The disturbing reality is that doctors and nurses were not only aware of what was happening, but, in cases too many to count, acted as first-line engineers of destruction in the name of spearheading a so-called genetic cleansing movement.

In her presentation delivered for the New York Genome Center, *Deadly Medicine: Nazi Eugenics and its Implications Today*, Dr. Mildred Solomon provides valuable insight into the historical context of these morally repulsive choices. World War I had left Germany decimated, with many of its youngest and most vital citizens having been injured on the battlefield, now becoming amputees or dependent on care in some other way. Young and healthy males debilitated to such an extreme seemed unjustly at odds with the sheltered lives of those who did not go to war, yet enjoyed the protection of social welfare. Physicians were among the first to rebel against what seemed to them an unjust allocation of resources. A misguided attempt to advocate for those who had been injured in the war that had stolen their youth and vitality, physicians decided to actively harm those they deemed less deserving of their care or of society's investment. Appalled, as we are, by their actions, Dr. Solomon awakens us to the uncomfortable truth that we, physicians today, may not be that different from our historical counterparts. In particular, in cases of genetic selection and manipulation, the precise lines between modern medicine and playing god in deciding who is worthy of living and who is not are alarmingly blurry [9].

Table 1 AAPS proposal

The *Association of American Physicians and Surgeons* adopted a list of patient freedoms in 1990, which was modified and adopted as a "patient's bill of rights" in 1995:

"All patients should be guaranteed the following freedoms:

To seek consultation with the physician(s) of their choice;

To contract with their physician(s) on mutually agreeable terms;

To be treated confidentially, with access to their records limited to those involved in their care or designated by the patient;

To use their own resources to purchase the care of their choice;

To refuse medical treatment even if it is recommended by their physician(s);

To be informed about their medical condition, the risks and benefits of treatment and appropriate alternatives;

To refuse third-party interference in their medical care, and to be confident that their actions in seeking or declining medical care will not result in third-party-imposed penalties for patients or physicians;

To receive full disclosure of their insurance plan in plain language, including:

CONTRACTS: A copy of the contract between the physician and health care plan, and between the patient or employer and the plan;

INCENTIVES: Whether participating physicians are offered financial incentives to reduce treatment or ration care;

COST: The full cost of the plan, including copayments, coinsurance, and deductibles;

COVERAGE: Benefits covered and excluded, including availability and location of 24-hour emergency care;

QUALIFICATIONS: A roster and qualifications of participating physicians;

APPROVAL PROCEDURES: Authorization procedures for services, whether doctors need approval of a committee or any other individual, and who decides what is medically necessary;

REFERRALS: Procedures for consulting a specialist, and who must authorize the referral;

APPEALS: Grievance procedures for claim or treatment denials;

GAG RULE: Whether physicians are subject to a gag rule, preventing criticism of the plan."

https://www.aapsonline.org/patients/billrts.htm

The rights of subjects in scientific human experiments, however, is not identical to the rights of patients. The concept of patient autonomy is only about 50 years old, and its degree, validity, and foundation are still the subjects of lively debates in bioethics today. Although to what degree patient autonomy reaches or should reach may be debated, it has clearly become a revered ethical principle in today's medicine [10]. Patient autonomy has replaced beneficence as the overriding factor in the clinical encounter, explicitly expressed in the Patients' Bill of Rights (see Table 1).

The paradigm for our current idea of the proper patient–doctor interaction in the United States can be traced to the late 1960s and early 1970s. These decades also mark the beginnings of bioethics along with the opening of the first institutions dedicated to its cause, such as the Hastings Center (1969) or the Kennedy Institute (1970). This was a time for celebrating the emergence of individuality both in matters of preference and in moral issues (e.g., the Vietnam War,

homosexuality, feminism). The celebration of personal choices grew rapidly and extended to include the medical sphere seamlessly. Parallel to the civil rights movement, with the evolution of scientific and technological advancements, a societal existential anxiety loomed over the meaning of progress. The fear was that by alienating the person from his or her body for the purposes of scientific inquiry, scientific progress may paradoxically endanger the very humanity it aims to protect [8].

The Surgeon–Patient Relationship: Legal Grounding

Prior to the emergence of individuality as an important social value, legal cases have paved the way to solidify the role of self-determination in the patient–doctor relationship and to underline the special moral obligations that physicians owe to patients. In 1901, in the case of *Hurley v. Eddingfield*, Dr. Eddingfield refused care to a pregnant woman in need of urgent medical attention who subsequently died. The Supreme Court of Indiana declared that doctors are different from other owners of establishments and, unlike innkeepers who may turn anyone away, physicians have a special obligation to help those in immediate need [11]. Decades later in Utah, the verdict in the case of *Ricks v. Budge* established the precedence for the duty to not abandon patients once treatment has begun [12]. Even when a surgeon–patient relationship is not explicitly stated, the surgeon becomes responsible for the patient's care when he makes an official medical recommendation. In *Mead v. Adler*, an on-call neurosurgeon was held liable for the permanent disability of a patient for whom he advised against surgery upon initial consultation. The court specified that "in the absence of an express agreement by the physician to treat a patient, a physician's assent to a physician – patient relationship can be inferred when the physician takes an affirmative action with regard to the care of the patient" [13].

The legal need for consent for surgery builds upon the 1914 case of *Schloendorff v. Society of New York Hospital*. The patient, Mary Schloendorff, had agreed to examination under anesthesia, but had explicitly forbade the performance of an hysterectomy, allowing only the excision of her fibroid tumor with which she was diagnosed. Her surgeon performed the operation against her consent, and, when she developed gangrene of her arm in the postoperative period, she blamed the surgery and filed suit. The court found that her operation constituted battery [14]. That a consent would only be valid if it was *informed* was delineated in the case of *Canterbury v. Spence* in 1972. Dr. Spence, a neurosurgeon in Washington, had performed a spinal operation on a patient who became paralyzed postoperatively. The surgeon had warned the patient of the possibility of weakness from the surgery, but had not informed him of the risk of paralysis. Although the jury found in favor of Dr. Spence, the court declared that all material risks must be disclosed to patients when asking for their consent to a procedure. Sufficient information would include what a reasonable person would need to know to make an informed choice [15].

The Surgeon–Patient Relationship: Bioethical Underpinnings

There is not a single system of bioethics that is more ethically appropriate to describe the governing forces of medical ethics than any other. Approaches to inform the ethics of the surgeon–patient encounter range from the theoretical to the purely practical. Virtue ethics espouses that to be a good surgeon, one must simply be a good person. It argues that the doctor as a value-neutral technician cannot be separated from the surgeon as a person and a moral agent [16]. As to whether it is possible to teach virtue, that has been a question for the ages ever since Meno put the question to Socrates [17].

Narrative ethics, another model for understanding the ethical underpinnings of the clinical encounter, focuses on nonverbal interactions in addition to the verbal ones and invites the formation of emotional connections between doctor and patient. It asks the surgeon to pay attention to the story told, rather than to the detached scientific problem. A story, it claims, is better suited to show the many shades of right and wrong [18]. Narrative ethics aims to remedy the rigid dichotomy and cool objectiveness of rational analysis. In its attempt at flexibility, however, narrative ethics risks becoming empirical rather than normative, or an excuse instead of an imperative. Feminist and relational ethics focuses on the patient as first and foremost a social being who, as the center of her social network, is not only influenced, but thoroughly defined by her relationships to others [19].

More recent times have seen a psychological account for moral choices, most successfully proposed by Jonathan Haidt. Haidt developed the social intuitionist model that explains moral judgment as mostly automatic and based on intuitions rather than reasoning. He names six moral "tastes" he claims are inherent and common to all: care, fairness, liberty, loyalty, respect for authority, and respect for sanctity. A significant body of empirical research supports his metaphor according to which we ought to visualize moral decision-making as a rider on an elephant. The rider represents our rationalization and the elephant the vastly more numerous nonrational causes of our choice [20]. Still others argue that American bioethics is impossible to separate from American Law. In the words of Annas: "In the United States, with its pluralism of beliefs and people, the law is what holds us together. There is no other ethos. Thus, the law – procedural, autonomy based and case focused – came into bioethics" [21].

While philosophically sound and conceptually interesting the above frameworks may be, perhaps for its simplicity and easy adaptability to clinical medicine, the most frequently taught framework for medical ethics is the one outlined by Beauchamp and Childress. In what has become known as the "Georgetown Mantra," the authors analyze ethical conflicts as the clash between two or more principles in their four-principle system of autonomy, nonmaleficence, beneficence, and justice [22]. Nonmaleficence is the imperative to do no harm and beneficence is the obligation of the physician to serve the best interest of the patient. Both principles have been accepted as the integral to the physician–patient relationship since its inception. The principles of autonomy and justice as ethical principles, however, as mentioned previously, are more modern. To better understand the meaning, importance, and limits of these concepts in the medical encounter, I will now examine these two principles in greater detail below.

The Principle of Autonomy in the Surgeon–Patient Relationship

Now, I say, man and, in general, every rational being exists as an end in himself and not merely as a means to be arbitrarily used by this or that will. *Immanuel Kant, Metaphysics of Morals*

Brother, let me ask you one more thing: can it be that any man has the right to decide about the rest of mankind, who is worthy to live and who is unworthy?

But why bring worth into it? The question is most often decided in the hearts of men not at all on the basis of worth, but for quite different reasons, much more natural ones. As for rights, tell me, who has no right to wish? *Dostoevsky, The Brothers Karamazov*

Above are two perspectives on autonomy. The former is that of Immanuel Kant, analytical philosopher, father of deontology. He defines reason as the foundation of morality and proceeds to deduce, as if proving a mathematical theorem, that ethical rules are inviolable and categorical [23]. The second example is from an interchange between Ivan and Alyosha, who are entertaining the possibility of killing their father. The passage later concludes with an agreement without any spoken words [24].

Autonomy, as a primary American value, is historical. We derive our bioethical concept of autonomy from Kant. With the heralding in of the Age of Reason starting from Western Europe and moving eastward, moral philosophers searched for a foundation for their reasoning independent of religion. Immanuel Kant put forth a system of morality in which the outcome had no effect on the moral verdict of the action. Each action would be judged in isolation as virtuous or not, depending only on whether the action in question did or did not follow a given moral categorical imperative and independent of its consequence. Morally correct directives are perceived as such "a priori" and uniformly by the human mind, and it is up to our practical reasoning to act in accordance with our innate moral mandates. Thus *autonomy*, in the Kantian sense, literally refers to moral guidance that is *autonomous*, or *self-legislative, based in the Greek political system.*

The principle of autonomy is so deeply embedded in our American values, and the appeal to its primacy in our everyday life is so pervasive that we accept its authority without questioning it, almost as a matter of faith. Over the twentieth century, autonomy has been defined in progressively more complex ways ranging from mere independence to self-mastery and detached reasoning, to speaking of it as a positive or negative right in the context of organized and law-based societies. While these notions are increasingly more refined, common to all these interpretations is the idea that autonomy is closely related to a rational and voluntary choice. An autonomous person is one who rules herself by her own reason and arrives at decisions in accordance with them.

Both the idea of voluntariness and the idea of rational are essential to autonomy's definition here [25].

The trouble begins when we apply the pure concept of autonomy to a human choice or decision that is necessarily bound by the messiness of the human condition, always complex and always contextual. The voluntary aspect of autonomy is often equated to freedom, but voluntariness is but one component of freedom. To be truly free to choose, one must have valid options to choose from; otherwise a choice is that in name only. Options and opportunity, in general, grow in proportion to one's education and wealth, among others. The individual who is well learned and has means above average has much more freedom of choice than her uneducated or impoverished counterpart [25]. Even if we were to grant the hypothetical existence of a truly autonomous choice – one that is both rational and given freely – we must discuss the value that we assign to autonomy and the context in which it remains valuable.

Constrains on patient autonomy are at least threefold. Foremost, transferring the knowledge, experience, and medical insight to patients that surgeons acquire from medical school and residency training often followed by subspecialty fellowships is simply not possible even in the most informed discussions. Even if all relevant information is thoroughly explained to the patient with its attendant risks and benefits, the experience, and, most importantly, the clinical judgment, of a trained surgeon cannot be summarized in percentages and data points for the untrained ear. In addition to the barriers posed by the lack of a thorough medical understanding, the patient tasked with a decision is ill, quite literally *diseased* [8]. To ask a person affected by psychological or physical pain to make a dispassionate, rational decision about his or her medical options is unrealistic. Empirical data shows that patients enjoy a greater sense of autonomy when actively advised and supported by their families and physicians [26].

Finally, patient autonomy is paradoxically limited by the societal constraints that protect it. Autonomy may be understood as either a negative or a positive right. In the healthcare setting, autonomy as a negative right simply means that the patient may refuse care. In contrast, when autonomy is understood as a positive right, it refers to the patient's ability to determine her own medical care within clinically appropriate options and, often, receive a course of treatment. Positive rights, however, are justified and secured by the constructs and arrangements of the social contract to which, in a democratic society, all citizens have implicitly agreed [27]. A key corollary is that the rights of any one person are equally important to those of any other persons [28]. The Kantian idea of autonomy evolved in a culturally homogeneous society, which stands in stark contrast with our American, pluralistic democracy that not only embraces, but positively idealizes diversity. Tolerance and acceptance are today's strengths and mutual respect the preeminent virtue among fellow citizens [8]. Therefore, patient autonomy must find its limits when it begins to encroach on the autonomy of other patients, such as in the case of the allocation or scarce resources.

The Principle of Justice in the Surgeon–Patient Relationship

The economics of medicine changed drastically in the 1980s and 1990s with the rise to dominance of third-party payers for physician-billed services. The transition of financial power to insurance companies away from individual patients who could no longer afford the exponentially rising price of advanced medical technology and pharmaceutical products created a conflict of obligations for physicians. This paradigm shift, in effect, transformed the doctor into a double agent [29]. Dual agency refers to the specific conflict of obligations physicians face in having to represent, on the one hand, the patient's best interest and, on the other hand, just allocation of scarce resources. On the one hand, the physician has a professional obligation to the patient to provide the best possible care for her; on the other hand, there is a societal call for the stewardship of scarce medical resources.

Several major medical societies explicitly call for their members to do both. The American Board of Internal Medicine's Physician Charter and the Professional Code of Conduct of the American College of Surgeons both define professionalism with the inclusion of these competing directives without providing any guidance on how to navigate the conflict (see Table 2). While bioethicists divide in their assessment of how to best manage this dual agency, doctors' responses to this conflict may be classified into three categories: "bunkering," "bailing," and "balancing" [30]. *Bunkering* physicians identify their primary professional obligation as serving the best interest of their patients. *Bailing* physicians consider themselves agents of society and make clinical decisions with public health concerns in mind. *Balancing* doctors attempt to do both, albeit unsystematically and without external guidance.

The most convincing case for *bunkering* comes from Marcia Angell. She argues that our underlying assumptions for cost containment are critically flawed. Most pertinent to her position is that our healthcare system is not a closed system akin to that of organ or other finite resource allocation systems that operate within well-controlled structures. There is no limit to the amount that we, as a society, spend on healthcare, often to the detriment of other social goods such as education. Additionally, there is no accounting for what happens to the money that is saved by cost-containing clinical decisions. She further points out that even if we were to set limits to spending, the open nature of the system does not force such limits to be universal. Consequently, those with less means suffer more, resulting in the loss of the equal rights promised in our constitution. Denying care to some and not to others breeds dishonesty and

Table 2 Code of Professional Conduct (Approved by the American College of Surgeons Board of Regents June 2003)

As Fellows of the American College of Surgeons, we treasure the trust that our patients have placed in us because trust is integral to the practice of surgery. During the continuum of pre-, intra-, and postoperative care, we accept the following responsibilities:

Serve as effective advocates of our patients' needs

Disclose therapeutic options, including their risks and benefits

Disclose and resolve any conflict of interest that might influence decisions regarding care

Be sensitive and respectful of patients, understanding their vulnerability during the perioperative period

Fully disclose adverse events and medical errors

Acknowledge patients' psychological, social, cultural, and spiritual needs

Encompass within our surgical care the special needs of terminally ill patients

Acknowledge and support the needs of patients' families

Respect the knowledge, dignity, and perspective of other health care professionals

Our profession also is accountable to our communities and to society. In return for their trust, as Fellows of the American College of Surgeons, we accept the following responsibilities:

Provide the highest quality surgical care

Abide by the values of honesty, confidentiality, and altruism

Participate in lifelong learning

Maintain competence throughout our surgical careers

Participate in self-regulation by setting, maintaining, and enforcing practice standards

Improve care by evaluating its processes and outcomes

Inform the public about subjects within our expertise

Advocate for strategies to improve individual and public health through communication with government, health care organizations, and industry

Work with society to establish just, effective, and efficient distribution of health care resources

Provide necessary surgical care without regard to gender, race, disability, religion, social status, or ability to pay

Participate in educational programs addressing professionalism

As surgeons, we acknowledge that we interact with our patients when they are most vulnerable. Their trust and the privileges we enjoy depend on our individual and collective participation in efforts to promote the good of both our patients and society. As Fellows of the American College of Surgeons, we commit ourselves and the College to the ideals of professionalism.

https://www.facs.org/about-acs/statements/stonprin#code

jeopardizes the morality of the medical profession. Ethics should not be driven by economics. Her solution is a single-payer system with a global cap. Such a transition would transform the system into a closed one that would engender universal rules, transparency, and accountability in health-care allocation decisions [29].

Bailing is a more common topic in the sphere of public health than in clinical medicine. Social iatrogenesis relates to this idea, a term coined by Ivan Illich and illustrated by Frey. Social iatrogenesis is the medicalization of a variety of societal issues previously thought to be independent of clinical medicine that now physicians find themselves confronting. Physicians become responsible for "certifying" people for activities outside of the medical encounter, such as for driving restrictions, school and work excuses, levels of disability, etc. "Our current society is driven more by individual demands than the collective social good. The response of the 'market' to patients as 'consumers' who, even if expressing unreasonable demands, need to be served cannot deal in a fair way with the problems of rationing of important resources like flu vaccine. If we as a society can't do it with flu vaccine, how will it deal with larger issues such as high cost-technology, screening for disease or Medicare costs?" [31].

Balancing is perhaps the most common. As Morreim, an avid advocate of physicians' obligations to balance the competing duties of direct patient care and societal responsibility, puts it: "(the) moral question is no longer whether to participate in cost containment (that would be rather like asking 'shall we abide by the law of gravity?'), but how to do so in morally credible ways." How to balance, she readily admits, is difficult. Guidelines may aid in general, but clinical cases tend to be evaluated on an individual basis and are vulnerable to exceptions. Consequently, guidelines may either be completely ineffective, or physicians will have to deny certain available treatment choices to patients. Despite the challenge of having to make what amount to bedside rationing decisions, the alternative would be to allow third parties to make allocation decisions without the advantage

of medical knowledge. This, in turn, would be equivalent to transferring the practice of medicine to those without a license [32].

The Surgeon–Patient Relationship: Other Conflicts of Interests and Obligations

The principles of bioethics often stand in opposition to one another and surgeons manage these conflicts often without even recognizing them as such [33]. The most immediately apparent example of how surgeons balance conflicting imperatives that is integral to the practice of surgery is the balancing of the principle to do no harm against the principle of beneficence. To perform an operation, even one as small as an incision and drainage of an abscess or as common as a laparoscopic appendectomy, the first thing we ask for after the local anesthetic is a knife. This is an easy equation to balance when the indications for the procedure are straightforward and the risks are relatively small. The less steady the ground is on which our decision to operate stands and the greater its potential unintended consequences, the shakier our feet feel on the balance beam of ethical decision-making.

In addition to balancing opposing ethical principles, surgeons navigate numerous other conflicts of interests and obligations. These include personal financial conflicts, mediating patient and family disagreements about treatment goals, leading research and innovation initiatives, and, in the academic setting, providing training to surgical residents at the potential expense of direct patient care. Perhaps it is because of the ubiquity of these ethical conflicts in surgical practice, or perhaps it is because of the way surgeons are trained to own their decisions with certainty that most surgeons appear be quite comfortable with making ethically charged decisions every day. Surgeons, for example, routinely make different end-of-life decisions from internists [34]. Even when the ethical choice is less obvious to the surgeon, they prefer to discuss their decision-making within their own group, that is, with other surgeons, rather than with so-called outsiders.

Science and Charity

After the invention of the stethoscope in 1816 by Rene Laennec [35] and paralleling the physical distance created by such technological tools, the documentation of most patient encounters evolved to use objective and clinical terms. Despite this trend to embrace the detached tone of clinical writing, British Medicine in the nineteenth century made an exception for the description of angina and heart disease. While other disease processes were described with the levelheaded coolness of a rational observer, encounters with heart disease elicited a romantic, passionate language from the physician-scientist. The patient was described as a collection of symptoms in the former scenarios, but as a person in the latter ones. In her article published in *Literature and Medicine* in 2014, Meegan Kennedy discusses the many examples and potential reasons for this phenomenon. In Western culture, the heart is identified as the seat of the soul. Diseases of the heart extract a special sympathy from the physician who is at a loss in his craft, unable to treat an organ so deeply hidden and inaccessible to him. The doctor's helplessness is reflected in the distress of his language when describing his patient's agony: vivid imagery replaces scientific objectivity [36].

Surgeons are often accused of not displaying adequate empathy for their patients. Advocates of a holistic approach to healing call for the inclusion of forms of communication in the doctor–patient relationship other than verbal discussion. Sharing music or other forms of art have been proposed as means of propagating the intimacy and trust in the patient's relationship with her doctor, especially when the patient is too ill for rational discourse [4]. Contemporary medicine does allow room for the inclusion of such less traditional methods of healing, although these practices tend to be limited to the domain of palliative care in cases of terminal diseases. When advocating for changing the goals of care to focus on comfort instead of a cure, physicians often do turn to vivid imagery to describe patient suffering to the family to gain their support in ending it. Interestingly, although many doctors admit to relying on vivid imagery for these discussions, in a study assessing the ethical appropriateness of end-of-life communication methods, most physicians have also judged the use of such language ethically inappropriate [34].

The Surgeon–Patient Relationship: Emerging Challenges

1. The surgeon as the leader of the surgical team
   Growing emphasis has been placed on the idea that patient outcomes are more closely tied to successful team dynamics than to the performance of the individual surgeon. Hospitals enforce team training exercises to improve communication and collaboration, and advanced care providers constitute an increasingly larger proportion of the treatment team. The empowerment of team members both inside and outside the operating room has not been accompanied by a similarly shared responsibility to establish and maintain patient trust [37]. It is still the attending surgeon whose name is marked on the patient's chart that remains ultimately responsible for the patient's care both legally and in our traditional notion of patient ownership. Attention to communication is only the start to address this discrepancy.

2. Social worth in a society of scarce resources
   Allocation of finite resources in closed systems necessitates prioritizing the interests of some patients over those of others. Transplant regulations of lifesaving organs such as livers or hearts follow carefully crafted algorithms based mostly on medical benefit, but ones that also take into account social factors such as reliability, trustworthiness, and lifestyle habits

[38]. The same rules have not yet been applied to resources that are also scarce but less demonstrably finite. Examples would include expensive treatments, access to care in rural areas, and the number of overall hospital beds. As healthcare spending becomes increasingly controlled and stewardship of resources not only encouraged but often institutionally rewarded, it seems as only a matter of time that social worth will enter the equation of all resource allocation decisions. To protect patient trust, it will be critical to keep these decisions to society and not up to the subjective interpretation of individual surgeons.

3. Telemedicine and surgery

Robot-assisted surgeries have become commonplace, especially when performing operations in restricted areas such as the pelvis, but are also used frequently for routine surgeries such as a hernia repairs or cholecystectomies. The console to the robot may or may not be in the same physical space as the arms of the device, and it is easily conceivable that, similarly to providing intensive care via tele-ICU to patients who are across the country from the intensivist, robotic surgeries may begin to take place across not only states, but possibly continents. What would the implications of such distance be on the surgeon–patient relationship? The American Medical Association (AMA) instructs that the responsibilities of the doctor–patient relationship must hold true even in instances of telemedicine [39]. It specifies that new relationships must include face-to-face contact and that anonymity is not acceptable. But does face-to-face contact imply in person contact, or can the specifications of the AMA be satisfied via electronic devices? More importantly, should they be? Ongoing dialogue must address these questions.

The Surgeon–Patient Relationship: Ethics in Practice

The many tentacles of professional expectations may appear overwhelming to the surgeon and feel like they detract from what we are really here to do: take care of our patients. To extract oneself from one's obligations as a doctor and "just operate" is tempting. Institutional learning modules on proper conduct, societal expectations of certain behaviors, and unrealistic patient demands based on the media are enough to drive one to focus on the single thing one can most readily control: the conduct of the operation. I submit to you that such an ostrich policy is not acceptable. As much as satisfying hospital policies and fulfilling social expectations may be necessary, ultimately it is a single trait that the surgeon must practice without compromise. The surgeon must be honest.

We owe it to our patients to be honest, not just with them, but also with ourselves and each other. When patients quite literally put themselves in our hands, they honor us with their trust, the cornerstone of the surgeon–patient relationship [40]. The honesty in response to this trust entails questioning why we do what we do; tailoring our practice to incorporate scientific evidence; examining and learning from our mistakes; and, most difficult perhaps, being transparent with our patients about our systems' and our own limits. And we must be honest with patients about them: their choices and chances with and without us and our recommendations to meet their goals.

Honesty should extend to dealing with the constraints of the system in which we operate. As surgeons, we sometimes tend to play Robin Hood and game the system for the benefit of our patients. In a survey assessing surgeons' willingness to deceive third parties in the interest of patient care, the more serious the disease process was and the greater the consequences of insurance approval for a given procedure, the larger number of us were willing to lie to secure care for the patient [41]. A better and more sustainable way to advocate for our patients is to help change the system, both by taking on institutional roles and as active citizens in our communities.

We may be proud of our specialty in that many of these aspirations we already practice routinely. We hold morbidity and mortality conferences to review our decision-making and inform our future judgments; we belong to professional organizations such as the American College of

Surgeons that provide lifelong education for us; and we have strict criteria for certification by our American Board of Surgery. To make the best use of these formal structures and processes, however, we must leave our egos behind and continually examine and re-examine our thought processes, motivations, and judgment. Formal ethics curricula in medical schools, residencies, and faculty members may aid in this goal [42]. The purpose of integrating formal ethics training in surgical programs is not to tell doctors what is right and what is wrong. It is to facilitate reflection by providing a framework for recognizing the ethical dimensions of our everyday decisions and a language with which to discuss them meaningfully [43]. Ethics, as applied to the surgeon–patient relationship, is best thought of as a means of contemplation [4].

Concluding Remarks
- The nature of the surgeon–patient relationship is historically determined and remains in evolution to reflect changing societal priorities.
- Patient autonomy and societal resource allocations weigh more heavily in the surgeon–patient relationship than before, and surgeons must devise a way to accommodate their conflicting professional obligations.
- Honesty and transparency with our patients, ourselves, and our colleagues fosters a reflective environment that is conducive to maintaining patient trust and improving care. Incorporating formal ethics education into surgical training may facilitate this practice.

References

1. Bosk CL. What would you do? Juggling bioethics and ethnography. Chicago: University of Chicago Press; 2008.
2. Langerman A, Angelos P, Siegler M. The "call for help": intraoperative consultation and the surgeon-patient relationship. J Am Coll Surg. 2014;219(6):1181–6.
3. Axelrod DA, Goold SD. Maintaining trust in the surgeon-patient relationship: challenges for the new millennium. Arch Surg. 2000;135(1):55–61.
4. Strebler A, Valentin C. Considering ethics, aesthetics and the dignity of the individual. Cult Med Psychiatry. 2014;38(1):35–59.
5. Locke J. Two treatises of government. Cambridge: Cambridge University Press; 1988.
6. Schwarze MLBC, Brasel KJ. Surgical "buy-in": the contractual relationship between surgeons and patients that influences decisions regarding life-supporting therapy. Crit Care Med. 2010;38(3):843–8.
7. Shah RD, Rasinski KA, Alexander GC. The influence of surrogate decision makers on clinical decision making for critically III adults. J Intensive Care Med. 2015;30(5):278–85.
8. Tauber AI. Historical and philosophical reflections on patient autonomy. Health Care Anal. 2001;9(3):299–319.
9. https://www.youtube.com/watch?v=ryqpYnlZMqk.
10. Ramsey P. The patient as person: medical and legal intersections. New Haven: Yale Univeristy Press; 1970.
11. Hurley v Eddingfield, 156 Ind 416, 59 NE 1058 (Ind 1901).
12. Ricks v Budge, 91 Utah 307, 64 P2d 208 (Utah 1937).
13. Mead v Adler, 231 Or App 451, 220 P3d 118 (Or 2009).
14. Schloendorff v. Society of New York Hospital, 105 N.E. 92, 93 (N.Y. 1914).
15. Canterbury v. Spence (464 F.2d. 772, 782 D.C. Cir. 1972).
16. Blake JH, Schwemmer MK, Sade RM. The patient-surgeon relationship in the cyber era. Communication and information. Thorac Surg Clin. 2012;22(4):531–8.
17. Plato: Meno.
18. Wilks T. Social work and narrative ethics. Br J Soc Work. 35(8):1249–64.
19. Donchin A, Scully J. Feminist Bioethics, The Stanford Encyclopedia of Philosophy (Winter 2015 Edition), Edward N. Zalta (ed.), https://plato.stanford.edu/entries/feminist-bioethics/.
20. Haidt J. The righteous mind: why good people are divided by politics and religion. New York: Pantheon; 2012.
21. Annas GJ, Grodin MA. The Nazi doctors and the Nuremberg code. Human rights in human experimentation. Oxford: Oxford University Press; 1992.
22. Beauchamp TL, Childress JF. Principles of bioethics. 4th ed. Oxford/New York: Oxford University Press; 1994.
23. Kant I. The metaphysics of morals. New York: Cambridge University Press; 1996.
24. Dostoevsky F. The brothers Karmazov. San Francisco: North Point Press; 1990.
25. Gaylin W, Jennings B. The perversion of autonomy: the proper use of coercion and constraints in a liberal society. New York: Free Press; 1996.
26. Gilbar R. Family involvement, independence, and patient autonomy in practice. Med Law Rev. 2011;19(2):192–234.

27. Dworkin G. The theory and practice of autonomy. Cambridge: Cambridge University Press; 1988.

28. Rawls J. A theory of justice. Cambridge: The Belknap Press of Harvard University Press; 1971.

29. Angell M. The doctor as double agent. Kennedy Inst Ethics J. 1993;3(3):279–86.

30. Tilburt JC. Addressing dual agency: getting specific about the expectations of professionalism. Am J Bioeth. 2014;14(9):29–36.

31. Frey JJ 3rd. Writing an excuse or educating the patient. Virtual Mentor. 2012;14(1):13–6.

32. Morreim EH. Fiscal scarcity and the inevitability of bedside budget balancing. Arch Intern Med. 1989;149(5):1012–5.

33. Kavarana MN, Sade RM. Ethical issues in cardiac surgery. Futur Cardiol. 2012;8(3):451–65.

34. Morrell ED, Brown BP, Qi R, Drabiak K, Helft PR. The do-not-resuscitate order: associations with advance directives, physician specialty and documentation of discussion 15 years after the Patient Self-Determination Act. J Med Ethics. 2008;34(9):642–7.

35. Wade NJ, Diana D. Binaural hearing—before and after the stethophone. Acoustics Today. 2008;4(3)

36. Kennedy M. Let me die in your house: cardiac distress and sympathy in nineteenth-century british medicine. Lit Med. 2014;32(1):105–32.

37. Wilk AS, Platt JE. Measuring physicians' trust: a scoping review with implications for public policy. Soc Sci Med. 2016;165:1–7.

38. Persad G, Wertheimer A, Emanuel EJ. Principles for allocation of scarce medical interventions. Lancet. 2009;373(9661):423–31.

39. Skinner S. Patient-centered care model in IONM: a review and commentary. J Clin Neurophysiol. 2013;30(2):204–9.

40. Pellegrini C. Trust: the keystone of the physician-patient relationship. Bull Am Coll Surg. 2017;102(1):58–61.

41. D'Amico TA, McKneally MF, Sade RM. Ethics in cardiothoracic surgery: a survey of surgeons' views. Ann Thorac Surg. 2010;90(1):11–13.e1–4.

42. Helft PR, Eckles RE, Torbeck L. Ethics education in surgical residency programs: a review of the literature. J Surg Educ. 2009;66(1):35–42.

43. Snelgrove R, Ng S, Devon K. Ethics M&Ms: towards a recognition of ethics in everyday practice. J Grad Med Educ. 2016; 8(3):462–4.

Suggested Readings

1. Blake V. When is a patient-physician relationship established? Virtual Mentor. 2012;14(5):403–6.

2. Daniels N. Accountability for reasonableness. BMJ. 2000;321(7272):1300–1.

3. Putman MS, Yoon JD, Rasinski KA, Curlin FA. Directive counsel and morally controversial medical decision-making: findings from two national surveys of primary care physicians. J Gen Intern Med. 2014;29(2):335–40.

4. Bernat JL, Peterson LM. Patient-centered informed consent in surgical practice. Arch Surg. 2006;141(1):86–92.

5. Burkle CM, Mueller PS, Swetz KM, Hook CC, Keegan MT. Physician perspectives and compliance with patient advance directives: the role external factors play on physician decision making. BMC Med Ethics. 2012;13:31.

The Surgical Decision-Making Process: Different Ethical Approaches

Christian J. Vercler and Sagar S. Deshpande

The interaction between patient and surgeon that results in the decision to proceed with an operation is one of the most sacred traditions of our profession. – Steven Charles Stain [1]

In my observation, doctors sometimes slip into the tempting trap of seeing the law of informed consent as stating the whole of the physician's duty to the patient's autonomy interests. – Carl Schneider [2]

The surgical decision-making process is the crystallization of the uniqueness of surgical ethics. This process, performed several times per day by surgeons all over the world, involves constantly weighing the prima facie duties of beneficence, nonmaleficence, justice, and respect for patient autonomy. It is in this way that the practice of surgery is inherently an ethical discourse—albeit not an explicit one.

Acute surgical decision making is often binary: "go/don't go to OR." A sign of surgical maturity in a trainee is when he or she can commit to this decision and start a presentation with, "This is a patient who needs to go to the OR. He is a 23-year-old male…." However, much of surgical decision making involves urgent, elective, or semi-elective operations where many possible options are available. The optimum surgical encounter is one where the right operation is being done on the right person at the right time for the right reasons and by the right surgeon. The goal of surgical education is to arrive a trainee to this point of excellence in decision making. Focusing on surgical decision making in this way can eclipse the patient's perspective from view. We may recognize that we live in a pluralistic multicultural society with no one dominant worldview and yet forget that the medical perspective is just one of many narratives clamoring for dominance. Hence there is the need to focus on the uniqueness of the patient in front of us during the surgical encounter. However, respecting that uniqueness does not simply collapse into doing whatever the patient requests in an attempt to "respect patient autonomy." In surgery, it is much more complex.

"Shared decision making" (SDM) is the current model of medical decision making that could be considered the "gold standard" [3]. At the core of this approach is the distinctly Enlightenment ideal of individual self-determination as a laudable and achievable goal. In this approach the role of the surgeon is to act in a way that facilitates the actualization of the patient's expression of her autonomous desires about her body and her life. The

C. J. Vercler (✉)
Department of Surgery, University of Michigan, Ann Arbor, MI, USA

Center for Bioethics & Social Sciences in Medicine, University of Michigan, Ann Arbor, MI, USA
e-mail: cvercler@med.umich.edu

S. S. Deshpande
University of Michigan Medical School, Ann Arbor, MI, USA

© Springer Nature Switzerland AG 2019
A. R. Ferreres (ed.), *Surgical Ethics*, https://doi.org/10.1007/978-3-030-05964-4_18

triumph of this model in the twenty-first century is held up against the horrors of paternalism evidenced throughout the twentieth century, where surgeons made decisions on behalf of patients with little to no involvement of the patients themselves. Shared decision making is most appropriate when there is uncertainty as to the best clinical option— when two or more approaches may reasonably address the patient's problem of concern [4]. This chapter will review the principles and techniques used in shared decision making, examine alternate approaches, and discuss some of the difficulties in implementing this approach in surgical cases of varying acuity. The concepts of "surgical buy-in" and the moral agency of the surgeon in declaring someone "not a surgical candidate" will also be discussed, as well as special considerations in the pediatric patient.

One of the oversimplifications made in shared decision making is to assume that the surgeon brings the facts to the equation and the patient brings the values [5]. The problem with this oversimplification is that it supposes that the surgeon has access to a set of value-free objective facts to be discussed. This is rarely the case. Institutional practices, regional variances, and training biases can affect the decision-making process, as well as heuristics and implicit biases [6–8]. These factors affect how the surgeon sees, interprets, and conveys the information about the patient's case and in turn affects how the "facts" are communicated, as well as how any uncertainty about the facts are discussed. Particularly challenging and prone to error are future predictions of quality of life for certain states of health [9]. The challenges to presenting value-neutral facts can seem insurmountable. However, recognition of the inherent uncertainty and fallibility of the assessment of the facts can also create space for allowing patient preferences to develop through a discussion of the uncertainties.

There is a long legal and ethical precedent for the rights of capacitated patients to refuse any proposed intervention, despite the outcome. One example is the debilitated patient who refuses a metastasectomy for an isolated hepatic recurrence of her colon cancer. This is a matter of informed *consent* or informed *dissent* and is discussed elsewhere in this book. Relative to surgi-

cal decision making though, following the principle of respect for patient autonomy does not inhere the opposite absolute right for a surgeon to provide a patient any procedure he/she requests. Some argue that the surgeon refusing to provide a requested intervention or even limiting the options presented to a patient entails a form of paternalism. This is not the case. Paternalism is defined by philosophers as "the interference with a person's liberty of action justified by reasons referring exclusively to the welfare, good, happiness, needs, interests, or values of the person being coerced" [10]. It is the professional duty of the surgeon to only offer or provide those operations that can achieve the goals of the intervention. Indeed, this is for the good of the patient, and so part of the definition of paternalism is fulfilled, but it does not amount to coercion. An individual person's right to liberty does not entail gaining access to an operation that is not indicated, or appropriate and second opinions should always be offered. While there are aspects of surgical decision making that seem irreducibly paternalistic, the following discussion is aimed at providing a more nuanced view.

Ezekiel and Linda Emanuel cogently summarized four different approaches to the patient–physician relationship: paternalistic, informative, interpretative, and deliberative [11] (Table 1). *Paternalism* is generally mentioned only to be condemned by bioethicists, as this model represents the "bad old days" where the "surgeon knows best" and the patient's only role is to accept the decisions that are handed down. At its worst, paternalism ignores the specific values and concerns of the patient in favor of pursuit of a goal that is informed solely by the values and determination of the surgeon. At its best, the paternalistic surgeon uses his/her knowledge, experience, and expertise to arrive at the decision that prioritizes the best interests of the patient

Table 1 Four models of patient–MD relationship

Model	Role of autonomy
Paternalism	Assent
Informative	Total control
Interpretative	Self-understanding
Deliberative	Self-development

over all other considerations. The paternalistic surgeon would never let a patient make a decision that would lead to an inferior outcome. The role of the patient is a passive one, as a child being guided and protected by a loving parent.

At the other end of the spectrum from paternalism is the *informative* model. The language of the informative model is prevalent in our contemporary system, which identifies patients as "consumers" and surgeons as "healthcare providers." This merchant–consumer dynamic is superimposed onto the doctor–patient relationship, which then obligates the provider to supply all the relevant information necessary to the patient/consumer so that he/she can make the best decision for himself/herself. Patient choices are maximized. Where the paternalistic model presumes that a surgeon would be able to unilaterally determine the best interests of a patient, the informative approach presumes that a patient would be able to perform the work required to process the information provided and be able to determine a course of action congruous with her goals and values.

The *interpretative* model recognizes that some patients may not have the ability to interpret the medical information for themselves and so requires the surgeon to understand the values of the patient and help him/her apply them to the medical facts and options available. This could be seen as potentially demonstrating the highest respect for patient autonomy, as it aims to assist the patient in elucidating his/her own goals and then offering options to achieve those goals in a nonjudgmental way. If the informative model is the most *laissez faire*, then this model provides more guidance: "Given that your primary goal is to get out of the hospital as quickly as possible and get back home to your family, the below-knee amputation would be the safest way for us to achieve that." In this model the surgeon would say that even if in his/her opinion the best option for the patient would be a femoral–popliteal artery bypass.

The *deliberative* model allows for the surgeon to persuade the patient to make the "right decision," such that in the case above, the surgeon would make a case that a fem-pop bypass is bet-

ter for the patient given many other considerations that the patient did not take into account. The surgeon is considered a teacher rather than a provider or technician. The idea is that the patient is open to growing in his/her understanding of what health-related values should be important to him/her and that both parties are morally engaged in choosing the "best thing, all things considered" for the patient. The general consensus of the received tradition in bioethics is that the deliberative or interpretative models are the ideal.

Whitney, McGuire, and McCullough proposed a further typology of decision making to help identify when shared decision making is most appropriate [4]. For situations of high risk and high certainty, for example, a GSW (gunshot wound) to the abdomen, the concept of shared decision making does not have much usefulness, and if the patient is conscious, he/she is informed of his/her situation and an operation is performed with presumed consent. It is generally clear in a case like this that without an emergent operative intervention the patient will have a poor outcome. However, in situations of high risk and high uncertainty, the model of shared decision making is the ideal. An example is a young woman with early-stage breast cancer, who has an option for mastectomy or lumpectomy with radiation and an additional myriad of options for breast reconstruction. There is almost certainly no *one* right decision for any given patient and the trade-offs between options are significantly preference-sensitive. There is no way that a surgeon can ethically navigate this without shared decision making.

How then does one actually *do* shared decision making? Elwyn et al. [12] suggest that apart from a foundation of a good relationship and good communication skills, the core of SDM is to confer agency to the patient by providing information and supporting the decision-making process. They offer a three-step approach to use as a technique for conducting a discussion that results in a truly shared decision: choice talk, option talk, and decision talk. Choice talk occurs after a diagnosis is communicated to the patient and involves letting the patient know that more than one treatment option exists. This sets the stage by introducing the idea that individual preferences matter and that uncertainties may exist about the outcomes. This phase also assures patients that they will not be abandoned to the choices but rather be guided through them. Option talk requires checking what the patient already knows about his/her options and then listing options and discussing the risks and benefits of each option. Decision-support aids (printed literature, graphics, videos, websites, etc.) can be useful during this portion. Before proceeding to the final step, having the patient "teach back" what they understand about their options is important to clear up misunderstandings or miscommunication. Decision talk elicits preferences by asking, "in your opinion, what matters most?" It also asks patients if they are ready to make a decision or not, with the goal of bringing them to a point where their initial preferences have matured into informed preferences. This process requires a deliberation between the patient and surgeon, with the surgeon checking that the patient's decision accords with those values elucidated. The ideal outcome is an intervention that is consistent with the patient's goals [13].

Decision Aids/Decision Support

Decision aids are tools available in a variety of media such as online, print, or video that help inform patients of their options from an evidence-based perspective, encourage active engagement with the decision-making process, and assist patients in thinking through their values so that they can make a choice consistent with those values [14]. Over the past several years, there has been increasing activity at the state and federal level to support the increasing use of decision aids as a part of shared decision making [15]. In 2007, the state of Washington passed legislation to encourage the use of certified decision aids in patients making preference-sensitive decisions about surgery [16]. Hence researchers have endeavored to measure the quality of decisions made using these tools in surgical decision making [17]. The ideal decision is one that is considered clinically appropriate, adequately informed,

and consistent with the patient's goals, concerns, and preferences [18]. The decision dissonance score is a survey instrument that has been developed and validated and in a large survey of Medicare patients who underwent CABG, prostatectomy, or lumpectomy or mastectomy for breast cancer showed patients who used decision aids reported being more informed about their decision and scored lower on the decision dissonance score. As more decision aids are developed, these types of instruments will be important to ascertain the effectiveness of these tools. The promise of the routine use of decision aids in surgical practice is that they can potentially standardize the process of shared decision making that is prone to a highly variable enactment by individual surgeons who have more or less time to spend with any one individual patient.

Emergency Patients

Except for trauma surgeons in the busiest of trauma centers, these situations comprise a minority of the patients that a surgeon encounters. The Acute Trauma Life Support algorithm suggests rendering definitive treatment for life-threatening conditions resulting from trauma in the "golden hour." Decision making in these cases is entirely unilateral, with the surgeon determining and performing the life-saving interventions under the aegis of "presumed consent"—that is, engaging the patient to the extent possible about the nature of the interventions being performed but also presumably proceeding despite voiced opposition by the patient. The emergent nature of the situation and the potential loss of life if the surgeon makes a false-positive determination of decision-making capacity in the patient justify the intervention. That is, in an emergency situation, incorrectly interpreting a dissenting comment from a patient as one that truly represents their goals and values and hence forgoing treatment and unnecessarily losing a life is the worse than saving the life of a dissenting patient. The first situation has no recourse due to the finality of death [19]. It is however the surgeon's duty to engage with a patient or surrogate decision making postoperatively, after the acute life-threatening situation is over, to discuss ongoing and further interventions and how those fit into the goals of care. At this point there is time to determine what the patient's goals and values are vis-à-vis the proposed treatments. For example, a patient in a motor vehicle collision who is post-op from an exploratory laparotomy to control bleeding, who is found to also have a devastating neurological injury, may have family who—using substituted judgment—determine to forego further life-sustaining interventions by refusing a tracheotomy and removing the patient from the ventilator.

Acute, Not Yet Emergent Patients

The patient who carries a life- or limb-threatening diagnosis but who does not require an emergent operation is often the most difficult situation for both the surgeon and the patient. Unlike the elective surgical patient, where a non-operative approach is generally acceptable and completely up to the patient, the patient with a diagnosis of a slow-growing tumor who wants to adopt a "watch and wait" approach can cause an incredible amount of anguish for the surgeon [20]. These comprise a large number of surgical practices and are ideal situations for shared decision making. Examples include the patient with claudication who is still smoking, the patient with CHF and COPD and a large abdominal aortic aneurysm, and the active person with a few hospitalizations for bleeding diverticulosis. All of these patients could benefit from immediate operations, some could be optimized with "preconditioning" preoperatively, and some could reasonably be observed. These are cases where the surgeon presenting the "one right answer" would be inappropriate. And while the patient may delegate his/her agency to someone else (even the surgeon), these decisions cannot be made without elucidation of the patient's values, hopes, fears, and goals and the surgeon dutifully interpreting the options for the patient. Once established, the nature of the patient–surgeon relationship requires that the surgeon not abandon the patient. Otherwise these

are situations when full-blown paternalism might inappropriately occur. For example, "you are at high risk for repairing your aneurysm, so you need to enroll in our pre-conditioning program and we should proceed with repairing this as soon as you are optimized. If you choose not to follow this recommendation, I will not see you when you return with worsening symptoms." This is clearly coercion, and yet it may be the case that the patient is not a surgical candidate when they return to the ER with symptoms from their ruptured aneurysm. However, the professional duty of the surgeon would be to still engage the patient and discuss what options may be left open to them. Ensuring that the patient knows this when he/she makes his/her decisions helps to ensure that the appropriate "nudging" of patients toward a decision does not become coercion. This is when the deliberative model is most appropriate, where the surgeon may not accept an initial refusal of an operation on its face but ask further questions, clarify the reasons for the refusal, and discuss frankly that options that achieve the patient's goals may not be available later.

Elective Patients

The very nature of elective operations is such that some surgeons perform more of them than others. There is a financial advantage to performing elective operations and one often wonders why that one surgeon in the hospital seems to perform more cholecystectomies for symptomatic cholelithiasis than everyone else. Elective cases seem to be a situation where the *informative* model may actually have a place; however even in cases where a non-operative approach may be equivalent to an operative approach, or where the results of the operation are primarily cosmetic, there is still a significant amount of work that has to be done on the part of the surgeon to uphold the professional responsibility that he/she has to the patient. Eliciting the patient's values and goals and discussing risks and benefits of the operation in light of those goals are the heart of the idealized shared decision-making process. A recent review of studies examining the use of SDM (including decision aids) in decisions for elective operations found that decisional conflict decreased with SDM and decisional quality increased [21]. Framing the discussion in a way that downplays the risks in an effort to nudge the patient toward an operation solely for the financial benefit of the surgeon is ethically suspect. Standardized decision aids for certain procedures may potentially mitigate some of these concerns.

Surgical Buy-In

The idea of "surgical buy-in" is one that has been recently developed and explored by Schwarze [22, 23]. This concept is aimed at describing more fully what non-surgeons have identified as surgeons' "difficulty giving up" on our patients who have a dismal postoperative outcome and require an extensive amount of intensive care. Specifically, the notion of "buy-in" relates to the idea that when a patient agrees to undergo an operation, he/she is also agreeing to all of the postoperative interventions aimed at prolonging life and facilitating hospital discharge. Schwarze has shown that patients do not often realize or understand all that the surgeon thinks have been agreed upon or discussed. Antidotes to this problem include a more thorough discussion preoperatively but also frank discussions postoperatively when a complication or physical deterioration may more tangibly weigh into the decision-making process from the patient and family's perspective. Unfortunately, when a patient is critically ill postoperatively, he/she may no longer be able to participate in the discussion, placing increasing importance on the preoperative discussion. This raises the question of whether a surgeon may rescind an offer to operate if the patient cannot agree to comply with the possible prolonged ICU course postoperatively.

Surgical Candidacy and Moral Agency

The decision of whether or not a particular patient is an appropriate surgical candidate can be con-

tentious. These include the decision of whether or not to accept someone as a living organ donor, whether or not a tumor is unresectable, whether or not to replace a reinfected valve on an active IV drug user, or whether to perform a surgical palliation on a child with trisomy 18 and hypoplastic left heart syndrome. None of these decisions are capable of being made with value-free medical facts, and yet a surgeon may reasonably refuse to perform any of these operations despite requests from the patients and families. In a society that is increasingly hostile to the idea of medical authority and conscientious objection, surgery remains a discipline where there is some finality to the decision that an operation is not warranted. Two concepts undergird this position: the moral agency of the surgeon himself/herself and the professional integrity of the practice of surgery.

"If a patient undergoes a harmful procedure, the moral responsibility for that action does not belong to the patient alone; it is shared by the doctor who performs it. Thus a doctor is in the position of deciding not simply whether a subject's choice is reasonable or morally justifiable, but whether he is morally justified in helping the subject accomplish it" [sic] [24]. Hence it is the case that surgeons infrequently (if at all) perform operations on patients against which they have recommended an operation. Respecting a patient's autonomous decisions about her health cannot induce an action that the surgeon would not offer. "Surgeons are not ethically obligated to provide treatments that they reliably judge will cause more harm than benefit or that will violate appropriate standards of care" [25]. The concern here is the finality of such a decision. Surgeons who refuse to perform a requested operation should encourage the patient to get a second opinion or transfer to another hospital if the patient and family persist in the request. Referring for a second opinion removes the surgeon from being the proximate cause of harm to the patient and is necessary because it recognizes the fallibility of human reason. Forcing a surgeon to operate when he/she feels that the operation is unindicated, futile, or technically impossible or will result in more harm than good is stultifying to surgical virtue.

Surgeon as Mere Technician

Surgeons sometimes face a situation in which the decision to operate has been purportedly made without the surgeon's involvement and he/she is being asked to be the motor end plate of the neuron. The trouble begins when the surgeon disagrees with the surgical decision or discussion that was completed without him/her. A common example is the otolaryngologist who is asked to place a tracheostomy in the neurologically devastated stroke patient with unclear goals of care, or the surgeon called in to "remove the dead bowel" from patient post-op from a complex cardiac operation who has thrombosed his/her SMA and necrosed the entire small bowel and appears completely moribund. Internists have written about the supposed illegitimacy of a surgeon refusing to perform operations in cases like these [26]. However, it is essential to the integrity of the profession of surgery that the surgeon can choose who to operate upon and what operation to perform, keeping her fiduciary responsibility to the patient primary. Some surgeons may find it easier to acquiesce and perform operations that other members of the team have decided upon and ones that he/she personally disagrees with; however this is problematic. A surgeon may decide to operate on someone despite thinking that the harms outweigh the benefits for the patient in order to collect on the billing or to keep the family or referring physicians happy. Both of these reasons are morally corrupt according to a Kantian framework that demands that the individual person be treated always as an end in himself/herself and never only as a means to an end [27]. If a surgeon thinks to himself/herself, "I know this patient is going to die immediately post-op, but I need the billing this month" or "It seems clear to me this patient never would have wanted this operation, but the family and referring MDs want to be able to say 'we did everything' so at least we will be keeping them happy," he/she has violated this fundamental concept of respect for persons.

Medicine and surgery have become so complex that inevitably sick patients have multiple teams that are involved in caring for them. The converse of the above scenarios of medical teams treating a surgeon as a mere "proceduralist" is the surgeon who makes surgical decisions in isolation from the rest of the care teams. Tumor boards and cleft teams are two examples where decisions about patient care are discussed in a multidisciplinary fashion and perspectives from more than just the patient and surgeon are considered. Most patients do not have a coordinated multidisciplinary approach to their care though, and myopia and miscommunication can plague the surgical decision-making process. An example is the patient with metastatic cancer who has developed a gangrenous leg from a thrombosed popliteal artery. The oncology team estimates that the patient has days to weeks to live, but the consulting surgeon performs an amputation because "she will die without an amputation" and the "family wants everything done." A discussion with the palliative care team could have better informed the decision for an operation, as they had multiple discussions about his/her goals of care around his/her end of life. These examples are meant to show that surgeons should embrace the idea of coordinated team decision making and be active participants when possible but should continue to resist the attempts of teams removing the surgeon from the process of applying surgical judgment and experience to the situation.

Pediatric Patients

Generally speaking, pediatric patients do not have legal control over their bodies until the age of 18. Until that time parents have legal authority to make medical and surgical decisions for their children. When a parent signs a consent form to authorize an operation, they are not giving informed consent as much as they are giving *permission* for the surgeon to proceed with the operation [28, 29]. Unlike the concept of "substituted judgment" that a surrogate decision-maker might use to weigh the risks and benefits of an intervention for an incapacitated adult, the classic standard applied to decision making in children is to follow what is in the child's "best interests." This places great moral authority in the standard of care, and often when a parent refuses an operation thought to be in the best interests of a pediatric patient (e.g., debridement of a full-thickness burn that is making the patient septic), child protective services can become involved, legal guardians can be put in place, and parental wishes overridden. Some have argued that "best interests" are too high of a standard to uphold and that the *harm principle* is a more practical and fair approach. The example of a burned patient requiring debridement is an example of when this principle also applies. It is not just that it is in the best interest of the patient to receive debridement; the patient will suffer harms if the debridement does not occur. John Stuart Mill articulated this principle as one that justifies state intrusion into the lives of citizens [30], and Doug Diekema established this as a dominant concept in pediatric ethics [31].

As a pediatric plastic surgeon, the author most frequently deals with requests for operations that may be unnecessary or not in the patient's best interests. Purely elective cases, that is, instances where there are little or no medical indication for the procedure, should involve the patient himself/ herself in the decision whenever possible. Most of these procedures address quality of life, which is best assessed by the pediatric patient himself/ herself, and about which we have not yet developed a gold standard for patient-reported outcomes [32]. The AAP states that patients 14 years old and up should be involved in the process and themselves *giving* consent (while parents sign the form that gives legal permission) and the younger than that children should be involved to the extent possible and giving *assent*. The complicated and unique circumstances of pediatric surgery are more fully explored in a subsequent chapter.

While it is clear that a shared decision-making approach is the ethical ideal, surgeons have been weighing the risks and benefits of cutting their patients since the beginning of the profession. The prudent surgeon understands that there is not one model that is appropriate in every scenario and that the good surgeon utilizes different approaches in different cases based on the particular context of the surgical scenario. Many

experienced surgeons tacitly understand this, but the challenge is training young surgeons in a way that they appreciate and develop the clinical wisdom to employ the appropriate model in every situation.

References

1. Stain SC. Informed surgical consent. J Am Coll Surg. 2015;222(4):717–8.
2. Schneider CE. The practice of autonomy: patients, doctors, and medical decisions. New York: OUP; 1998. p. xiv.
3. Barry MJ, Edgman-Levitan S. Shared decision making–pinnacle of patient-centered care. New Engl J Med. 2012;366(9):780–1.
4. Whitney SN, McGuire AL, McCullough LB. A typology of shared decision making, informed consent, and simple consent. Ann Intern Med. 2003;140:54–9.
5. Brock DW. The ideal of shared decision making. Kennedy Inst Ethics J. 1991;1(1):28–47.
6. Kelly ML, Sulmasy DP, Weil RL. Spontaneous intracerebral hemorrhage and the challenge of surgical decision making: a review. Neurosurg Focus. 2013;45(5):1–7.
7. Kahneman D. Thinking, fast and slow. New York: Farrar, Straus, and Giroux; 2011.
8. Tversky A, Kahneman D. Judgment under uncertainty: heuristics and biases. Science. 1974; 185:1124–31.
9. Ubel PA, Loewenstein G, Jepson C. Whose quality of life? A commentary exploring the discrepancies between health state evaluations of patients and the general public. Qual Life Res. 2003;12:599–607.
10. Dworkin G. Paternalism. In: Sartorius R, editor. Paternalism. Minneapolis: University of Minnesota Press; 1987. p. 19–34.
11. Emanuel EJ, Emanuel LL. Four models of the physician-patient relationship. JAMA. 1992; 267:2221.
12. Elwyn G, Frosch D, Thomson R, et al. Shared decision making: a model for clinical practice. J Gen Intern Med. 2012;27(10):1361–7.
13. Fowler FJ Jr, Gallagher PM, Drake KM, Sepucha KR. Decision dissonance: evaluating an approach to measuring the quality of surgical decision making. Joint Comm J Qual Patient Saf. 2013;39:136–44.
14. International patient decision aids standards collaboration. *Criteria for judging the quality of patient decision aids*. 2005. www.ipdas.ohri.ca/IPDAS_checklist. pdf.
15. Kuehn BM. States explore shared decision making. JAMA. 2009;301(24):2539–41.
16. University of Washington. Shared decision making project at the University of Washington. 2009. http:// depts.washington.edu/shareddm/waleg.

17. Fowler FF Jr, Gallagher PM, Drake KM, Sepucha KR. Decision dissonance: evaluating an approach to measuring the quality of surgical decision making. Jt Comm J Qual Patient Saf. 2013;39(3):136–44.

18. Collins ED, Moore CP, Clay KF, et al. Can women with early-stage breast cancer make an informed decision for mastectomy? J Clin Oncol. 2009;27(4):519–25.

19. Mattox KL, Engelhardt HT Jr. Emergency patients: serious moral choices with limited time, information, and patient participation. In: McCullough LB, Jones JW, Brody BA, editors. Surgical ethics. New York: Oxford University Press; 1998. p. 78–96.

20. Shuman AG. Contemplating resectability. Hastings Cent Rep. 2017;47:3–4.

21. Boss EF, Mehta N, Ngarajan N, et al. Shared decision-making and choice for elective surgical care: a systematic review. Otolaryngol Head Neck Surg. 2016;154(3):405–20.

22. Schwarze ML, Bradley CT, Brasel KJ. Surgical "buy-in": the contractual relationship between surgeons and patients that influences decisions made regarding life-supporting therapy. Crit Care Med. 2010;38(3):843–8.

23. Nabozny MJ, Kruser JM, Steffens NM, Pecanec KE, Brasel KJ, et al. Patient reported limitations to surgical buy-in: a qualitative study of patients facing high risk surgery. Ann Surg. 2017;265:97–102.

24. Ross LF, Glannon W, Gottlieb LJ, Thistlethwaite JR Jr. Different standards are not double standards: all elective surgical patients are not alike. J Clin Ethics. 2012;23(2):118–28.

25. McCullough LB, Jones JW, Brody BA, editors. Surgical ethics. New York: Oxford University Press; 1998. p. 91.

26. Wicclair MR, White DB. Surgeons, intensivists, and the discretion to refuse requested treatments. Hastings Cent Rep. 2014;44(5):33–42.

27. Kant I. Groundwork of the metaphysics of morals. Cambridge: Cambridge University Press; 2005.

28. American Academy of Pediatrics Committee on Bioethics. Informed consent, parental permission, and assent in pediatric practice. Pediatrics. 1995;95(2):314–7.

29. American Academy of Pediatrics. Informed consent in decision-making in pediatric practice. Pediatrics. 2016;138(2):e20161484.

30. Mill JS. On liberty. In: John Stuart Mill, on liberty and utilitarianism. New York: Bantam Books; 1993. p. 12.

31. Diekema DS. Parental refusals of medical treatment: the harm principle as threshold for state intervention. Theor Med Bioethics. 2004;25(4):243–64.

32. Ranganathan K, Vercler CJ, Warschausky SA, MacEachern MP, Buchman SR, Waljee JF. Comparative effectiveness studies examining patient-reported outcomes among children with cleft lip and/or palate: a systematic review. Plast Reconstr Surg. 2015;135(1):198–211.

The Surgical Informed Consent Process: Myth or Reality?

Miguel A. Caínzos
and Salustiano Gonzalez-Vinagre

Key Points

1. Informed consent does not just consist of the patient's authorization to be operated on. It represents a process which may be complex.

2. The paramount aspect of this process requests the surgeon providing the adequate information to the patient so that he or she can make a free and not coercive decision based on what they know and in accordance with their own interests and values.

3. It is necessary to disclose information about the potential risks, including the risks that are common to all types of surgery and the specific risks of the proposed surgery.

4. Some tools can improve the information disclosed to the patient.

5. The decision-making process affects both parties – the patient and the surgeon – and both play an active role.

6. The ultimate goal of the informed consent process must be the development of trust in this dyadic relationship.

M. A. Caínzos (✉) · S. Gonzalez-Vinagre
Department of Surgery, Hospital Clínico
Universitario, Santiago de Compostela, Spain
e-mail: miguelangel.cainzos@usc.es;
salustiano.gonzalez.vinagre@sergas.es

The Informed Consent

According to Hanson and Pitt [1], "Informed consent for surgery has become a critical component of surgical practice. Informed consent for surgery entails what surgeons communicate to their patients about the proposed surgery and is a key element in the trust patients have in surgeons." Wheeler [2] believes that as surgeons, standards become increasingly globalized, and many countries will need to adapt existing practices related to the process historically covered by the term "informed consent." Grant et al. [3] believe that "surgeons must not only empower patients to make autonomous decisions that agree with physicians' recommendations but also respect the autonomy of patients who disagree with these recommendations by not ignoring, insulting, or demanding them or by being inattentive to alternative choices."

Surgical informed consent is a complex process and not just an event and is a very important reality in the current surgical setting [4]. The Canadian Medical Protective Association reported in 2016 that over a recent 5-year period, 65% of the medical legal actions involving informed consent referred to surgery, and only 21% of these cases were decided in favor of the surgeon [5].

The year 2014 marked the hundredth anniversary of the famous verdict by Justice Benjamin Cardozo in "Schloendorff vs. New York Hospital" [6]. His decision, finding the surgeon liable for

Table 1 Main rulings in the development of informed consent during the twentieth and twenty-first century

Year	Sentence
1905	*Mohr* vs. *Williams* (USA) [7] "Need to proceed according to the preoperative agreement"
1914	*Schloendorff* vs. *Society of New York Hospital* (USA) [6] "Every human being of adult years and sound mind has a right to determine what shall be done with his own body"
1957	*Bolam* vs. *Friern Hospital Management Committee* (UK) [8] "Bolam principle"
1957	*Salgo* vs. *Leland Stanford, Jr. University Board of Trustees* (USA) [9] Introduced the term "informed consent"
1972	*Canterbury* vs. *Spence* (USA) [10] "Reasonable patient standard"
1980	*Truman* vs. *Thomas* (USA) [11] The risk of "not acting or postponing"
1990	*Moore* vs. *Regents* of the University of California (USA) [12] "Anyone can take someone's garbage and sell it"
1992	*Roger* vs. *Whittaker* (Australia) [13] "The reasonable doctor standard"
1993	*Arato* vs. *Avedon* (USA) [14] "Legal duty to disclose to the patient all material information that would be regarded as significant by a reasonable person"
2001	*Duttry* vs. *Patterson* (USA) [15] "Physicians do not have to disclose experience when obtaining informed consent, but ethically they should do so"
2015	*Montgomery* vs. *Lanarkshire Health Board* (Scotland, UK) [16] "The doctors must ensure the patient is aware of any material risk involved in any recommended treatment, and of any reasonable alternative or variant treatments"

the removal of an abdominal fibroid tumor without the express consent of the patient, paved the way together with other sentences for the development of the defense of patients' rights, in accordance with major developments in the fields of surgery and anesthesia during the twentieth and twenty-first centuries [6–16] (Table 1).

In the context of twenty-first century surgery, it is fully accepted that the traditional paternalist relationship between the patient and his or her physician has been superseded by a new type of relationship in which the patient detents a major participative role, with the aim of a more interactive dialogue that makes it possible to provide the patient with the information they need before the operation and to allow the surgeon to receive feedback from the patient. This new concept of the patient-physician relationship has come about due to several facts: the preeminence of the patient's autonomy, and that patients are not only more mature but also more informed by different sources and wish to be actively engaged in their

care. For this reason, it is essential to establish a more fluid dialogue between both sides of this dyadic relationship.

Within this new context of the patient-physician relationship, the informed consent process plays a major and significant role. For patients waiting to undergo surgery, the whole process of informed consent requires a free, competent, autonomous, and willing patient, which is not always the case. The legal principle emphasizes the fact that the patient is an independent adult who has the capacity to authorize what is going to be done to their body and mind. Therefore, any operation that may infringe this principle is not only considered illegal and liable to result in a claim for unlawful injury caused to the patient but also ethically unacceptable.

In Spain, the law 41/2002 defines informed consent as "the free, voluntary and conscious agreement of a patient, stated in the full use of their faculties after being suitably informed, so that an action may be undertaken that affects their health" [17]. The general concept of this Law is similar in all western countries. It is a legal situation that underlines the voluntary authorization granted by a patient who has fully understood the risks involved in the diagnosis and performance of a medical or surgical treatment. Thus, the former paternalistic relationship with the physician gave way during the late twentieth century to the current situation whereby patients are able to make decisions characterized by patient autonomy, shared decision-making, self-determination, and patient value system. The aim of the patient-physician relationship is to respect the patient's interests and system of values and, in general, to guarantee the patient's rights.

Surgery without consent can be done only in emergency situations when the patient is unconscious and not capable and no substitute decision-maker is available [1].

Experts increasingly agree that informed consent is a process in which the surgeon plays a truly important role. This role has been defined as "the art of obtaining informed consent in the clinical setting" [18]. All surgeons, including those in training, must know and excel in this art; in the opinion of Childers et al. [18], "the informed consent must be an integral part of every surgeon's daily practice." This process has significant ethical and legal aspects, but it also plays an important role in modern surgical practice. The surgeon is the protagonist together with the patient where they can bring all of their experience and knowledge into play in order to obtain the informed consent of a wide range of patients in very different clinical situations.

Weaver [19] has proposed using the term *operative request* instead of *operative consent* and *operative permit*, as the latter terms come from the legal field, which is not conductive to an appropriate physician-patient relationship: "when we use legal terminology we place ourselves in a venue that is not appropriate." In Weaver's opinion, "labeling this document as an operative request is a more appropriate terminology for medicine, and extricates it from the legal arena." It would also be beneficial to young surgeons since "viewing this surgical document with the terminology of *informed request* rather than

informed consent is helpful to young physicians in training, because it will influence their understanding of their relationship with the patient throughout their surgical career."

As a principle, the informed consent is personal for the physician to whom it is granted, although in today's team medicine, as is the case of surgery, it seems to be accepted that this authorization is valid for all of the physicians involved in their surgical treatment, unless the patient grants this consent exclusively to a specific surgeon.

The Components of Informed Consent

It is generally accepted that the process of informed consent comprises two components derived from the rights that affect the patient: the right to receive adequate information that allows them to make the best decision and the right to give their consent [18, 20]. The review carried out by Leclercq et al. of the surgical informed consent shows that the three cornerstones of the informed consent process are the following: preconditions, information, and consent [21].

Preconditions

The first component of informed consent – the preconditions – includes the patient's competence and willingness [8]. The patient needs to have the full use of their faculties to be able to provide informed consent. In situations of impaired capacity due to situations such as cognitive dysfunction or psychiatric illness or legal incompetence, the final decision to accept the surgical proposal will have to be performed by a legally appointed representative who is capable of making decisions for the patient [14].

Information Provided to the Patient

Without information it is impossible for the patient to make a grounded decision and grant their consent for a surgical procedure. The information the patient receives from the physician must be accurate, simple, transparent, and intelligible. The surgeon can use his or her skills to provide the necessary information to the patient so that they can make a correct decision according to their own set of values and interests.

Patients can be informed either in a single session or progressively. Regarding the information given in a single encounter, the patient can receive either too much or too little information to be able to immediately grant consent. In general, this type of information is lacking in detail and is aimed at obtaining the patient's consent quickly, usually because of the pressure exerted by the healthcare system. Because of this pressure, it is difficult to find enough time to provide detailed information to the patient and gauge it to the real needs of the patient and to establish a dialogue with the patient, allowing them to actively participate in the process of granting their informed consent. Providing information gradually, from the first consultation until all of the diagnostic tests have been carried out, is ideal so that the patient can progressively assimilate the information received and forward any questions and doubts they may have about their illness and surgical treatment, risks, and prognosis.

For Childers et al. [18], the information process consists of three stages: physician disclosure, patient understanding, and patient decision-making.

Physician Disclosure

The first stage, physician disclosure or information disclosure, involves providing the patient with relevant, genuine, and truthful information. The physician should have the skill to adjust the information to each patient, taking into account their own set of values and interests. The information must be clear and include a discussion of the diagnosis, therapeutic options, and possible alternative surgical and nonsurgical treatments. The physician's decision on how much information to provide is always a complex problem with an important ethical component. In general, the physician must provide information on the proposed treatment's benefits, risks, potential complications, and alternative procedures and always, and most importantly, in a language that the patient can understand and assimilate. From the different models of informed consent that have arisen over the years – the professional, the reasonable, and the subjective models – Childers et al. [18] consider that perhaps the best approach to information disclosure uses a model that combines elements of both the reasonable standard and the subjective standard: *the balanced model* where the disclosure and discussion are based on the most important and relevant interests, values, and goals of the patient, as identified by both the patient and the physician. In order to achieve this goal of providing the correct information, "the specific choice of words used by the physician is critical" [18].

The importance of the surgeon disclosing his or her own volume of surgery performed and the outcomes continues to be a highly debatable matter. Char et al. conducted surveys to determine which types of information, including a surgeon's volume/outcomes, are essential for a patient to decide whether to have surgery or to compare patients' and surgeons' attitudes [22]. For patients, the most important piece of information was whether the surgeon was performing the procedure for the first time: 79% considered this information essential to decision-making. The second most important information for patients was their surgeon's volume of surgery and outcomes, after the risk/benefits. About two-thirds of this information are essential to decision-making, while only one-quarter to one-third of surgeons considered this information essential ($p < 0.001$).

In this regard, Heneghan and Walter [23] have detailed the elements of a well-designed informed consent process according to the Centers for Medicare and Medicaid Services [24] (Table 2), which are the requirement for informed consent according to the Joint Commission [25] (Table 3).

Skowron and Angelos [26] have highlighted the fact that informed consent in its current fashion does not meet the needs of the shared decision-making model. They believe that in practice, surgeons do not provide patients with

Table 2 Elements of a well-designated informed consent

A description of the proposed operation, including the anesthesia to be used
The indications for the proposed procedure
Material risks and benefits for the patient related to surgery and anesthesia
Treatment alternatives
The probable consequences of declining recommended or alternative therapies
Who will conduct the surgical intervention and administer the anesthesia
Whether physicians other than the operating practitioner, including but not limited to residents, will be performing important tasks related to the surgery
Residents performing surgical tasks will be under supervision of the attending surgeon
Important surgical task include: opening and closing, dissecting tissue, removing tissue, administering anesthesia, implanting devices, and placing invasive lines
Whether, as permitted by the state law, qualified medical practitioners who are not physicians will perform important part of the operation or administered the anesthesia, and if so, the types of tasks each type of practitioner will carry out

Heneghan and Walter [23]

Table 3 Requirements for informed consent discussion

The nature of the proposed care, treatment, services, medications, interventions, and procedures
Potential benefits, risks, or side effects, including potential problems that may occur during recuperation
The likelihood of achieving goals
Reasonable alternatives
The relevant risks, benefits, and side effects related to alternatives, including the possible results of not receiving care, treatment, and services
When indicated, any limitations on the confidentiality of information learned from the patient

Heneghan and Walter [23]

all of the possible information and leave out what may be critically relevant information, and as such, the current informed consent process is necessary but not enough to help a patient reach a decision regarding surgery: "Surgeon-patient expectations are typically well aligned in the case of routine, low-risk surgery. However, when a patient is faced with a high likelihood of death if they do not proceed with a high-risk operation, expectations for the postoperative course may differ dramatically." This is especially important

when discussing surgery with a frail, elderly patient; in these cases, they recommend training surgeons to use the "Best Case/Worst Case" scenario to improve their ability to communicate with elderly patients and enhance shared decision-making [27], as has been demonstrated by Kruser [28] and Taylor [29]. This is a novel communication tool for difficult surgical decisions. Patients are provided with a visual diagram of the available options. The surgeon writes the "best-case" and "worst-case" outcomes of each option on a linear continuum, as well as how likely they are for the patient in question. This includes an explicit explanation of the quality of life that the patient may expect in each scenario. The patient can then visually determine how their personal preferences fit into the best or worst possible outcomes.

Concurrent surgery is a controversial topic and its practice should be forbidden [30].

Surgeons also have to take into account to what extent the patient wants to be informed. Keulers et al. [31] found that "surgeons generally underestimate their patients' desire for extensive information prior a surgical procedure of any complexity." The authors administered a questionnaire to a group of surgeons and a group of patients. The questionnaire comprised 80 topics with fields of information on disease, physical examination, preoperative period, anesthesia, operation, postoperative period, self-care, and general hospital issues. Both groups were asked their opinion on what they considered to be important and useful preoperative information for patients. There were significant differences in the responses between the patients and surgeons. Patients scored the following items higher ($p < 0.001$): preoperative period, anesthesia, operation, postoperative period, self-care and hospital information. The patient group had greater interest in information on the operation: specific questions on the procedure, operating time, location of the operation room, waiting list, immediate postoperative contact with their family, and complication rate ($p < 0.007$). Interestingly, women demonstrated a significant higher need for information than men, while the surgeons thought that their patients desired more extensive information on the cause, effect, and prognosis of the disease itself ($p < 0.001$). It is clear that the patient can request more information or can restrict the information that they receive. The patient continues to be the final decision-maker, even though they do not want to know by listening or reading the information [32].

Patient Understanding

The second stage of the information process described by Childers et al., patient understanding, is very important in the whole informed consent process. The aim is to identify how thoroughly the patient understands the information provided and disclosed by the surgeon. It is important to know if the patient fully understands everything that has been explained to them, stressing that when they ask questions they identify their degree of cognitive understanding. If the patient does not has a suitable understanding, it will be necessary to explain the situation again and make sure that there are no misunderstandings, and that the patient's values and interests are being well respected and placed in high consideration.

Despite all of the efforts made to ensure a correct process of information and understanding, it is interesting to note that the assessment of patient's recall and understanding is generally poor. Scheer et al. [33] carried out a study at the Ottawa Hospital Cancer Assessment Center in adult patients with rectal cancer treated with low anterior resection or abdominoperineal resection. The aim of the study was to find what the needs of these patients were when deciding on the surgical treatment of their disease. The primary outcomes measured were the patients' knowledge and understanding of decision and their decisional needs. When questioned about the main outcomes of rectal cancer surgery, 47% could not recall a preoperative discussion of risk about sexual function, and 57% could not recall a preoperative discussion of the risk to urinary function. Nearly half of the patients could not recall having a discussion regarding postoperative bowel function. Only 20% could recall specific aspects

regarding their probability of survival. The majority of patients (73%) could recall a discussion regarding overall quality of life following surgery, such as body image, functional outcomes, and the appearance of stomas and scars. For most, the discussion was centered on returning to activity, work, and regular day-to-day life. These results, which have also been discussed by others [34, 35], show the contrast between the outcomes patients value most and those that surgeons value. The conclusion is that patients retain little of the content of the informed consent discussion.

Faced with this situation, it is necessary to consider whether patients adequately understand the information provided. Fink et al. [36] studied 576 patients to identify independent factors that could influence patient understanding, finding that total consent time was the strongest predictor of patient comprehension. Their study also revealed that comprehension during informed consent discussions may be limited in individuals with potential language difficulties due to ethnicity or education and that being 70 years of age or older is another important factor that may reduce patient comprehension (Table 4). However, in that study gender, marital status, the SF-12 (the Short Form Health Survey-12) physical and mental scales, anxiety, and the REALM (Rapid Estimate of Adult Literacy in Medicine) score were not significant in the bivariable or multivariable analyses. Most authors consider that advanced age, low education level, and ethnicity are important factors, obliging the surgeon to make a greater effort in order to correctly transmit the information. In the cases with these factors, the design of the informed consent document given to the patient is extremely important and should avoid techni-

Table 4 Factors that may affect patient understanding

Older age: >70 years ($p < 0.02$)
African-American race ($p < 0.01$)
Hispanic ethnicity ($p < 0.05$)
Operation type ($p \ll 0.01$)
Lower levels of education ($p < 0.0002$)
Total consent time: <15 minutes ($p < 0.0001$)

Fink et al. [36]

cal terms and long sentences, so that the documents are easy to read and understand [37].

An important part of the surgeon's responsibility is assessing the patient understanding of the information presented. To do so, they must ask the patient about the information they have been provided and evaluate their real level of true understanding. Braddock et al. [38] have developed a model that makes it possible to evaluate nine key factors in the decision-making process, including an assessment of the patient's understanding.

The main goal of the informed consent process is to gain the patient's confidence and trust. In this case, the surgeons must use all the available tools and strategies to effectively inform (Table 5). Mulsow et al. [39] recently carried out a review using MEDLINE and PubMed for articles on different techniques that may be used to improve patients' understanding, e.g., patient information leaflets, multimedia intervention, patient decision aids, the Internet, structured informed consent platforms, and repeat-back.

Leaflets provide written information with or without illustrations. They are short and generally include information on the operation and postoperative course, together with details of any possible complications and results. This type of information has been shown to be valid for educational in screening programs. However, the validity of the pamphlet for the informed consent process is questionable because often they do not provide the basic information required in order to make decisions and, sometimes, are difficult to read.

Table 5 Strategies and techniques for improving patient understanding

Increase discussion time: 15–30 minutes
Patient information leaflets
Multimedia interventions
Patient decision aids
The Internet
Structured informed consent platforms
Repeat-back
Guidelines and evidence-based standardized material provided by specialized organizations

Mulsow et al. [39] and Fink et al. [36]

Multimedia intervention uses a combination of interactive computer programs, videos, and animation. In the studies reviewed by Mulsow et al. [39], the use of multimedia was associated with a significant improvement in patient recall as assessed by questionnaire. The mean improvement in knowledge score per study was 13.6% when compared with patients who received information in the standard way.

Decision aids contain detailed evidence-based information on medical conditions and their treatment. This type of tool can help patients make up their mind by helping them recognize the relative importance and value of different therapeutic options, as well as their potential risk and outcomes, and the aids usually offer structured guidance on decision-making. For this reason, this type of assistance is especially effective in cases where the choice of treatment is based on the patient's preferences. The few studies that are available on the use of decision aids focus on breast cancer, and the aids have shown to be effective in increasing the knowledge scores.

The Internet as a new technology in the field of healthcare has become increasingly accessible and is used by patients of all ages and social conditions to obtain information about medical issues and by a significant proportion of patients undergoing common surgical procedures. This was clearly shown by Tamhankar et al. [40]. They studied patients who were undergoing elective abdominal hernia repair (epigastric, paraumbilical, incisional, and groin) and cholecystectomy; 98% of the patients were provided with printed leaflets regarding their operation, and over 95% of them considered this information to be good or very good. Nevertheless, out of the 59% of the patients with Internet access, 31% used it to acquire additional information about their operation, and 58% used Internet search engines to acquire this additional information. In this study, of the patients who searched the Internet regarding their operation, 79% considered the information to be either very good or good, while 26% were confused and/or worried by the information they received. The main criticisms raised against the use of medical information found on the Internet is that it is often deficient, unreliable, suboptimal, and of poor quality [41, 42]. With respect to surgical treatment options and surgical complications, there is much variability [43]. Also, websites found by search engines can be confusing, subject to bias, and some are commercial [40].

To address these deficiencies of information on the Internet, one solution might be the active collaboration of governments or specialized non-government organizations which could guarantee patients high-quality web-based medical information with evidence-based standardized material and present the ethical and legal aspects of the information they provide. This has been done by the governments of the USA (healthfinder) or the UK (National Health Service (NHS) Direct Online) [44] and through the definition of national guidelines by the UK Department of Health [45] or surgical associations such as the British Orthopaedic Association (http://www.orthoconsent.com), which offers consent forms that are accessible to patients and physicians [46]. In Spain, the Spanish Association of Surgeons (www.aecirujanos.es) and the Spanish Neurosurgery Association (www.senes.es/.), among others, produce informed consent forms, although these are only available to members of the specific association. It has also been proposed that hospital websites provide patients with regulated, easy-to-understand information [40], and the website could provide information about the data from the hospital itself.

Another tool for delivering information to the patient is the structured informed consent platforms. This tool allows for a structured conversation and standardized processes. The system is widely used in the USA as automated structured consent tools (e.g., iMedConsent). However, the impact this tool has on patient understanding is not clear [39]. Fink et al. [47] demonstrated that there was a slight improvement in patient understanding of information when the computer-based consent program iMedConsent followed by the "repeat-back" technique, in which the patient repeats what they have understood during the discussion with the physician, which making it possible to identify and solve any problems that may arise.

On the website of the American College of Surgeons (ACS), the largest organization of surgeons in the world, with more than 82,000 members (www.facs.org), it is recommended that patients should seek the answers to questions such as [48]:

– What are the indications that have led your doctor to the opinion than an operation is necessary?
– What, if any, alternative treatments are available for your condition?
– What will be the likely results if you don't have the operation?
– What are the basic procedures involved in the operation?
– What are the risks?
– How is the operation expected to improve your health or quality of life?
– Is hospitalization necessary and, if so, how long can you expect to be hospitalized?
– What can you expect during your recovery period?
– When can you expect to resume normal activities?
– Are there likely to be residual effects from the operation?

These are the basic questions the surgeon should be prepared to answer and explain to the patient, together with any other specific questions the patient may ask.

One of the most complex questions the surgeon has to answer is: What is the risk of the operation? The American College of Surgeons National Surgical Quality Improvement Program (ACS NSQIP®) Surgical Risk Calculator (SRC) was presented in 2013 as a new tool that estimates patient-specific postoperative complications risk for 1557 surgical procedures across all surgical subspecialties with the exception of transplant and trauma (http://riskcalculator.facs.org). The risk calculator tool predicts the chance that patients will have any of nine different outcomes within 30 days following surgery. The outcomes included are mortality, morbidity (any of the following intraoperative or postoperative events: surgical site infection, wound disruption, pneumonia, unplanned intubation, pulmonary embolism, on ventilation >48 hours, progressive renal insufficiency, acute renal failure, urinary tract infection, stroke/cerebral vascular accident, cardiac arrest, myocardial infarction, deep venous thrombosis, systemic sepsis), pneumonia, cardiac event (cardiac arrest or myocardial infarction), surgical site infection, urinary tract infection, deep venous thrombosis, and renal failure (progressive renal insufficiency or acute renal failure). The universal risk calculator is based on 1,414,006 patients [49]. The goal of the ACS NSQIP risk calculator is to provide patient-specific risk information to guide surgical decision-making and informed consent.

The ACS NSQIP risk calculator has been widely adopted as an aid in the decision-making process and an informed consent tool by surgeons and patients. Because lack of calibration can lead to systematic errors in assessing surgical risk, the Surgical Risk Calculator (SRC) model calibration has recently been evaluated from nearly 3 million patients and for 1887 operations defined by the CPT code. The non-recalibrated Surgical Risk Calculator performed well, although there was a slight tendency for predicted risk to be overestimated for the lowest- and highest-risk patients. After recalibration, this distortion was eliminated, and the performance of NSQIP Surgical Risk Calculator models was shown to be excellent and improved with recalibration [50]. The SRC has been shown to exhibit good calibration and discrimination in large-scale investigations without design limitations, which facilitate its intended purpose, which among others is to support surgeon-patient and family engagement, shared decision-making, and informed consent, providing a general purpose risk calculator, which is applicable across many surgical domains, using easily understood and generally available predictive information [51].

A pediatric Surgical Risk Calculator was recently developed based on 181,353 cases covering 382 CPT codes across all specialties using standardized data from 67 hospitals in the USA. It had excellent discrimination for mortality, morbidity, and seven additional complications. The Hosmer-Lemeshow statistic and graphic representation also showed excellent calibration. It can be used as a tool in the shared decision-making process by providing clinicians and families with useful information for many of the most common operations performed on pediatric patients [52].

The ACS NSQIP risk calculators for adults and children are now considered to be very effective tools for decision-making during the informed consent process. However, a recent study by Lubitz et al. [53], in colorectal surgery, demonstrated that the Surgical Risk Calculator accurately predicts outcomes for elective operations; predicted and actual outcomes were significantly better in patients undergoing elective operations compared with those undergoing emergency procedures. They recommended to use this tool with caution in emergency cases, as it has the potential to underestimate serious complications that were higher than predicted (47% vs 63%; $p < 0.05$) and length of stay because it was longer than estimated (14.4 vs 19.2 days; $p < 0.05$). So far this tool is only available in English, and it would be desirable to have the Surgical Risk Calculator available in other languages.

For some authors, the fact that current clinical practice with regard to incidental findings is suboptimal is a cause for concern, proposing that a more robust informed consent process would be necessary in order to enable patients to correctly anticipate incidents and perceive the associated risks [54]. It is necessary to consider *the material risks*, which include risks that are common to all surgery and risks that are specific for the proposed surgery, even if they are rare. In this case, Hanson and Pitt [1] propose using the surgeon's checklist based on the Consent to Treatment Policy of the College of Physicians and Surgeons of Ontario [55], which includes eight questions that have to be answered:

– Date
– Who was involved?
– Material risk
– Unique risks
– Special circumstances of the patient
– Risks of not undergoing intervention
– Consent given or refused?
– Findings of incapacity and identity of substitute decision-maker

Special consideration must be given to informed consent processes in clinical research, as it is more complex than the conventional informed consent because the language of medical research uses the most technical terminologies. Informed consent is essential for the development of biomedical research. In an study related to clinical research carried out in Brasov with 68 patients, the authors found that 35.3% of the patients did not ask any questions, and from those who did, 20.6% of the questions referred to general aspects of the clinical trial, suggesting a lack of interest and a failure to understand the information presented in the informed consent form [56]. In Korea, Kim and Kim [57] evaluated the effect of a simplified informed consent form in a clinical trial with a total of 150 patients who were randomly assigned to one of two groups and provided with either standard or simplified consent forms for a clinical trial for cancer treatment. This simplified informed consent form included plain language, short sentences, diagrams, pictures, and bullet points and proved to be effective in enhancing the participants' subjective and objective understanding, regardless of their health literacy. In an observational study of adult women seeking surgical treatment for pelvic floor disorders in which 150 participants were enrolled, the association between decisional satisfaction and knowledge persisted after controlling for demographic and clinical variables, including education level, health literacy, race/ethnicity, age, surgeon years after completing fellowship, diagnosis, number of visits in the past 6 months, and number of days between the informed consent discussion and the survey. The study found that patient knowledge and understanding of surgery were important components of the patient's satisfaction with her decision to proceed with pelvic floor surgery [58].

Grady et al. [59] believe that the traditional prototype and the classic interaction in which informed consent is obtained for research, asking the participants to read and sign detailed written consent documents, are becoming outdated, as consent forms are increasingly longer and more complicated, obscuring important details. Digital technology has transformed how people communicate, learn, and work. Technological changes in information practices offer new opportunities for the innovative implementation of informed consent. In their opinion, apps, tablets, video, interactive computers, robots, personal digital assistants, mobile phones and smartphones, as well as wearable technology could all help to modernize, alter, and improve methods of informed consent. Electronic and digital informed consent can improve the disclosure and understanding of information, as well as voluntariness and authorization. Informed consent using electronic devices (e-consent) often includes multimedia, such as graphics or videos about essential study features that may increase the understanding of the study; participants can sign documents electronically, using individual passwords. When e-consent is provided remotely, the identity of the person who is giving their consent can be confirmed by a digital signature, username, and password, or biometrics. E-consent does have some disadvantages (increasing time for videos and quizzes) and some problems in international trials, as countries may have different requirements for e-consent. Using mobile devices, people can participate and contribute their data to research more easily, and researchers can have wider access to populations, as more detailed information on individual activities. Big data can come from this research. Video-informed consent is another tool, but which also suffers from a number of problems: setting up video recording of the informed consent process is not simple in a busy clinical environment, and each recording has to be reviewed for quality, increasing the need of time for the study team and the need for backup equipment. In the authors' opinion, two ethical issues have to be considered: first, only participants who consent to video recording may participate in a clinical trial and second, the question of confidentiality, as there is no clarity with regard to the control the participant has over the process.

The goal of these new tools should be to provide the participants and researchers with better tools to enhance medical research. However, other authors, such as Goh and Shin [60], believe that "true informed consent needs to go beyond symbolic measures such as clicking blocks electronically or supplying a signature" and that "in-person discussion is an imperative step in obtaining informed consent in most clinical trials."

Regardless of the tools used during the disclosure of information, all of the experts coincide in the need to take the proper amount of time to suitably explain the information to the patient and their family, proposing that this time should be extended to 30 minutes (Table 5). However, this is not normally the case, and it is not even possible to suggest it in many healthcare systems, in which the total time allocated to dealing with patients is normally less than 15 minutes.

There can be no doubt that informed surgical consent is a complex process that calls for the surgeon to be competent and for the information to be transmitted in a correct manner. For this reason, and because of its complexity, this process should not be carried out by residents [61].

Patient Decision-Making

The third stage of the information process is the decision-making. Giving informed consent is clearly a complex and delicate decision for the patient, although this will be much easier if the second informative stage has been carried out correctly. This stage is when the patient contemplates the information provided by the surgeon, discusses it with their family and friends, or even with other physicians if they consider it appropriate, and makes the decision they consider to be the most suitable according to their interests and values.

The decision-making process also has a significant effect on the surgeon. The patient and the surgeon both depend on mutual trust and commitment in order to ensure that the surgical procedure will be carried out successfully. Winning the patient's trust has a major influence on the decisions the surgeon will make and the actions they will perform.

A study by McKeally et al. [62] focused on how this decision-making process is seen from the surgeon's perspective. A series of interviews were carried out with thoracic and general surgeons. The surgeons reported that in the decision-making process, they routinely disclosed to patients the diagnosis, prognosis, treatment alternatives, and inherent risks and benefits and were prepared to follow the patients' preferences for treatment. As part of this process, surgeons give their own consent to accept the risks and responsibilities of performing the proposed operative treatment. Important factors that influence the surgeon's decision to operate on an individual patient were described in terms of objective, "hard" findings (e.g., tumor stage, pulmonary function, etc.) and affective "soft" findings (e.g., courage and determination to survive). The surgeon significantly values the determination of the patient to recover from treatment or having a strong will to live. These factors can affect the decision to perform a more aggressive operation, if intraoperatively the illness is found to have spread more widely than previously thought.

In a questionnaire given to surgeons and anesthetists in the UK, the majority of respondents agreed on the need to clearly explain the possibility of dying to patients if this risk was present [63].

In the opinion of McKeally et al. [62], during the decision-making process there is "a foundation of empathy and respect and trust, without which nothing else happens…." and "trust creates a psychological contract, an implied promise to succeed, that binds surgeons to persist in their remarkably relentless pursuit of the best outcome for their patients." "This psychological contract transcended the ritual of written consent."

Consent

The third component of the process of informed consent is the consent itself. This consists of the patient signing a legal document that authorizes the surgeon and surgical team to carry out the operation.

Childers et al. [18] clearly indicated the criteria that must be met by the informed consent document. For these authors, the essential components are a clear description of the planned procedure and its risks and benefits; possible alternative treatment, therapies, including the option of no treatment, and their risks and benefits; documentation that shows that the patient had chance to ask questions; authorization with signature of the patient or surrogate decision-maker (family or legal representative); and signature of physician.

With regard to the risks of complications described in the informed consent document, some guidelines, such as those of the Association of Anaesthetists of Great Britain and Ireland recommend that even "rare but serious complications should be included in written information, as should the very small risk of death" [64].

Most experts in the field of informed consent agree that if this process is carried out correctly, it satisfies the patient, gives the patient a sense of security, may reduce their anxiety, makes the patient more committed to the treatment, and makes it possible to forge stronger bond between the physician and patient, reducing the tendency to solicit legal claims for medical errors [65, 66], and we agree with Skowron and Angelos [26] that "when the informed consent process is optimally undertaken, surgeons can be satisfied that we have done our very best for our patient."

Concluding Remarks

- The surgical informed consent process is the corollary of an intense and adequate patient-surgeon relationship grounded upon trust.
- The ethical need for the informed consent process lies on the principle of respect for patient's autonomy.
- The major factor underlying autonomy is the decision-making capacity of the patient.
- The information provided by the surgeon and/ or surgical team should be truthful, trustworthy, and loyal, taking into consideration the patient's set of values and beliefs.
- Although there may seem to be many limitations to an adequate informed consent process, this process should be improved and enhanced.

References

1. Hanson M, Pitt D. Informed consent for surgery: risk discussion and documentation. Can J Surg. 2017;60:69–70.
2. Wheeler R. The evolution of informed consent. BJS. 2017;104:1119–20.
3. Grant SB, Modi PK, Singer EA. The surgical informed consent process. In: Ferreres AR, Angelos P, Singer EA, editors. Ethical issues in surgical care, vol. 2017. Chicago: American College of Surgeons Division of Education; 2017. p. 137–63.
4. Caínzos M, González-Vinagre S. Informed consent in surgery. World J Surg. 2014;38:1587–93.
5. Canadian Medical Protective Association. Risk Fact Sheet CMPA. Available: www.cmpa-acpm.ca/documents/10179/300031190/informed_consent-e.pdf. Accessed 1 Mar 2016.
6. *N.E. Schloendorff v. Society of New York Hospital* 1914, 211 N.Y. 125 (106 N.E. 93) (N.Y. 1914).
7. Supreme Court of Minnesota. *Mohr v Williams*, 1905 (104 N.W. 12 Supreme Court of Minnesota).
8. *WLR Bolam vs. Friern Hospital Management Committee*, 1957 (1 WLR 583).
9. *Salgo v. Leland Stanford Jr. University Board of Trustees*, 1957 (317 P.2d 170).
10. *F. Canterbury vs. Spence*, 1972 (464 F.2d 772 (d.c. 1972)).
11. *P. Truman v. Thomas*, 1980 (611 P.2d 902 (Cal 1980).
12. *Moore v. Regents of the University of California*, 499 US 936 (1990).
13. Skene L, Smalwood R. Informed consent: lessons from Australia. BMJ. 2002;324(7328):39–41.
14. *Arato v. Avedon*, 858 Pad 598 (Cal 1993).
15. *Duttry v. Paterson*, 771 A2d 1255 (Pa 2001).
16. *Montgomery v Lanarkshire Health Board* (Scotland), (2015); UKCS 11.
17. Ley 41/2002 de Noviembre, básica reguladora de la autonomía del paciente y de derechos y obligaciones en materia de información y documentación clínica. Boletín Oficial del Estado, 15 de Noviembre de 2002. Págs. 40126–32.
18. Childers R, Lipsett PA, Pawlik TM. Informed consent and the surgeon. J Am Coll Surg. 2009;208: 627–34.
19. Weaver JP. A problem with informed consent. J Am Coll Surg. 2009;209:286–7.
20. Acea B. Informed consent in the surgical patient. Reflections on the basic law of patient autonomy. Cir Esp. 2005;77:321–6.
21. Leclercq WKG, Keulers BJ, Scheltinga MRM, et al. A review of surgical informed consent: past, present, and future. A quest to help patients make better decisions. World J Surg. 2010;34:1406–15.
22. Char SJL, Lo B, Kirkwood KS. How important is disclosing surgeon experience when obtaining informed consent? J Am CollSurg. 2011;213(3S):S112–3.
23. Heneghan K, Walter KR. Legislative activities and informed consent. Bulletin American College of Surgeons. 2016;101:61–5.
24. Centers for Medicare & Medicaid Services. State Operations Manual. Appendix A – Survey protocol, regulations and interpretive guidelines for hospitals. Rev. 151. Nov 20, 2015. 482.51 (b). www.cms.gov/Regulations-and-Guidance/Guidance/Manuals/downloads/som107ap_a_hospitals.pdf. Accessed 5 July 2016.

25. The Joint Commission 2009 Requirements Related to the Provision of Culturally Care Hospital Accreditation Program (HAP). Hospital Accreditation Standards. Oakbrook Terrace, IL: The Joint Commission. 2009. Standard RI.01.03.01. www.jointcommission.org/assets/1/6/2009_CLASRelatedStandardsHAP.pdf. Accessed 5 July 2016.

26. Skowron KB, Angelos P. Surgical informed consent revisited: time to revise the routine? World J Surg. 2017;41:1–4.

27. Angelos P. The evolution of informed consent for surgery using the best case/worst case framework. JAMA Surg. 2017;152:538–9.

28. Kruser JM, Nabozny MJ, Steffens NM, et al. "Best case/worst case": qualitative evolution of a novel communication tool for difficult in-the-moment surgical decision. J Am Geriat Soc. 2015;63:1805–11.

29. Taylor LJ, Nabozny MJ, Steffens NM, et al. A framework to improve surgeon communication in high-stakes surgical decisions: best case/worst case. JAMA Surg. 2017;152:531.

30. Largerman A. Concurrent surgery and informed consent. JAMA Surg. 2016;151:601–2.

31. Keulers BJ, Scheltinga MRM, Houtermann S, et al. Surgeons underestimated their patients' desire for pre-operative information. World J Surg. 2008;32:964–70.

32. Rastogi P. Through the looking glass – understanding informed consent. Clinical Ethics. 2015;10:41–3.

33. Scheer AS, O'Connor AM, Chan BPK, et al. The myth of informed consent in rectal cancer surgery: what do patients retain? Dis Colon Rectum. 2012;55:970–5.

34. Solomon MJ, Pager CK, Keshava A, et al. What do patients want? Patient preferences and surrogate decision-making in the treatment of colorectal cancer. Dis Colon Rectum. 2003;46:1351–7.

35. Masya LM, Young JM, Solomon MJ, et al. Preferences for outcomes of treatment for rectal cancer: patient and clinician utilities and their application in an interactive computer-based decision aid. Dis Colon Rectum. 2009;52:1994–2002.

36. Fink AS, Prochazka AV, Herderson WG, et al. Predictor of comprehension during surgical informed consent. J Am Coll Surg. 2010;210:919–26.

37. Ruíz Lopez R. Informed consent in surgery. Distance between theory and practice. Cir Esp. 2013;91:551–3.

38. Braddock C 3rd, Hudal PL, Feldmann JJ, et al. Surgery is certainly one good option: quality and time-efficiency of informed decision-making in surgery. J Bone Joint Am. 2008;90:1830–8.

39. Mulsow JJW, Feeley M, Tierney S. Review. Beyond consent – improving understanding in surgical patients. Am J Surg. 2012;203:112–20.

40. Tamhankar AP, Mazari F, Everitt NJ, Ravi K. Use of the internet by patients undergoing elective hernia repair or cholecystectomy. Ann R Coll Surg Engl. 2009;91:460–3.

41. Berland GK, Elliott MN, Morales LS, et al. Health information on the internet: accessibility, quality, and readability in English and Spanish. JAMA. 2001;285:2612–21.

42. Murphy JO, Sweeney KL, O'Mahony JC, et al. Surgical informatics in internet: any improvement? Surgeon. 2003;1:177–9.

43. Murphy MA, Joyce WP. Information for surgical patients: implications of the world wide web. Eur J Surg. 2001;167:728–33.

44. Evans R, Elwyn G, Edwards A. Making interactive decision support for patients a reality. Inform Prim Care. 2004;12:109–13.

45. Department of Health. Good practice in consent implementation guide: consent to examination or treatment. London: Department of Health; 2001.

46. Atrey A, Leslie I, Carvell J, et al. Standardised consent forms on the website of the British Orthopaedic Association. J Bone Joint Surg Br. 2008;90:422–3.

47. Fink AS, Prochazka AV, Henderson WG, et al. Enhancement of surgical informed consent by addition of repeat back: a multicenter, randomized controlled clinical trial. Ann Surg. 2010;252:27–36.

48. American College of Surgeons. www.facs.org. Public information from the American College of Surgeons, 2013.

49. Bilimoria KY, Liu Y, Paruch JL, et al. Development and evaluation of the universal ACS NSQIP surgical risk calculator: a decision aid and informed consent tool for patients and surgeons. J Am Coll Surg. 2013;217:833–42.

50. Liu Y, Cohen ME, Hall BI, et al. Evaluation and enhancement of calibration in the American College of Surgeons NSQIP surgical risk calculator. J Am Coll Surg. 2016;223:231–9.

51. Cohen ME, Liu Y, Ko CY, Hall BL. An examination of American College of Surgeons NSQIP surgical risk calculator accuracy. J Am Coll Surg. 2017;224:787–95.

52. Kraemer K, Cohen ME, Liu Y, et al. Development and evaluation of the American College of Surgeons NSQIP pediatric surgical risk calculator. J Am Coll Surg. 2016;223:685–93.

53. Lubitz AL, Chan E, Zarif D, et al. American College of Surgeons NSQIP risk calculator accuracy for emergent and elective colorectal operations. J Am Coll Surg. 2017;225:601–11.

54. Kole J, Fiester A. Incidental findings and the need for a revised informed consent process. AJR. 2013;201:1–5.

55. College of Physicians and Surgeons of Ontario. Consent to treatment. Dialogue. 2015;11:50.

56. Purcaru D, Preda A, Popa D, et al. Informed consent: how much awareness is there? PLOS ONE. www.plosone.org. 2014;9:1–6.

57. Kim EK, Kim S. Simplification improves understanding of informed consent information in clinical trials regardless of health literacy level. Clin Trials. 2015;12:232–6.

58. Halloch JL, Rios R, Handa RL. Patient satisfaction and informed consent for surgery. Am J Obstet Gynecol. 2017;217:181.e1–7.

59. Grady C, Cummings SR, Rowbotham MC, et al. Informed consent. N Engl J Med. 2017;369:856–67.

60. Goh HG, Shin SW. Informed consent. N Engl J Med. 2017;376:e41.
61. Angelos P, Darosa DA, Bentram D, et al. Residents seeking informed consent: are they adequately knowledgeable? Curr Surg. 2002;59:115–8.
62. McKneally MF, Martin DK, Ignani E, D'Cruz J. Responding the trust: surgeons' perspective on informed consent. World J Surg. 2009;33:1341–7.
63. Jamjoom AAB, White S, Walton SM, et al. Anaesthetists' and surgeons' attitudes towards informed consent in the UK: an observational study. BMC Med Ethics. 2010;11:1–7.
64. The Association of Anaesthetists of Great Britain and Ireland Consent for anaesthesia. Revised ed. London: AAGBI; 2006.
65. Benjamin DM. Reducing medication errors and increasing patient safety: case studies in clinical pharmacology. J Clin Pharmacol. 2003;43:768–83.
66. Baum N. Informed consent – more than a form. J Med Pract Manag. 2006;22:145–8.

Informed Consent and Disclosure of Surgeon Experience

Surgical Ethics: Principles and Practice

Sabha Ganai

Key Points
- Informed consent is part of the daily practice of surgeons.
- Informed consent aims to respect patient autonomy and limit patient harm.
- Informed consent is a process, not a document.

Informed Consent

The act of surgery is a transaction wherein intentional physical harm is directed to a person for the prospect of a net benefit to that individual. Hippocrates, a founder of the ethical standard of nonmaleficence, suggested that physicians swear to "not use the knife," leaving "to such as are craftsmen therein" [1]. Medieval European practice kept barber surgeons stratified within merchant guilds professionally distinct from physician healers, who otherwise originated from a clergy class, prioritizing delivery of sacraments at the end of life, favoring soul over body [2]. While surgical science later expanded through apprenticeship, battlefield practice, and human anatomical dissection, cadavers were often pro-

cured in questionable ways, even by ethical and legal standards of the time [2, 3]. As the healing arts have rapidly evolved over the last two centuries with scientific progress, so has the prospect of benefit from surgery, particularly with the advent of anesthesia and antisepsis [4], leading to surgery becoming a legitimate and fundamental component of modern medical care, with strong ethical foundations.

An essential component to the act of surgery is the unique relationship between surgeon and patient, where the patient must trust the surgeon enough to submit all or part of their body to the care of the surgeon, and the surgeon then must then claim a fiduciary duty to act in their benefit and interest [5]. The modern practice of anesthesia allows procedures to be done while the patient is completely unconscious, leading to a power shift where the patient must submit themselves completely but temporarily to the care of medical practitioners. While the framework of patient-physician determination has transitioned from the age of paternalism, where the doctor decided what is best, to the age of patient autonomy, where the patient decided what is best, to the age of shared decision-making, where a negotiation between the doctor and patient must occur, so has the process of communication and comprehension that encompasses informed consent [6].

Nearly every encounter in the daily practice of surgeons requires informed consent from the patient or their surrogate. While the process of informed consent is meant to expand and pro-

S. Ganai (✉)
Southern Illinois University School of Medicine, Department of Surgery, Springfield, IL, USA
e-mail: sganai@mail.harvard.edu

© Springer Nature Switzerland AG 2019
A. R. Ferreres (ed.), *Surgical Ethics*, https://doi.org/10.1007/978-3-030-05964-4_20

tect patient autonomy, it should ideally also provide the patient the opportunity to properly weigh the risks and benefits of an intervention, so they can make a decision on whether or not to receive a specific treatment based on their own perception of benefit, after being properly informed. While the ethical principle of autonomy becomes a fundamental part of discourse relevant to informed consent as it recognizes that each patient has an intrinsic right to determine what happens to their body, the principle of nonmaleficence still must be respected as part of the fiduciary duty of the surgeon to the patient through an attempt to balance benefits over risks. Furthermore, transparency of disease process and outcome must be provided for the patient to make an informed and autonomous decision. Nonetheless, one can question whether informed consent can ever be achieved, as thorough disclosure of all relevant information by a clinician is not always feasible or practical and does not always translate to comprehension by the patient [7]. In addition, the concept of surgical "buy-in" suggests that a complicated relationship often occurs between patients and their surgeons, where agreement to surgery creates a commitment to keep the operation and postoperative management as a package deal that may otherwise require renegotiation of goals of care in response to complications [8].

For the novice clinician, informed consent is often perceived to be "just a signature," as it becomes interpreted as a mundane but necessary task that gets marked off on the preoperative checklist of nurses and house staff [9]. Minimum elements of an informed consent document include the name of the procedure; the hospital where the procedure will take place; the responsible clinician; a statement of benefits, risks, and alternative therapies; as well as a signature with date and time (Box 1) [9]. While "disclosure of the indications, risks, benefits, and alternatives of a procedure" is often considered a requisite definition of informed consent, this alone does not imply true from the patient's side to make an autonomous decision from an ethical context. Moreover, the patient's signature on a consent form is hardly sufficient as legal protection

against litigation. Informed consent should ultimately be a conversation between the patient and physician within the framework of shared decision-making with the document simply being a record that this process took place. Informed consent is ultimately the process, not a document.

> **Box 1 Minimum elements of an informed consent document** [9]
> - Name of the procedure
> - Name of the hospital where the procedure will take place
> - Name of the responsible clinician(s)
> - Statement of benefits, risks, and alternative therapies
> - Signature of the patient with date and time and, if appropriate, a legal representative
> - Signature of the witness to consent

Canterbury v. Spence established the legal standard of informed consent as an "objective" duty to disclose, otherwise known as the reasonable patient test [10]. In this case, Jerry Canterbury sued his physician for negligence after complications ensued in that he was not properly informed of the risks involved with an elective procedure, a laminectomy, performed for back pain. While "subjective" standards rely on risks disclosed that are pertinent to an actual patient's decision to accept therapy, and "community practice" standards rely on what other local practitioners deem appropriate for disclosure, the "reasonable person" standard established the importance of disclosing what a reasonable patient would want to know under given circumstances (Box 2). The court ultimately felt that "full" disclosure was a norm that was prohibitive and unrealistic to demand from physicians, so it was favored to require disclosure of risk as "material when a reasonable person, in what the physician knows or should know to be the patient's position, would be likely to attach significance to the risk or cluster of risks in deciding whether or not to forego the proposed therapy" [10]. Material risks may have a "high degree of likelihood but a low degree of severity" or a "very low degree of likelihood but high degree of severity" [9]. After elective thyroidectomy, for example, recurrent laryngeal nerve injury leading to permanent vocal cord paralysis would be considered a material risk, even if this complication is exceedingly rare in expert hands.

> **Box 2 Standards of practice for disclosure of risk**
> - Subjective standard – risks disclosed are pertinent to the actual patient's decision to accept therapy.
> - Community practice standard – risks are disclosed based on what other local practitioners deem appropriate.
> - Reasonable person standard – risks disclosed should be what a reasonable person would want to know under given circumstances.

Current legal precedent requires the disclosure of information that would be relevant to the ability of a patient to make a decision under given circumstances. This almost always includes discussion of the risks and benefits, potential alternatives, and expected postoperative course relevant to a procedure or disease process in order for it to be a truly informed decision. Unfortunately, the objective legal standard focusing on information that a "reasonable person" would want to know is still ambiguous and subjective as what can be defined as a material risk may vary widely between different patients.

From an ethical perspective, in order to fulfill duties to limit patient harm and honor patient autonomy, it thus becomes a responsibility of the clinician to determine what information is relevant to a particular patient. In *The Nicomachean Ethics*, Aristotle discusses *nous* (comprehension) as a virtue based not simply on the acquisition and synthesis of *episteme* (knowledge), *techne* (craft), or *phronesis* (practical wisdom) but as applicable to the exercise of opinion for the purpose of rendering a decision or judgment (Box 3)

[11, 12]. Using the framework of Aristotelian virtue ethics, "good" informed consent requires the provision of sufficient information that will allow for judgment at a personal level [13]. The process of disclosure of relevant information can be extensive, as this discourse may not only be procedure-specific and disease-specific but also patient-specific, and must be taken in context with a patient's preferences and values.

> **Box 3 Three types of knowledge that inform comprehension (Aristotle)**
> - Scientific knowledge (episteme)
> - Art or craftsmanship (techne)
> - Practical wisdom (phronesis)

In the process of informed consent for an elective procedure, a conversation between the surgeon and patient may begin with the surgeon learning what the patient knows about their disease, followed by instruction about the natural history of their disease process, anatomical considerations, and what therapeutic options are available to manage relevant symptoms or pathology, including observation. A surgeon, who often leads this conversation by providing information, should ideally break their discussion periodically with "Before I go on, what questions do you have?", in order to explore the level of comprehension of the patient, as well as determine what additional information is relevant for the patient. While the informed consent process for bariatric surgery may happen over the course of a year and may be assisted by a multidisciplinary team to ensure that adequate disclosure and understanding have taken place, it can be argued that a majority of surgical procedures cannot be practically undertaken with the time and effort needed to fully educate the patient, particularly related to the time frame and urgency of many relevant surgical indications, including malignancy, obstruction, perforation, ischemia, and hemorrhage.

Even with elective surgery, up to 13% of patients have major deficits in the informed consent process, such as not knowing the major risks or even the procedure being performed, and another 33% had less serious deficits in the process, such as not having their values, preferences, or goals assessed [14]. Patient understanding and recall are poor after 4–5 days after obtaining consent for an open inguinal hernia repair, with two out of three patients understanding that they would have mesh in their groin and less than 3% aware of the potential to develop chronic pain [15]. While a "reasonable person" would most likely want to know these details, the recall and understanding of patients who were presented this information are often limited less than 1 week after consenting to treatment. These studies were on benign surgery procedures and certainly do not include the potential psychological shutdown a patient may have after just being told they have cancer, only to be followed by having to decide on surgical options that same visit. Even given more time to contemplate, patients with rectal cancer on a comprehensive, several-week multidisciplinary clinical pathway including chemoradiation therapy followed by surgery with multiple preoperative visits by the surgical team still retain very little of informed consent discussions and often did not perceive decisions surrounding surgery as being reflective of a true choice [16].

It is challenging to know when a patient truly does understand all of the relevant information, even if they verbalize understanding. While two-thirds of patients in a study felt that they were extremely to moderately well-informed about their procedure, there was no relationship between perceptions of being informed and actual knowledge scores [17]. While these findings may lead to the conclusion that surgeons do a poor job of informing patients, they also highlight challenges in the process of disclosure of information to patients. Conversely, while surgeons are considered an important source of information for the consent process, a majority of patients seeking elective surgery may have already decided on whether they want the procedure done prior to even meeting their surgeon [18].

The content and methodology of informed consent have led both patients and physicians to be dissatisfied with the process of consent [19,

20]. It has been argued that the eras of paternalism and patient autonomy have led to a general dissatisfaction of the doctor-patient relationship by both stakeholders and that the current era of "bureaucratic parsimony," or "shared decision-making," is appealing because it fosters both autonomy and collegiality in the decision-making process [6]. This newer paradigm appears to require clinicians to relinquish their role as a sole authority, but rather than give up their expertise, they must train to become more effective coaches for their patients [21]. To further explore the process of informed consent in the context of a shared decision-making framework, the process of coaching the patient will be examined as two parts: (1) informing the patient and (2) getting consent [20].

Informing the Patient

Informing the patient can be divided into three areas that need to be met during the informed consent process: (1) surgeon disclosure, (2) assessing patient understanding, and (3) shared decision-making [22]. In clinical practice, these stages may overlap and not be distinct and can happen over time and over multiple encounters (Box 4).

Box 4 Process of informed consent [20, 22]
- Informing the patient:
  - Surgeon disclosure
  - Assessing patient understanding
  - Shared decision-making
- Obtaining consent

The first part of informing the patient is surgeon disclosure. One manner of disclosure, particularly for emergent procedures, may involve a rote listing of indications, risks, benefits, and alternatives, all while the patient is reviewing and signing a consent document. Ideally, the act of disclosing information to the patient should take place as early as possible in the clinical setting in order to allow the patient time to reflect on the given information, allowing the patient to develop and ask pertinent questions. Certainly, this ability to properly disclose may be influenced by the severity and acuity of the disease process, as well as confounding medical issues on mental capacity that can pose a requirement for a surrogate decision-maker.

The American College of Surgeons recommends an informed consent discussion to include the contents of Box 5 as conducted by the surgeon and without exaggeration, promises, or guarantees [23]. The act of disclosure by the surgeon can be challenging to the surgeon and overwhelming to the patient because of the breadth and complexity of the information that may need to be delivered. Whether in an elective or acute setting, it may be impossible to discuss every facet of the procedure with the patient, so the focus has to be on the most important values and interests as determined by the patient and the physician together. The discussion should almost always include the diagnosis, an explanation of the procedure, a discussion of the major risks and benefits of the procedure, and alternatives including nonsurgical management or nonintervention. Disclosure can also include detail on other topics that could be relevant to the patient, such as prognosis depending on treatment choice, change in functional status after treatment, side effects of treatment, and expected postoperative course.

> **Box 5 American College of Surgeons recommendations for informed consent discussions**
> - The nature of the illness and the natural consequences of no treatment.
> - The nature of the proposed operation, including the estimated risks of mortality and morbidity.
> - The more commonly known complications, which should be described and discussed. The patient should understand the risks as well as the benefits of the proposed operation. The discussion should include a description of what to expect during the hospitalization and posthospital convalescence.
> - Alternative forms of treatment, including nonoperative techniques.
> - A discussion of the different types of qualified medical providers who will participate in their operation and their respective roles.

The language used during disclosure is important, and if the physician is not thoughtful, they can unintentionally coerce the patient into making a specific decision. The goal is to be as objective as possible while delivering information to the patient and to try to avoid personal opinions until after disclosure is complete or unless the patient asks for them specifically. As an extreme example, telling a patient "you may die horribly without an operation" can be construed as manipulative and may negatively impact the decision-making process, particularly by not providing enough information for a decision to be made relevant to the patient's values. As detailed by Dr. Schwarze and colleagues, it may be of preferable to provide the best-case and worst-case scenarios for both surgical decisions and alternative options, disclosing a range of uncertainty with anticipated good and bad outcomes for each treatment choice [24]. Providing contextual details of how a surgical procedure may very likely lead to survival with associated disability and dependency, versus observation leading to a large possibility of death in the comfort of family, and a smaller possibility of continued independent status, a patient may elect to choose the nonsurgical option if they value independence over quantity of life.

Disclosure of objective findings, such as "a third of people may have complications that affect sexual function and urinary continence, however choosing this type surgery ultimately leads to a cure in a greater number of patients than the other options," may provide information relevant to making an autonomous decision depending on if a patient values quantity of life over sexual function. However, it can be helpful to put additional context and significance to numbers by explaining "10–15% of patients may develop a pancreas leak, but the majority of times that happens it means a drain may need to be left in a couple weeks after you are discharged from the hospital. It usually does not mean another operation." In addition, contrasting short-term and long-term outcomes may be of value for patients to make a decision: "while overall survival is identical at 20 years, the recovery from choosing breast conserving surgery combined with radiation is typically less involved than mastectomy with reconstruction. A lumpectomy is a relatively quick and lower-risk outpatient procedure…." It may also be necessary to counteract framing bias by presenting the data in both directions: "one out of five people develop a recurrence within five years after this procedure; that means that four out of five do not."

The word "doctor" is derived from the Greek *docere*, meaning "to teach." An essential component of disclosure requires the surgeon to teach their patient as much as practical about their disease process and how surgery and other therapies may influence their disease course. A common pitfall comes from the overuse of medical jargon. It becomes crucial that simple and easy to understand language is used to ensure comprehension, especially because patients do not typically stop physicians to ask for clarification. Drawings and illustrations may be more effective methods of describing surgical procedures and educating the patient than using medical terminology. Decision-making tools, videos, and pamphlets may also assist with provision of relevant information to make an informed choice depending on the complexity of the disease process and procedure.

The second stage of informing the patient is assessing patient understanding, where beyond disclosure of information to the patient, patients should understand the information that is provided to them. One of the best ways to assess this is by reflecting on the types of questions the patient is asking [22]. When patients ask questions suggesting an incorrect understanding of the information, or are reluctant to ask questions, the physician should ask probing questions to clarify any misunderstanding. The goal is to encourage patient participation in an open dialogue about the current situation and the choices available.

A useful way to assess patient comprehension is the repeat-back method [25]. This requires the patient to use their own words to tell you what they understand about the procedure and can simply be performed by asking the patient to explain the procedure to you or one of their family members. A recent multicenter, randomized controlled trial showed that adding the repeat-back method significantly improved patient comprehension with no differences in patient anxiety or satisfaction, and it only added about 2.5 minutes to the time spent by the provider [25]. Another recent study examined which factors predicted comprehension during informed consent and found that total consent time and use of the repeat-back method were both strong predictors of patient comprehension [26].

The last stage of informing the patient is shared decision-making. Here, the patient must analyze the information and discuss their goals with the physician in accordance with their preferences and values. There are three primary components of shared decision-making: (1) the sharing of information between parties, (2) the surgeon offering options and then describing their risks and benefits, and (3) the patient expressing his or her preferences and values [20]. Shared decision-making is best facilitated when the physician acts less like an authority and more like a coach or partner in making the decision. Physicians should ideally do their best to avoid making therapeutic decisions for the patient whenever possible, unless specifically requested to do so.

While surgeons are at liberty to make modifications in technique during a procedure based on operative findings and use an approach based on their training, sometimes disclosure of differences in approach such as anterior versus posterior spinal fusion, or the use of minimally invasive and robotic techniques as opposed to open, may be of material value to the patient in their choice of pursuing an operation and even a particular surgeon. It is also important to note that patients often need time to reflect, process information, and make an informed choice and that they may also value the opinion of their primary care doctor in addition to the perspective of the surgeon. While this is certainly not realistic in emergencies or cases in which urgent care is required, physicians should ideally limit having extensive discussions on the same day that the consent document is being signed in order to allow adequate processing of information. Patients can sometimes be reassured by intentionally reminding them that they are at liberty to change their mind after thinking about things, even on the day of their operation.

Obtaining Consent

The second part of the informed consent process is obtaining consent for treatment. As previously described, having a patient sign a legally required form does not imply that they understand the decision or truly underwent informed consent, but fulfills a secondary requirement for documentation that consent has taken place. The form must be filled out correctly and signed after the patient has been informed. Ideally, the surgeon should confirm on the day of surgery that the patient has no additional questions prior to proceeding with surgery.

There are several minimum components that the informed consent document must have to be valid from a legal standpoint [9, 22]. The document must have a clear description of the planned procedure, anticipated benefits, material risk, and alternative therapies, including the options of nonintervention or observation. In addition, there should also be accompanying documentation supporting the conversation(s) with the patient, including notation that the patient had the opportunity to ask any questions they might have. This may be described in a clinic note stating "all the patient's questions were answered to their satisfaction" [22]. The last component of the consent document is authorization with the signature of the patient or the surrogate decision-maker and confirmation of patient authorization with the signature of the physician and/or a witness.

The process of informed consent is an important contract establishing an essential fiduciary relationship between the provider and patient. It promotes patient autonomy while fostering the doctor-patient relationship and advancing that unique bond. Providers will not only fulfill the important ethical and legal requirements for obtaining informed consent by putting recommendations into practice, but by placing respect toward the process of informed consent, patients may understand their treatments better, have greater satisfaction in their care, and put more trust in their physicians.

Challenges in Informed Consent

There are several situations in which the process of obtaining consent remains challenging. One situation is when a previously consented patient decides that they no longer want treatment. This can occasionally be perceived as a frustrating situation, and care must be taken in how surgeons respond to patients in such circumstances. Patients have the right to refuse treatment, even if they have previously given their consent to treatment [23]. It may be helpful to explore the reasons behind the patient changing their decision as it can offer insight into the patient's thought process or on occasion may confirm that they do not have capacity for medical decision-making. It is important to articulate that refusing treatment does not imply that the patient will lose care and support, but efforts should be made to inform the patient of any expected implications of treatment refusal. Depending on the circumstance, the physician should also make sure that the patient understands that refusing treatment now may or may not necessarily preclude having the procedure done at a later time.

Added complexities are present when treatment refusal occurs from the parents and/or guardians of minors. Here, a best interest standard may be considered if parental refusal negatively impacts child welfare. In these situations, courts may exercise power under the doctrine of *parens patriae*, which allows state interference to protect a child's interests over parental autonomy [27]. However, courts have been conflicted on how to approach adolescent refusal of care, ranging from allowing a teenage Jehovah's witness with leukemia to refuse blood transfusions based on religious belief to requiring chemotherapy after a teenager with leukemia refused it based on side effects, because of a high prospect of overall benefit [28]. The "mature minor" doctrine where minors may have common-law rights to refuse medical treatment has only been accepted by a few states so this is subject to local practice [27].

Another challenge of provision of informed consent is where the patient is simply misinformed. Sometimes a patient presents with a preconceived notion on what is the right clinical option for them, often not supported by data, but sometimes supported by the opinions of friends, family, or even other clinicians. In these scenarios, anxiety may also play into the decision-making process. For example, in the case of patients committed to contralateral prophylactic mastectomy for non-hereditary breast cancer, sometimes the choice is related to a false belief that breast cancer can spread from one breast to another, or that mastectomy is a low-risk procedure that can completely eliminate risk of cancer recurrence, or that patients with mastectomies no longer require surveillance. Occasionally, spending time to inform the patient can help educate the patient about the risk-benefit profile of surgical options, but sometimes the surgeon and patient have to negotiate a treatment plan that is satisfactory to both, with the surgeon often accepting the patient's choice for reasons of "symmetry" or "peace of mind" [29]. Indeed, a surgeon is not obligated to manage care if the patient agrees to an operation conditionally or makes demands that are unacceptable to the surgeon [23]; however, in such situations, it is probably best to refer the patient to another surgeon who will be willing to provide a second opinion on the case.

Other commonly encountered situations are when the patient does not have the capacity to make medical decisions. The legal standards for having decision-making capacity can vary based on jurisdiction, but generally follow four criteria

where the patient is able to (1) communicate a choice, (2) understand the relevant information, (3) appreciate the medical consequences of the situation, and (4) reason about the treatment options (Box 6) [30]. All of these criteria must be met for a patient to be considered competent to consent for medical treatment.

> **Box 6 Requirements for establishing decision-making capacity** [28]
> - Patient must be able to:
>   – Communicate a choice.
>   – Understand the relevant information.
>   – Appreciate the medical consequences of the situation.
>   – Reason about the treatment options.

In many instances, the acute and chronic medical problems of our patients can suddenly or gradually result in diminished capacity for decision-making. A study of medical inpatients with acute conditions estimated that as many as 48% were incompetent to consent to medical treatment [31]. Of interest was not how many patients had diminished capacity but that the clinical team responsible identified impaired decision-making capabilities in only a quarter of this cohort [31]. While the vast majority of people are capable of making their own decisions, diminished capacity can be very common in surgical patients. When a patient has diminished capacity, it is important that we do not default to calling them incompetent, but that we should engage patients to determine their level of understanding. A sliding scale approach was endorsed by the President Commission for the Study of Ethical Problems in Medicine and Biomedical and Behavioral Research, where, in practice, only patients with impairment at the far left end of a performance bell curve should be considered incompetent [32]. Ultimately, the stringency of the test used to determine capacity should correlate with the severity and consequences of the likely outcomes of the patients' decisions.

Assessment of Competence

Physicians can determine if patients have the capacity to participate in shared decision-making, but the determination of competency is a legal issue that requires either psychiatry consultation or a judicial process. If this is required to clarify ambiguity or inconsistency in a patient's thought process, the physician should be as candid as possible with the patient and let them know that a psychiatry consult will be considered. With that being said, the goal of a psychiatry consult should ultimately be to improve the patients' capacity and not to simply affirm that the patient needs a proxy [32]. If possible, providers should always first try to identify and correct reversible causes of impairment. If a patient is deemed incompetent, then a surrogate decision-maker must be found, with the ultimate goal to respect the patients previously described wishes and values through a process of substituted judgment. First the provider should look for any advance directives the patient might have made displaying choices on treatment or prior documentation of a proxy. The hierarchy of possible decision-makers can vary based on state, but they generally follow the same guidelines. A durable power of attorney for healthcare decisions as previously designated by the patient when they were competent will always take priority in this hierarchy. If the incompetent patient did not previously appoint a proxy, then responsibility typically goes to the spouse, then adult children, then siblings, and finally to other family members or close friends. If none of these are available to make decisions for the patient, then the courts can appoint a surrogate for them, or a best interest standard can be followed if lifesaving decisions must be made emergently.

There are instances when the provider and proxy disagree on what the patient would have wanted. While many informed consent discussions using surrogate decision-makers often focus on acute and emergent indications that harbor a threat for life, there are several elective procedures where discussions may become particularly contentious as they may not improve quality of life but may otherwise be considered life-extending, including tracheostomy for prolonged ventilator support and gastrostomy for long-term feeding support. If thorough discussion cannot achieve resolution of conflict, then the physician should consider ethics consultation to help clarify goals of care and gain understanding of family dynamics relevant to making decisions that respect a patient's previously expressed wishes. There is evidence that ethics consultation services can effectively build consensus in disagreements regarding perceived nonbeneficial treatments and are a valuable resource [33]. Unfortunately, physicians with the least training in ethics are also the least likely to have access to an ethics consultation service [34].

Disclosure of Surgeon Experience

Disclosure of surgeon experience and procedure-specific outcomes is an area of ambiguity, as there is marked variability in the performance characteristics of procedural learning, as well as possible distributive justice issues related to distance and scarcity of surgeons, particularly in rural regions. The question of whether surgeons are legally required to disclose their experience for certain procedures is unclear. Legal precedent on this issue remains unsettled, as to date there have been two state Supreme Court cases that examined issue, with both courts coming to opposite conclusions [35, 36]. In *Johnson v. Kokemoor* (1996), the legal standards of informed consent were expanded to include providing a surgeon's performance data if considered material to the decision-making process, while *Duttry v. Patterson* (2001) indicated that a physician's prior experience is outside the scope of an informed consent claim. While it remains unclear whether there is an ethical obligation toward blanket disclosure of surgical experience, if prompted by the patient, it should be considered fundamental to uphold professional standards with respect to truth-telling [13]. This obligation to disclose becomes even more important in highly complex procedures with greater associated risks where the surgeon has limited experience. In these cases it may be considered appropriate and material to disclose this information and offer the patient a referral to a surgeon with more experience if it is desired.

The issue of disclosing surgeon-specific performance ratings, or public "report cards," is even more contentious, as these report cards have been criticized for being overwhelming, confusing, and misleading to health consumers [37]. Disparate outcomes from surgeons may not be controlled for patient comorbidities, referral patterns, and team characteristics and may not be reflective of the expertise and technical skill of an individual surgeon [13]. At this time many experts believe that since surgeon-specific performance data are currently inaccurate and misleading, there is no ethical obligation to disclose them as part of the informed consent process [38–40]. If the accuracy and applicability of these statistics improve in the future, then a case could be made that we do have an ethical duty to disclose it in order to minimize patient harm. Even if we do improve our data on surgeon-specific performance to the point of relevance, one can question if we should put the burden of selecting the "best" surgeon on our patients who may not have complete autonomy in making all decisions. Competing with autonomy is justice, and patients still may have to judge the relative degree of importance of surgeon skill with issues with access related to distance, as well as containing the cost of their deductible by staying within "in-provider" insurance networks.

While the goal of reporting surgeon-specific performance ratings is ultimately to improve the quality of surgical care and enhance patient autonomy, it is questionable whether this actually enhances autonomous decision-making [39]. While sharing of data within peer groups to improve performance rates is currently justifiable from a quality and process improvement stance, as well as increasing transparency of data provided at an institutional level, it is unclear of the benefit unadjusted surgeon-specific outcomes data provides to patients. Schwarze argues that disclosure of performance ratings would be similar to mandated disclosure of flight disasters faced by individual pilots, and would provide an excess burden on the consumer to make a deci-

sion that may not be feasible or even relevant to their future, and would certainly question the ability of the airline industry to self-police [39]. Using comparative data in an unadjusted fashion, it becomes unclear if a heart surgeon has a high complication rate because of poor technique, or because this particular surgeon takes the hardest cases, or because he or she is readily available as backup to salvage complications from a particularly aggressive interventional cardiology group. An expert pancreas surgeon may have a higher complication rate than a novice surgeon who has merely gotten away with a few well-selected cases without any morbidity. Individual statistics unadjusted for volume or stratified by comorbidities may not tell the whole story in a straightforward fashion.

Surgeons cannot know what their true outcomes are in order to compare to others with a reasonable degree of certainty, as point estimates of outcomes like mortality rates are simply estimates of a true value, which actually lies somewhere within a very wide confidence interval for any individual [40]. Furthermore, application of prediction models developed at a population level cannot reliably be applied to individual surgeons – this is otherwise considered an ecological fallacy. Of interest, patients may not even be influenced by performance data if available, but instead may be swayed more by the good opinion of their referring physician [38].

For disclosure of surgeon experience to be material, it is essential that there is a potential for an autonomous decision that can be made [39]. The disclosure of surgeon-specific data is probably most relevant for highly complex procedures and rare or unusual disease processes, but because the procedures are less frequently performed, the error bars on point estimates will be great, so there will be greater uncertainty in the data. Unfortunately, while there is data supporting relationships of hospital and surgeon volume and in-hospital mortality for pancreatectomy and esophagectomy [41, 42], it is unclear if socioeconomically deprived patients in rural locations can always choose to go to a high-volume provider as opposed to a medium- or low-volume provider, which underscores a conflict of autonomy with access to care. There are many regions where there may only be one expert surgeon within a several hour radius that frequently performs such complex procedures. It may then be material for a surgeon offering complex procedures to disclose to a patient their own level of training and case experience, along with outcomes if they are even known (or knowable), and allow the patient to decide if he or she wish to travel to another center or stay closer to home. These are very challenging questions, and it is unclear if there is a right answer, but a duty to respect the patient's ability to decide for themselves should prevail.

Forecasting

One can argue that the ability to adequately provide informed consent requires data, experience, wisdom, and an uncanny ability to look into the future. In making good predictions, contemporary forecaster Nate Silver endorses gaining a solid understanding of the accuracy, honesty, and value of a forecast [43]. During the process of informed consent, it may be appropriate for the surgeon to not only disclose risk but also to articulate the level of uncertainty surrounding risk estimates, especially if there is greater system complexity surrounding disease process and technique [13]. As advocated by Schwarze and colleagues, providing estimates and error bars or ranges for outcomes of both the surgical intervention and alternative using the best-case/worst-case model may be an effective tool to inform patients of their options and help them synthesize a plan as these can then be aligned with personal goals and values [24]. In the process of prognostication and disclosure, it is essential for the surgeon to be intellectually honest and self-reflective; understand their own limitations, biases, and conflicts of interest; and potentially seek assistance of those with greater experience or the counsel of multidisciplinary teams, especially when risk is high and personal experience is low. It is also important to provide information that will be of clear value to the patient, including personal performance data if that information may help them lead to a more informed decision.

Humility and intellectual honesty in the process of disclosure may help engender trust in the surgeon and can only help foster the surgeon-patient relationship for the better.

Concluding Remarks

- The process of informed consent within a framework of shared decision-making requires a surgeon to effectively communicate with patients the indications, risks, benefits, and alternatives of a procedure and ensure that the patient comprehends and synthesizes salient issues in light of their own goals and values [23].
- The reasonable person standard established the importance of disclosure of material risks or what a reasonable person would want to know given similar circumstances [10].
- The best-case/worst-case model may be an effective tool to inform patients of their options and help them synthesize a plan as these can then be aligned with personal goals and values [24].
- Surgeons are not currently obligated to provide patients with raw numbers of operations done or how their results compare with those of others, as doing so could misinform and mislead patients [39, 40].

Glossary

Best-case/worst-case model Tool to inform patients of estimates and ranges for outcomes of both a surgical intervention and an alternative in order to help them synthesize a plan in alignment with personal goals and values.

Best interest standard Process of making healthcare decisions with an intention to minimize harm and maximize benefit to a patient when there is no available surrogate decision-maker to allow for substituted judgment.

Community practice or professional standard Relies on what other local practitioners deem appropriate for disclosure.

Decision-making capacity Requires the patient to be able to (1) communicate a choice, (2) understand the relevant information, (3) appreciate the medical consequences of the situation, and (4) reason about the treatment options.

Doctrine of parens patriae Allows state interference to protect a child's interests over parental rights to refuse care.

Durable power of attorney for healthcare decisions Surrogate decision-maker who was previously designated by the patient when they were competent. Takes priority in the hierarchy of possible decision-makers.

Fiduciary duty Highest standard of care, where a person holds a legal or ethical relationship of trust and responsibility to act on the behalf of another party.

Informed consent A process of disclosure of risks, benefits, and alternatives of treatment decisions.

Material risk Risk when a reasonable person, in what the physician knows or should know to be the patient's position, would be likely to attach significance to the risk or cluster of risks in deciding whether or not to forego the proposed therapy.

Mature minor doctrine Situation where minors may have common-law rights to refuse medical treatment.

Medical paternalism Attitude and practice where a physician decides what is best for the patient; may compete with autonomous decision-making by the patient.

Reasonable person standard Disclosure of what a reasonable patient would want to know under given circumstances.

Repeat-back method Requires the patient to use their own words to tell you what they understand about the procedure; assesses patient comprehension during informed consent.

Shared decision-making Framework of doctor-patient relationship requiring (1) the sharing of information between parties, (2) the clinician offering options and then describing their risks and benefits, and (3) the patient expressing his or her preferences and values.

Subjective standard Rely on risks disclosed that are pertinent to an actual patient's decision to accept therapy.

Surrogate decision-maker Has authority to act on behalf of a patient's previously described wishes and values when a patient lacks decision-making capacity.

Substituted judgment Process of acting on behalf of a patient's previously described wishes and values.

References

1. Hippocrates of Cos. Hippocrates, Jones WHS, Withington ET, Potter P, Heraclitus of Ephesus. Cambridge: Harvard University Press; 2015: 299–301.
2. Bagwell CE. Respectful Image: Revenge of the Barber Surgeon. Ann Surg. 2005;241:872–8.
3. Ghosh SK. Human cadaveric dissection: a historical account from ancient Greece to the modern era. Anat Cell Biol. 2015;48(3):153–69.
4. Gawande A. Two Hundred Years of Surgery. N Engl J Med. 2012;366:1716–23.
5. Pellegrini CA. Trust: The Keystone of the Patient-Physician Relationship. J Am Coll Surgeons. 2017;224(2):95–101.
6. Siegler M. The progression of medicine: from physician paternalism to patient autonomy to bureaucratic parsimony. Arch Int Med. 1985;145:713–5.
7. Schmitz D, Reinacher PC. Informed consent in neurosurgery – translating ethical theory into action. J Med Ethics. 2006;32:497–8.
8. Schwarze ML, Bradley CT, Brasel KJ. Surgical "buy-in": The contractual relationship between surgeons and patients that influences decisions regarding life-supporting therapy. Crit Care Med. 2010;38(3):843–8.
9. Cocanour CS. Informed consent – it's more than a signature on a piece of paper. Am J Surg. 2017;214:993–7.
10. Canterbury v. Spence, 464 F.2d 772 (D.C. Cir. 1972).
11. Aristotle. The Nicomachean Ethics, 2nd ed. Irwin T., (transl.) Indianapolis: Hackett Publishing Co.; 1984.
12. Aristotle. Aristotle's Nicomachean Ethics, Bartlett RC, Collins SD (transl.) Chicago: The University of Chicago Press; 2011.
13. Ganai S. Disclosure of surgeon experience. World J Surg. 2014;38:1622–5.
14. Ankuda CK, Block SD, Cooper Z, Correll DJ, Hepner DL, Lasic M, Gawande AA, Bader AM. Measuring critical deficits in shared decision making before elective surgery. Patient Educ Couns. 2014;94:328–33.
15. Uzzaman MM, Sinha S, Shaygi B, Vitish-sharma P, Loizides S, Myint F. Evaluation of patient's understanding and recall of the consent process after open inguinal hernia repairs. Int J Surg. 2012;10:5–10.
16. Scheer AS, O'Connor AM, Chan BP, Moloo H, Poulin EC, Mamazza J, Auer RC, Boushey RP. The myth of informed consent in rectal cancer surgery: what do patients retain? Dis Colon Rectum. 2012;55:970–5.
17. Sepucha KR, Fagerlin A, Couper MP, Levin CA, Singer E, Zikmund-Fisher BJ. How does feeling informed relate to being informed? The DECISIONS survey. Med Decis Mak. 2010;30:77S–84S.

18. Hall DE, Morrison P, Nikolajski C, Fine M, Arnold R, Zickmund SL. Informed consent for inguinal herniorrhaphy and cholecystectomy: describing how patients make decisions to have surgery. Am J Surg. 2012;204:619–25.
19. Faden RR, Becker C, Lewis C, Freeman J, Faden AI. Disclosure of information to patients in medical care. Med Care. 1981;19:718–33.
20. Katz J. Reflections on informed consent: 40 years after its birth. J Am Coll Surg. 1998;186:466–74.
21. Barry MJ, Edgman-Levitan S. Shared decision making— the pinnacle of patient-centered care. N Engl J Med. 2012;366:780–1.
22. Childers R, Lipsett PA, Pawlik TM. Informed consent and the surgeon. J Am Coll Surg. 2009;208:627–34.
23. American College of Surgeons Statements on Principles. Section IIA. Revised 12 Apr 2016. Accessed at https://www.facs.org/about-acs/statements/stonprin#iia
24. Taylor LJ, Nabozny MJ, Steffens NM, Tucholka JL, Brasel KJ, Johnson SK, Zelenski A, Rathouz PJ, Zhao Q, Kewekkeboum KL, Campbell TC, Schwarze ML. A framework to improve surgeon communication in high-stakes surgical decisions: Best Case/Worst Case. JAMA Surg. 2017;152(6):531–8.
25. Fink AS, Prochazka AV, Henderson WG, et al. Enhancement of surgical informed consent by addition of repeat back: a multicenter, randomized controlled clinical trial. Ann Surg. 2010;252:27–36.
26. Fink AS, Prochazka AV, Henderson WG, et al. Predictors of comprehension during surgical informed consent. J Am Coll Surg. 2010;210:919–26.
27. Wooley S. Children of Jehovah's Witnesses and adolescent Jehovah's Witnesses: what are their rights? Arch Dis Child. 2005;90:715–9.
28. Blake V. Minors' Refusal of Life-saving Therapies. Virtual Mentor. 2012;14(10):792–6.
29. Angelos P, Bedrosian I, Euhus DM, Hermann VM, Katz SJ, Pusic A. Prophylactic mastectomy: Challenging considerations for the surgeon. Ann Surg Oncol. 2015;22(10):3208–12.
30. Appelbaum PS. Assessment of patients' competence to consent to treatment. N Engl J Med. 2007;357:1834–40.
31. Raymont V, Bingley W, Buchanan A, David AS, Hayward P, Wessely S, Hotopf M. Prevalence of mental incapacity in medical inpatients and associated risk factors: cross-sectional study. Lancet. 2004;364:1421–7.
32. President's Commission for the Study of Ethical Problems in Medicine and Biomedical and Behavioral Research. Making health care decisions: a report on the ethical and legal implications of informed consent in the patient-practitioner relationship, vol. 1. Washington, DC: Government Printing Office; 1982.
33. Nelson R. Nonbeneficial Treatment and Conflict Resolution: Building Consensus. Perm J. 2013;17:23–7.
34. DuVal G, Clarridge B, Gensler G, Danis M. A national survey of U.S. internists' experiences with ethical dilemmas and ethics consultation. J Gen Intern Med. 2004;19:251–8.
35. Johnson v. Kokemoor, 545 NW 2d 495 (Wis. Supreme Court 1996).
36. Duttry v. Patterson, 771 A.2d 1255 (Pa. Super., 2001).
37. Schneider EC, Lieberman T. Publicly disclosed information about the quality of health care: response of the US public. Qual Health Care. 2001;10:96–103.
38. Burger I, Schill K, Goodman S. Disclosure of individual surgeon's performance rates during informed consent: ethical and epistemological considerations. Ann Surg. 2007;245:507–13.
39. Schwarze ML. The process of informed consent: neither the time nor the place for disclosure of surgeon-specific outcomes. Ann Surg. 2007;245:514–5.
40. Sade RM, Boan A. Accounting for outcomes: lies, damned lies, and statistics. J Thoracic Cardiovasc Surg. 2017;153:1212.
41. Birkmeyer JD, Siewers AE, Finlayson EV, Stukel TA, Lucas FL, Batista I, et al. Hospital volume and surgical mortality in the United States. N Engl J Med. 2002;346:1128–37.
42. Birkmeyer JD, Stukel TA, Siewers AE, Goodney PP, Wennberg DE, Lucas FL. Surgeon volume and operative mortality in the United States. N Engl J Med. 2003;349:2117–27.
43. Silver N. The signal and the noise: why so many predictions fail – but some don't. New York: The Penguin Press; 2012.

Suggested Literature

Appelbaum PS. Assessment of patients' competence to consent to treatment. N Engl J Med. 2007;357:1834–40.
   A case vignette that provides strategies on how to assess decision-making capacity.
Childers R, Lipsett PA, Pawlik TM. Informed consent and the surgeon. J Am Coll Surg. 2009;208:627–34.
   An overview of ethical requirements for provision of informed consent.
Pellegrini CA. Trust: The Keystone of the Patient-Physician Relationship. J Am Coll Surgeons. 2017;224(2):95–101.
   A framework for communication between patients and physicians that that help establish a bond of trust.
Taylor LJ, Nabozny MJ, Steffens NM, Tucholka JL, Brasel KJ, Johnson SK, Zelenski A, Rathouz PJ, Zhao Q, Kewekkeboum KL, Campbell TC, Schwarze ML. A framework to improve surgeon communication in high-stakes surgical decisions: Best Case/Worst Case. JAMA Surg. 2017;152(6):531–8.
   A framework to improve surgeon communication in high-stakes surgical decisions.

The Pediatric Patient as a Self-Individual and Decision-Maker

Rosa Angelina Pace, Susana Ciruzzi, and Alberto R. Ferreres

Key Points

- Decision-making is the process and logic rationality through which individuals arrive to a decision.
- The novelty of the minor as a participant in the health and surgical decision-making comes from the fact of considering the minor as an active citizen and a subject of rights.
- The medical decision-making process is a multilateral process that is shared and mutually discussed. In the pediatric universe, this dialogue presupposes the active participation and a major role of the minor and his or her parents.
- Every minor should be offered the possibility to make decisions regarding the health surgical care.
- A new category has been developed in the health arena, which is that of the mature minor.
- Nonetheless there are moral and practical reasons for displaying caution when the minor and parents disagree.

R. A. Pace (✉)
Ethics Committee Italian Hospital,
Buenos Aires, Argentina
e-mail: rosina.pace@hospitalitaliano.org.ar

S. Ciruzzi
Ethics Committee "Prof Juan P Garrahan" Hospital and "Dr Alfredo Lanari" Institute, University of Buenos Aires, Buenos Aires, Argentina

A. R. Ferreres
Department of Surgery, University of Buenos Aires, Buenos Aires, Argentina

Department of Surgery, University of Washington, Seattle, WA, USA

Decision-making is the process and logic rationality through which individuals arrive at a decision [1]. In recent years there has been a great interest in the decision-making processes in the health-care area, including the field of childhood. The increasing interest in giving children a voice in decisions and services for them has accompanied the emergence of a new conception of the child as an active citizen. The participation of the sick minor in this process is more than just asking them for their ideas and views. It is about listening to them, taking them seriously, and turning their ideas and suggestions into reality. It is also about providing them with the ability to influence some of the issues and circumstances that affect them and at the same time helping adults understand children's issues through their lens [2]. This movement toward respect for the decisions of minors has had a strong impact on medical care. Generally speaking, a pediatric patient or

© Springer Nature Switzerland AG 2019
A. R. Ferreres (ed.), *Surgical Ethics*, https://doi.org/10.1007/978-3-030-05964-4_21

minor is an individual who has not reached the legal age of majority (in most countries, 18 years of age), an adolescent is usually defined as that individual between 13 and 18 years of age, a child refers to ages 1–12 years, and an infant is the one in his or her first year of life.

The Complexity of the Medical Decision-Making

Health outcomes are probabilistic; most decisions are made under conditions of uncertainty [3]. The medical decision-making is a multilateral process, shared, discussed, and dynamic, in which two fundamental actors participate: the health team and the patient. This interaction between those who hold scientific knowledge and those who hold the right to life, health, and death is not without tensions [4]. On the one hand, the physician and the health team as a whole must recognize that the patient is the one who holds the final decision in this process: he or she accepts or refuses the therapeutic proposal, and on the other hand, the patient must admit that the doctor is the one who is better prepared to assist and guide him or her in taking the most appropriate, suitable, and correct decision in their own benefit.

However, it is in the field of pediatrics that this relationship becomes even more complicated. This patient is no longer an individual who is fully capable from the legal standpoint and whose full autonomy is debatable. The minor is a vulnerable patient, sometimes immature, and the patient-physician relationship is no longer between two individuals, but the role of a third party, whether their parents or those with parental responsibility, carries a preponderant role.

In this fashion, the medical and surgical decision-making process generates a permanent tension among adults who are inclined toward the protection and care of the pediatric patient, not taking into consideration the autonomy of the minor, who, in some cases, is not willing or anxious regarding the medical decision concerned with his or her own well-being and issues not addressed by the adults. There may be also ten-

sion between the parental responsibility, which involves the set of rights and obligations tending to the protection, care, and development of their children, which implies making decisions on their behalf, and, on the other hand, the duty to create the space for the exercise of the autonomy of the child. Ross highlights the fact that the parents should facilitate not only the present-day autonomy but also the long-term autonomy, in the sense to bear in mind the full well-being and survival of the minor [5].

Health in the Pediatric Universe

A multidimensional concept has been added to the traditional definition of health endorsed by the World Health Organization ("the state of complete physical, mental and social well-being, and not only the absence of disease," Alma Ata, 1946) when pediatric patients are treated. Health outcomes in children must consider the ability of them to fully participate in the appropriate physical, social, and psychosocial activities for each age [6].

The adequate social and emotional development includes the minor's experience, expression, and management of emotions as well as the ability to establish positive and rewarding relationships with others. It encompasses both intra- and interpersonal processes. The core features of emotional development include the ability to identify and understand one's own feelings, to accurately detect and comprehend emotional states in others, to manage strong emotions and their expression in a positive way, to regulate one's own behavior, to develop empathy for others, and to establish and maintain relationships [7].

Infants and children experience, express, and perceive emotions before they are fully capable of understanding them. In learning to recognize, label, manage, and communicate their emotions and to perceive and attempt to understand the emotions of others, children build skills that connect them with family, peers, teachers, and the community. These progressive and growing capacities help children to become competent in negotiating increasingly complex social interactions, to participate effectively in relationships and group activities, and to reap the benefits of social support crucial to healthy human development and functioning.

Therefore, a series of dimensions are included related to the ability to perform daily activities (mobility and personal care) and cognitive acquisitions (memory, recall, concentration, and learning, among others) and develop emotions (positive and negative), self-perception, and interpersonal relationships with friends and family and with the environment [7]. McCormick defines it as the capacity for human relations, associated with the condition of the child. Following the Judeo-Christian tradition, life is not a value to be preserved in and for itself: it is a relative good – a value to be preserved insofar as the higher and spiritual purposes of life are attainable. The duty to preserve physical life, as well as the limits of that duty, is based on the possibility of achieving these values. Life is a value to be preserved only insofar if it keeps some potentiality for developing and nurturing human relationships. McCormick applies this concept and value theory even to defective newborn children, being the criteria to decide which efforts should be done and which should be precluded to sustain the child's survival and well-being and the potential for human relationships associated with the infant's condition [8].

Health care is an area where the welfare and development of the minor's potential are maximized even when suffering from chronic and/or serious diseases. In order to achieve these goals, the focus is not placed in prolonging life in its biological expression but on the so-called biographical life: the meaning that life has for a particular individual, adult, or minor, placing content and value on it. Quality of life is not just a measure of the concept of health, but refers to a dimension collecting a unique and very personal perception of all the elements making up the personality and reflecting the individual state experienced by the patient about their health and the medical aspects of life. The factors that affect it are many, but the state of consciousness is unique and must collect all the others connecting them

with the personal identity, which is the one that will provide the global dimension. Good health is, therefore, a complex and harmonious result of a set of parameters and conditions, which can turn into bad health by the single failure of one of its elements. Each patient is one and different, making real the old aphorism "there are no diseases but patients " [9].

Therefore, the quality of life associated with the right to health is a primarily evaluative concept and as such has a double perspective: objective and subjective. Objective determination focuses on what the individual can do, while subjective assessment includes the perception or personal estimation of living conditions, which translates into positive or negative feelings [10].

The Ethical and Legal Status of the Minor

In 1989 the United Nations adopted the Convention on the Rights of the Child (CRC), which changed the approach of childhood and adolescence [11]. The convention defines a child as any person under 18 years (otherwise a minor) and requires that childhood is recognized as a developmental period and that the laws must be developed in a manner consistent with the evolving capacities of the child (Article 5). As children grow and develop in maturity, their views and wishes must be given greater weight, and their development toward adulthood must be respected and promoted, making him or her a subject of rights. In that sense, the child is not only the holder of rights but is able to exercise them for him- or herself.

Those rights belong to the personal individual and are crossed by the concept of dignity. Dignity represents a supreme, absolute, and irreducible value, particular to the human and personal condition. Dignity is expressed by the individual freedom, nondiscrimination, and equality. In the particular case of a minor, it accounts for him or her to be valued and respected for their own sake as well as to be treated ethically [12].

For some, dignity is just a concept reducible to respect for autonomy; nonetheless the concept of dignity has played a significant role in the ethics world. It is worth remembering that the word has not either a Christian or Jewish tradition and was not used by Greek philosophers, such as Plato or Aristotle. Its use can be traced firstly to the Roman stoics, like Cicero and Seneca. Cicero defined it as "the honorable authority of a person which merits attention and honor and worthy respect" [13]. The Kantian concept of dignity exercises a powerful influence until the present times, his view of dignity is interlaced to one's humanity [14].

Sulmasy summarizes the three historical uses of the word "dignity" in the following classification [15]:

– Attributed dignity refers to the value that human individuals place upon others by acts of attribution, which represents a convention way of valuing others.
– Intrinsic dignity is linked to the value that human individuals possess by virtue and the sole fact that they are human beings.
– Inflorescent dignity means the value of other(s) in consequence of the human excellence and virtue.

These three facets of dignity suit the dignity of the minor. Since birth, the minor progress in the emerging of self-awareness as well as the perception of the outer world, which allows to know and relate to others. This human being is temporal and spatial and, in due time, acquires the needed skills to help him or her to face and handle the interrelation with the real world, according to the own interests.

The development of consciousness gives answer to the different questions presented in the evolution of an individual: who I really am, what is my identity, what my true personality looks like, and what is that makes me a singular subject and so different from others. The minor finds answers to these questions in a naturally spontaneous and progressive way, through the interaction between nature and upbringing, between the own innate abilities and what is perceived through the interaction with the environment. For this reason, the minor – as an individual in

constant evolution – possesses the ethical status of a person. This ethical status is the one which entails the possibility of being able to express his or her needs and desires and to act by his or her own [16].

The legal competence means that a subject may hold legal acts on their own and that they can perform binding actions according to their rights, duties, and obligations. In that sense a minor will endure restrictions depending upon the age; nonetheless in the health arena, regulations in some countries recognize the autonomy of those elder than 16 years. The legal competence to accept or refuse medical treatment requires the mental capacities to reason, think, rationalize, develop a set of values and goals, have a full insight of the internal and external circumstances, understand the information that is provided, and communicate a choice. When someone's right to accept or refuse treatment is jeopardized, it has to be decided if that individual's mental capacity is enough to hold legal competence and so their wishes should be respected [17]. But competence is a subject for the law as well as for morality [18]. In that sense, the ethical competence is a concept inserted in the domain of the exercise of personal rights and demands the necessary capacity to make those personal rights to life and health effective, allowing decisions regarding their care in their best interest. Ethical competence is not reached at a specific age or time, but it develops and evolves overtime and with the gradual acquisition of maturity; this situation represents the competence of a minor.

According to most western legal frameworks, the general assumption is that all adults are fully competent, both from the legal and ethical point of view. Below the age of majority (18 years in most countries), these assumptions are reversed. The notion of a valid informed consent is linked to the quality of judgment and, consequently, to that of competence.

The American Association of Pediatrics has endorsed the fact that older children and adolescents should be involved in their medical decision-making and consent process, favoring the concept of pediatric assent or refusal. Of course, the conflict will probably arise when they do not make what is considered the right choice [19]. It should be highlighted that the informed consent process in pediatrics involves two different, although linked, steps: providing information and obtaining proper consent, with the patients and their families and/or surrogates. It is interesting to stand out the fact the complexity of parents and surrogates' decisions regarding the care of a minor, which are influenced by many factors [20]. Parental decision-making should primarily be understood as the parents' responsibility to support the interest of their child, rather than being focused on their rights to express their own autonomous choices [5].

Ross focuses on the principle of respect to persons evolving from the Kantian tradition. When applied to the context of children, it is about respect for the person a child is, even if not a full Kantian person, and also about the person the child is becoming [21]. The ethical obligation of the health team is to ensure that the decision-making process in questions pertaining the health and well-being of a minor takes into account his or her voice in the sense expressed above and also to promote the progressive autonomy of the minor, adjusting to the understanding and competency.

The Informed Consent Process in Pediatrics

The doctrine of informed consent, specifically in the surgical field, has developed as a consequence of the supremacy of the ethical principle of respect for autonomy. It has roots within both law and ethics. The respect for autonomy conveys the possibility of self-determination, which represents a limit to the present medicalization of life. The shift from the preeminence of paternalism toward respect for autonomy was due to different events in the twentieth century, such as the distrust in the medical profession, ethical violations in research, and society's trend toward individualism.

The progressive maturation of the child precludes the increasing inclusion of the child's participation in the process. The 1995 American Academy of Pediatrics statement promoted the pediatric assent in the decision-making and informed consent process [American Academy of Pediatrics. Committee on Bioethics. Pediatrics 95:314–317, 1995]. Assent may not be assimilated to a full consent, but it should bear the following characteristics:

– Useful to develop an awareness regarding the disease status of the one affected
– Offering the pediatric patient a full view of his or her situation and cover the expectations
– Allow the health-care provider and/or the surgeon an evaluation of the patient's comprehension of the information that has been provided
– Requesting the will of the patient to accept the therapeutic proposal

The first step in this process is the determination whether the patient and his/her family and/or surrogates are able to understand the information disclosed by the physician or the health agent. It is generally understood that the lack of information represents a source of legal responsibility, since it prevents the patient and relatives from making a free choice in terms of treatment's acceptance or refusal.

In this way, the "disclosure of information, the evaluation and understanding of that information" (with reference to the patient's life experience and his value system) represents the central core of the doctrine of informed consent, to which it adds the freedom of the subject that decides, and the competence to consent.

The terms capacity and competence are frequently overlapped in clinical practice. As mentioned before, capacity refers to a clinical condition addressing the mental abilities; meanwhile competence is a legal determination, addressing society's interest in restricting decision-making when capacity is in question.

Another concept which originated in England and used in medical law is that of Gillick competence, which is used to decide whether a child under 16 years is able to consent to his or her own medical treatment without the need for parental permission or knowledge [22]. The standard is based on the 1985 decision of the House of Lords in "Gillick v West Norfolk and Wisbech Area Health Authority," where a mother of girls under 16 objected to the Department of Health advice that allowed doctors to give contraceptive advice and treatment to children without parental consent. The House of Lords held that a child under 16 had the legal competence to consent to medical examination and treatment if they had sufficient maturity and intelligence to understand the nature and implications of that treatment. This case is binding in the United Kingdom and has been adopted to different extents in Australia, Canada, and New Zealand.

This key principle is reflected in consent law applied to children, since children pass through three developmental stages of their journey to becoming an autonomous adult [23]:

(a) The child of tender years who relies on a person with parental responsibility to consent to treatment.
(b) The Gillick competent child under 16
(c) Minors of 16 and 17 years old who are able to consent to treatment as if they were full of age

The degree of maturity and intelligence needed and requested depends upon the gravity, seriousness, and necessary impact of the decision. A relatively young child would be considered to be sufficiently mature and intelligent and be competent to consent to a plaster on a small cut. In the same fashion, a child competent to consent to a dental treatment or the repair of a broken bone may lack competence to consent to more serious, invasive, or risk-prone procedures in order to treat more serious and life-compromising conditions. The Gordian knot behind the issue is if they are able to understand the treatment implications and future prognosis in case of refusal, because they feel overwhelmed by the burden of the decisions and lack the maturity to make them.

Since the Gillick competence is a functional ability to make a decision, it is task-specific: more complex procedures require greater levels of competence. When assessing Gillick competence, the physician will decide in each particular case whether the child is or not competent to make that particular decision in each particular case. It is not just an ability to choose where the child recognizes that there is a choice to be made and is willing to perform it (Fig. 1). Rather it should be considered as an ability to understand that there is a choice to be made, that that choice has consequences, and that they must be willing, able, and mature enough to make that choice. Health professionals and pediatricians should be satisfied that the child has a full understanding of the following facts:

- The need for a medical and/or surgical treatment and the reasons for it
- The risks and outcomes of the proposed treatment
- The alternative options to that therapy
- The full understanding of the consequences, risks, and prognosis of the refusal, delay, or nonacceptance of the proposal

Fig. 1 Gillick competency

Gillick competency

For a particular decision, a child under 16 years:

- Understands the health problem and its implications

- Understands the risks and benefits of the proposed treatment

- Understands the consequences if the treatment is refused

- Understands the alternative treatments

- Understands the implications on his/her own and the family

- Is able to retain (remember) the information

- Is able to consider the pros and cons

- Is able to make and communicate a reasoned and grounded decision regarding their wishes, goals, expectations and future quality of life

Conclusions

The minor is a person with own dignity and, in addition, is a subject of rights, even when confronted to situations of increased vulnerability, which require protection and care from adults and surrogates. Not only the minor achieves ownership of rights, but under specific conditions and circumstances, he or she can exercise them by himself or herself.

The decision-making and informed consent process is a continuum that traverses different stages and in the particular case of minors, intimately related to their specific status of maturation. Thus, the minor should be informed accordingly at any age, with appropriate language and adapted to their needs and abilities. When the minor is old enough, he or she should be listened to and invited to give their opinion. This opinion must be taken into account and respected whenever and wherever possible, and the parents need the legal and moral space within which to make decisions that will facilitate their child's long-term autonomy and not only their present-day autonomy [5].

Gracia reminds us that "the maturity of a person, whether older or younger, must be measured by the formal ability to judge and assess situations, not by the content of the values the individual assumes or handles. The typical mistake has been to consider immature or uncapable all those who had a set of values different from ours" [24].

Concluding Remarks

- The minor is an individual with own dignity and a subject of rights.
- There is a greater recognition of the role that the minor plays in society, a greater demand in terms of the information to be provided and the role that the child is given in decisions attaining his or her health.
- There is a duty to create the adequate frame for the exercise and empowerment of the autonomy of the minor.
- Competence and capacity are not just attributes the minor either own or not; much will depend upon the building of a relationship of trust and fidelity between the surgical team, the minor patient, and the parents and family. Minors should be encouraged, whenever possible, to develop autonomy and a role in health decisions regarding their own care.
- Nonetheless the parents should be reassured to supervise the minor decisions to warrant the minor well-being and benefit as well as the minor's long-term autonomy.

References

1. Nitta K. Decision making. Encyclopedia Britannica. Available in https://www.britannica.com/topic/decision-making. Accessed January 12, 2018.
2. Grootens-Wiegers P, Hein IM, van den Broek JM, et al. Medical decision-making in children and adolescents: developmental and neuroscientific aspects. BMD Pediatr. 2017;17:120–30.
3. Kaplan RM, Frosch DL. Decision making in medicine and health care. Ann Rev Clin Psychol. 2005;1:525–56.
4. Barry MJ, Edgman-Levitan S. Shared decision making: the pinnacle of patient-centered care. N Engl J Med. 2012;366:780–1.
5. Friedman Ross L. Health care decisionmaking by children. Is it in their best interest? Hast Cent Rep. 1997;27:41–5.
6. Starfield B, Bergner M, Ensminger M, et al. Adolescent health status measurement: development of the Child Health and Illness Profile. Pediatrics. 1993;91:430–5.
7. National Scientific Council on the Developing Child. Young children develop in an environment of relationships. Working paper no. 1. Http://www.developingchild.net. Accessed 2 Feb 2018.
8. McCormick R. Les soins intensifs aux nouveau-nés handicapés. Etudes. 1982;49:493–502.
9. Eiser C. Children's quality of life measures. Arch Dis Child. 1997;77:350–4.
10. Thompson HL, Reville MC, Price A, et al. The quality of life scale for children (QoL-C). J Child Serv. 2014;9:4–17.
11. https://www.ohchr.org/en/professionalinterest/pages/crc.aspx. Accessed 17 Jan 2018.
12. Arnold R. Human dignity and minority protection. Some reflections on a theory of minority rights. In: Elósegui M, Hermida C, editors. Racial justice, policies and courts' legal reasoning in Europe. Ius Gentium: Comparative Perspectives on Law and Justice, vol. 60. Cham: Springer; 2017. p. 3–14.
13. Cicero. On invention (translation Hubbell HM). Loeb Classical Library 386. Cambridge: Harvard University Press; 1949.

14. Kant I. Groundwork for the metaphysics of morals (Wood AW, editor and translator). New Haven: Yale University Press; 2002.

15. Sulmasy DP. The varieties of human dignity: a logical and conceptual analysis. Med Heatlh Care Philos. 2013;16(4):937–44.

16. McCabe MA. Involving children and adolescents in medical decision making: developmental and clinical considerations. J Ped Psychol. 1996;21:505–16.

17. Buchanan A. Mental capacity, legal competence and consent to treatment. J R Soc Med. 2004;97: 415–20.

18. Baumgarten E. The concept of competence in medical ethics. J Med Ethics. 1980;6:180–4.

19. Duncan RE, Sawyer SM. Respecting adolescents' autonomy as long as they may the right choice. J Adolesce Health. 2010;47:113–4.

20. McDougall RJ, Notini L. Overriding parents' medical decisions for their children: a systematic review of normative literature. J Med Ethics. 2014;40:448–52.

21. Friedman Ross L. Theory and practice of pediatric bioethics. Perspect Biol Med. 2015;58:267–80.

22. Griffith R. What is Gillick competence? Hum Vaccin Immunother. 2016;12:244–7.

23. Kennedy I, Grubb A. Principles of medical law. Oxford: OUP; 1998.

24. Gracia D, Jarabo Y, Martín N et al. Toma de decisiones en el paciente menor de edad. En: Gracia D, Júdez J (Ed.). Ética en la práctica clínica. Fundación Ciencia de la Salud. Madrid, 2004 (pp 127–160).

Suggested Literature

American Academy of Pediatrics. Committee on Bioethics. Informed consent, parental permission and assent in pediatric practice. Pediatrics. 1995;95:314–7.

Katz AL, Webb SA, AAP Committee on Bioethics. Informed consent in decision-making in pediatric practice. Pediatrics. 2016;138:e20161485.

End of Life Issues

Karen Brasel and Mary Condron

Key Points
- Identifying the goals and core values that are most important to your patient is essential to providing ethical end of life care.
- Families of dying patients often feel profound powerlessness.
- Removing life-sustaining therapies when and if they become inconsistent with your patients' goals is ethically justified.
- Navigating differences in culture and medical literacy can be difficult; overcoming these obstacles with frequent open communication is the key to providing excellent end of life care.

Death is the enemy. But the enemy has superior forces. Eventually, it wins. And in a war that you cannot win, you don't want a general who fights to the point of total annihilation. You don't want Custer. You want Robert E. Lee, someone who knows how to fight for territory that can be won and how to surrender it when it can't, someone who understands that the damage is greatest if all you do is battle to the bitter end. –Atul Gawande, Being Mortal: Medicine and What Matters in the End [1]

K. Brasel (✉) · M. Condron
Department of Surgery, Oregon Health
and Science University, Portland, OR, USA
e-mail: Brasel@ohsu.edu;
Mecondron@stcharleshealthcare.org

Articles and chapters about medical ethics often start with a cursory review of ethical principles: beneficence, non-maleficence, respect for autonomy, justice, and fidelity. These intuitive sounding principles are worth careful consideration in the context of caring for patients at the end of life. When encountered in this context, their application is often not immediately obvious, and the principles can seem to conflict with one another. Many of the most commonly encountered ethical dilemmas related to end of life surgical care, organized around a fitting ethical principle will be examined. Individual cases and complex circumstances may bring other principles to the forefront.

Beneficence

Beneficence is defined as the moral obligation to act for the others' benefit, helping them to further their important and legitimate interests, often by preventing or removing possible harms. Stated more simply, it is to doing what is good or beneficial for the patient. This pillar of patient care underlies several critical components of high-quality end of life care, including both effective symptom control and reasonable attempts at cure. Defining when the effort required for an attempt at rescue has become too heroic (and death is imminent) requires not only an excellent knowledge of the expected natural history of your patients' conditions but also a deep understanding of what your patient would consider a good outcome. Approximately 60–70% of seriously ill patients are unable to speak for themselves by the time their medical team initiates discussions about limiting treatment [2]. Because this makes getting to know your patients more difficult, inviting family to share stories about your patients' personality and values, not just their wishes regarding DNR (do not resuscitate) status, may provide critical guidance. For example, many patients and families care deeply about where death happens (hospital vs. home). A recent study showed that nearly three-quarters of recently deceased inpatients would have wanted an out of hospital death [3]. Their family members cared deeply about the patient getting to die in their preferred location; however this is rarely prioritized. Showing compassion and simply listening may be tremendously beneficent acts, but understandably these often do not get prioritized with our myriad service obligations.

The traditional model of the principle of beneficence was that routine, standard of care treatments are compulsory and unable to be withheld regardless of circumstance [2]. This thinking was heavily influenced by Roman Catholic moral theology, which forbids euthanasia. According to this thinking, life is a good that has been given as a gift from God and is a means for achieving a stronger communion with God by loving others; this imparts a duty to protect and preserve life [4]. A very influential series of writings by Kelly popularized the distinction between ordinary: "medicines, treatments, and operations, which offer a reasonable hope of benefit and which can be obtained and used with-out excessive expense, pain, or other inconvenience" and extraordinary: "which can not be obtained or used without excessive expense, pain, or other inconvenience, or which, if used, would not offer a reasonable hope of benefit" [5]. However in reality, the distinction between ordinary and extraordinary treatments is exceptionally hard to define, with nearly all interventions falling into both categories depending subjective interpretation of context. The difficulty of applying this principle was illuminated by the famed debate

regarding nutritional support spurred by the Terri Schiavo case, which proved divisive even among prominent theologians [6, 7].

Well supported by both case law and ethical theory, terminal sedation is grounded in the principle of beneficence. This approach involves aggressive symptom management in the imminently dying patient with analgesia and sedatives, as they expire from their underlying disease. Often this will result in rendering the patient unconscious. This is most appropriate in patients with otherwise difficult to control symptoms. Double effect is the framework that supports providing patients with treatments to control symptoms, acknowledging these treatments will likely hasten death, but are nonetheless justified given the importance of symptom control. Terminal sedation is a separate idea from that of the "double effect," though these cases do often overlap. Terminal sedation does not necessarily hasten death, but in many cases it may.

While the topic is nothing if not controversial, it is worth addressing the issue of physician-assisted suicide. If the purpose of medical treatment is the alleviation of suffering, rather than only the prolongation of life, one can construct a coherent ethical justification for physician-assisted suicide as a type of beneficence. First let us be clear about what physician-assisted suicide is and is not. Decidedly different from withdrawal of life support, where the ultimate cause of death is the underlying disease process, physician-assisted suicide is where the ultimate cause of death is the physician's intervention, not the underlying disease. That is to say, with nonintervention the patient would not die. There are many concerns about this approach. A frequently cited concern is that judgment-clouding depression that may be reversible is not uncommon among those with diagnoses of progressive, terminal illness. Some have argued that medicine should focus on developing better ways to palliate disease, such that pursuing a premature end of life wound no longer have any appeal.

Watching a family member die is always difficult, particularly in a setting where you are "out of your element" and feel all the more powerless. It is perhaps not surprising, therefore, that many family members display signs of post-traumatic stress disorder (PTSD) after the death of a loved one in the ICU. Your patient's family is a part of the unit to which you as the surgeon have an ethical duty; in this case it is worthwhile to understand some of the tools that have been found to protect and help heal family members. One such model is family-centered care which, among other interventions, emphasizes family engagement during rounds. This approach invites family members to participate in team rounds, giving them a structured, predictable opportunity to ask questions and raise concerns. Despite worry that this could diminish student learning or that poor situational awareness on the part of a presenting trainee could traumatize family members, family satisfaction and student learning have both been seen to improve with this approach [8, 9]. Another innovative way of caring for families of end of life surgical patients is post-ICU storytelling [10]. This approach is still being actively developed and studied in the population of end of life care decision-makers, but self-disclosure and narrative construction have been shown to significantly mitigate other types of trauma [11].

Non-maleficence

The principle of non-maleficence is critical to any discussion of invasive, potentially harmful interventions. As surgeons, this describes much of what we do. This most basic idea, primum non nocere, can be difficult to apply in cases of diagnostic or prognostic uncertainty. It can be quite difficult to prospectively determine which interventions will prove non-therapeutic. This is demonstrated by the common experience of wildly divergent options about the utility of a specific treatment between teams caring for the same patient (e.g., the operative and intensive care teams).

Beyond differences in perspective between providers with different roles in patient care, there are measurable differences in perspective based on practice environment when it comes to defining the line between reasonable and heroic treatments. There is significant variabil-

ity between the treatment approach to patients who are dying at "high intensity" as compared to "low intensity" hospitals; Barnato and colleagues observed that in the former patients were deemed to be dying when clinical status worsened in the face of maximal life support, whereas in the latter this determination was made at the time of diagnosis of a terminal illness [12, 13].

Determining when a procedure or intervention becomes non-therapeutic is highly dependent on the ability to accurately predict prognosis. Patients often have different views about what they would be willing to undergo if the cause of their illness is a temporary problem amenable to a "quick fix" or the irreversible progression of an underlying disease [14]. This can be difficult to navigate early in a patient's clinical course when there are many unknowns. For a patient "in extremis," awaiting additional medical records or diagnostic results for prognostication prior to decisions to proceed with care is not advisable. Being direct with patients and families about the unknowns in the patient's future, while clarifying the patient's goals and values is of critical importance. These early conversations serve as the foundation for your relationship with the patient and their loved ones: knowing more about who the patient is and what they value can help you guide the decision-makers through difficult decisions and avoid care inconsistent with the patient's goals. It is disheartening to hear during the first formal family conference that "he would say we've gone too far and done too much already" after difficult surgical heroics. Because early prognostication is imperfect and patients may be willing to tolerate aggressive intervention in the short term if a good outcome is expected, but not in the long term or if a good outcome is not expected, some interventions may be reasonable initially but with time it may become clear de-escalation is indicated.

Some have questioned the ethicality of removing life-sustaining therapies once they have been initiated. It is legally and ethically justifiable to consider removal of support devices equivalent to withholding them. Therefore a trial of a therapy does not require that this therapy continue; it may be discontinued if either the patient or their surrogate decides that it is not consistent with goals of care. It may also be discontinued if the provider determines that it is no longer medically indicated.

Moral distress arises when providers feel that they know the morally correct course of action but are powerless to enact it. Within current practice, this often means that the provider feels forced to continue treatments that they feel are harmful or at least not beneficial. For surgeons this is most commonly encountered when a patient with a protracted course is felt to have inconsistent, unrealistic goals of care, or when there are widely disparate opinions about prognosis between a given patient's providers [15]. This is almost always in the setting of a patient who can no longer express his/her wishes. Moral distress may be experienced by any of the members of the care team. This is a situation where the adage "it is better to stay out of trouble than to get out of trouble" rings particularly true. Moral distress is best avoided by early effective communication, as resolving entrenched differences of opinion about how best to proceed can be very difficult. Ethics consultation and structured team debriefings may be important adjuncts to navigating out of these cases and recovering working relationships afterward.

There is significant potential for harm with communication errors. This makes conversations with patients and surrogates across language barriers particularly fraught with risk. The process of translation can lead to insertion of values and judgments by the translator into medical decisions, not just the information the healthcare team wanted to transmit. This makes the use of an appropriately trained medical translator critical, despite the tempting convenience of relying on family members – particularly for uncommon languages where obtaining translation services can be challenging. Medical Orders for Life-Sustaining Treatment (MOLST/POLST) programs are increasingly available online in many languages. This can serve as a good starting point for clarifying end of life wishes, but does not adequately replace in-depth conservation [16]. Often where there are differences in language, there are differences in cultural attitudes toward medicine, family structure, and death. This increases the risk not only of miscommunication but also of mistrust and breakdown of the therapeutic alliance. It may be possible to avoid these issues by explicitly asking early on for preferences regarding the communication of "bad news," the locus of decision-making, and attitudes toward advance directives – prior to diving into goals of care discussion [17]. Online tools like CultureVision that allow providers to briefly familiarize themselves with common cultural views in their patients' communities may be useful for setting a platform for cross-cultural discourse, but care must be taken to avoid using allowing tools to propagate stereotypes.

Futility is a commonly used but hard to define term. Early attempts at defining futility were based on trying to assess either qualitative or quantitative futility. Qualitative futility required a judgment about quality of life expected. Quantitative, or physiologic, futility required judgment about the probability of treatment success, usually based in the ability of individual treatments to meet their physiologic goals – ignoring the patient's overall condition. Because these definitions were so value laden, conflict with families about continuing care that had been deemed futile was not infrequent. These conflicts lead to conflation of futility ("will this work") and rationing (is this expenditure of resources "worth it"). Because of these limitations, the concept of futility has fallen out of favor. The concept of futility can only be meaningful in regard to an end – assessing futility therefore always requires a value judgment. Because of this shortcoming, there is an emerging preference for discussing when interventions are "not medically indicated" rather than when they are "futile." The SCCM (society of critical care medicine) ethics committee noted that "treatments that are extremely unlikely to be beneficial, are extremely costly, or are of uncertain benefit may be considered inappropriate and hence inadvisable, but should not be labeled futile" [18]. This limits the judgment to whether or not there can be a reasonable expectation of medical benefit, instead of judging if the benefit would be "worth it."

In cases where there is a difference of opinion between family and the surgeon about if specific interventions are indicated, the surgeon has an obligation to at least obtain a second opinion and perhaps attempt to transfer care of patient to another provider whose views align with the family. While this addresses the family's concerns, it does keep the surgeon facilitating a service that they believe is of no benefit to the patient, potentially with a cost to society or inflicting suffering on the patient [2]. This places them at risk of moral distress. Many of these conflicts are actually rooted in faulty communication, rather than irreconcilable differences of opinions; most can be overcome with sustained attempts at improving communication to find common ground. Patients and families may be swayed by news reports of "miracles" that likely represent initially misdiagnosis combined with sensationalism, rather than true reversal of, for example, persistent vegetative state [19]. Non-indicated therapies may be suggested by family members with medical experience, but without expertise on the patient's specific condition. While this can be frustrating for the care team, it is helpful to consider how we would handle being in their role. Many of us would find it difficult to passively sit back instead of attempting to be helpful – and we have all seen healthcare teams that have missed things and made mistakes. Many of the negative behaviors seen in these cases can be sublimated by directly naming and addressing the fear of medical errors, as well as empowering the family member to assist in other ways (coordinating travel for distant family, taking lead in family learning about wound care/drain care, etc.)

Respect for Autonomy

Within surgical practice, respect for autonomy is the principle that the patient has a right to determine their own goals of care and to accept or decline offered treatments. This right is transferrable to a surrogate if the patient is unable to express his or her own wishes. Unfortunately, only one in three US adults completes any type of advance directives, and 80% of these were done by patients age 65 years or older [20]. Those who do complete advance directives were generally unable to predict the exact circumstances of their medical crisis, and in situation where the underlying condition is terminal (e.g., advanced ovarian cancer) but the acute problem (e.g., bowel obstruction) is likely reversible, these directives frequently provide little guidance [14]. Often they are completed without support in identifying relevant risks and expected outcomes – for example, it is unlikely to be common knowledge for those completing advance directives that 20% of patients over age 65 who undergo urgent or emergent abdominal surgery die within 30 days [21], even as they declare if they would want emergent surgery at an age over 65.

One of the critical difficulties in helping patients and families navigate making medical decisions during crisis or around end of life is that most people lack personal experience with the range of treatment options facing them, not to mention any medical crisis at all, so it is unlikely that they developed informed preferences outside of the medical setting. Because of this, their preferences can be influenced heavily by small changes in how options are presented. This issue is referred to as "preference construction." Two particularly important concepts in preference construction are context effect and framing effect, which describe how seemingly insignificant changes (such as whether "full code" or "DNR" is listed first on an advance directive form) can have a significant impact on patient and family preferences [13]. These patterns suggest that the choices being made are not in fact inviolable, but are constructed and reconstructed ad hoc to provide an answer for the question at hand. This raises the problematic question of whether patient and surrogate lead decision-making is actually giving them meaningful autonomy.

Further complicating the practical application of patient autonomy is a significant knowledge gap about many of the features of end of life care. Although many decision-makers have taken CPR (cardiopulmonary resuscitation) classes, it is common for their understanding of the outcomes of CPR to be based on depictions in TV or movies. In a recent study, more than 70% of surrogate decision-makers thought that survival after CPR was >75%, only 20% understood that brain damage could be present afterward, and merely 2% understood that the patient could be dependent on life support afterward [22]. This gap in understanding often leads to significant miscommunication or unrealistic expectations.

Another interesting observation about the role of surrogate decision-makers in end of life care is that more than 75% of them express a wish to limit or eliminate their role in medical decision-making. This suggests not only that being put into this position is emotionally difficult but also that their engagement in this role could easily fatigue [23]. Additionally, surrogate decision-makers have shockingly high measured levels of learned helplessness, with more than half reporting helplessness levels equivalent to alcoholics starting 12-step programs [24]. Learned helplessness can influence cognition and decision-making such that one believes that outcomes are independent of their actions. This results in decreased interest in participation and increased defensive behavior [24]. A finding worthy of further study is the observation that this helplessness is associated with lower educational level, but not with race or zip code. This may suggest that communication strategies that make medical decision-making more accessible across educational levels could mitigate this effect.

Risk of burdening family may influence patients' decision-making: in people age 65 or greater, those who died within the following year required 61 hours of care per week, which had a mean duration of 10 years if caregiver was spouse [25]. While these exact figures may not be widely known, many people over 65 have observed others become sources of significant care taking burdens after they became ill. When asking spouses to describe their experience care taking, 66% reported that it was significantly physically difficult [25] – suggesting that the exact people we will depend on for decisions when they are already struggling with the illness of a loved one come to this situation already fatigued.

Often our patients are not familiar with the vocabulary that is used during family meetings. Over two-thirds of older, dialysis dependent patients reported being unable to define "prognosis" or accurately define "hospice" [26]. This suggests not only that we are not sufficiently laying the foundation for these conversations but also that patients and their families may not be comfortable speaking out when we lose them in our jargon.

Shared decision-making is often cited as best practice [27], but details on how to enact this can be sparse. Describing the details of the complex decision tree and range of potential successes and complications that define prognosis is formidable. The traditional informed consent conversation outlines the risks of specific complications; it may even follow the "PARQ conference" (procedure, alternatives, risks, and questions in plain language) model as required in the state of Oregon [27]. However, to a patient without an understanding of what it would be like to live with renal failure, knowing that they have a 15% risk of becoming dialysis dependent is not the same as being truly informed. Enabling patients and families to imagine themselves experiencing these scenarios can help with comprehension much more than simply forecasting. The Best Case/Worst Case framework is one way of enacting this approach. This combines a narrative description and hand written graphic aid to illustrate the best case, worst case, and most likely outcomes sorted by different treatment strategies. This has been shown to shift the focus of decision-making conversations toward the treatment options, rather than the acute surgical problem at hand [28]. Ongoing studies are looking at family and surgeon experience and satisfaction with this particular tool; however decision aids in general were recently found in a Cochrane review to have a positive impact, with patients and families feeling that they were more knowledgeable, more clear about their values, and more accurate in their risk perceptions while only adding 2.6 minutes to consultation time [29].

Justice

One approach to the concept of justice is to divide it into three subcategories: distributive, rights based, and legal. Distributive refers to the fair distribution of scarce resources, rights based reflects respect for peoples' rights, and legal justice demands morally acceptable laws and application of those laws [30]. Each of these types of justice is based equality and equity. The right to be treated equally, and to have equal access to medical treatment, is codified in many groups' bylaws and even in some countries constitutions. In practice, there are many factors that can influence an individual's ability to access treatment, for example, age, place of residence, social status, ethnic background, culture, sexual preferences, disability, and insurance coverage. Some of these factors are more easily mitigated than others. The debate around the Affordable Care Act has at times focused on how to address distributive justice in the US healthcare system. Systems with limited access to insurance have the potential to exacerbate distributive conflicts. Systems that rely on the government to provide access to treatment have the potential for rights-based conflicts. For example, the Swiss Academy of Medical Sciences recently reported that doctors and other medical staff are increasingly refusing to administer potentially useful treatment for economic reasons [31] and there has been considerable debate in the UK over the refusal of expensive treatment to patients who could potentially benefit from it [32, 33].

With regard to equality in the provision of care, there can be a perception that some people are not treated with the same degree of respect as is accorded to others. This can impact the likelihood that patients in marginalized communities will seek care, as well as how easily they will be able to engage in open and trusting communication with the healthcare team. Disparities in the care for people with disabilities as well as members of the lesbian, gay, bisexual, and transgender communities related to these barriers are widespread and well documented [34–36]. The impact of race on end of life care is complex. In addition to the potential for issues of trust and communication when the patient and their family have a different background than then healthcare team, there have been well-documented differences in the rate of completion of advance directives, provision of aggressive end of life therapy, and pursuit of organ donation across racial groups [37, 38].

Additional factors that influence just distribution of end of life resources are logistic and regulatory barriers. To qualify for the Medicare hospice benefit, a patient must have a physician certify that they are expected to live 6 months or less. This 6 month limit was designed with the ability for Medicare to pay in mind, rather than anything intrinsic to 6 months of remaining life. Only half of eligible Medicare beneficiaries ever use their hospice benefit. This may be a function of late referral: approximately one-third of Medicare hospice patients dying within their first week in hospice and nearly two-thirds within the first 30 days [39]. This suggests that many of these patients could have benefitted from earlier referral. Access to hospice may be biased toward those with the education and resources necessary to be aware of, request, and advocate for these services themselves.

One of the root causes behind the limited discussion of end of life issues is that this activity had traditionally not been reimbursed. The argument against paying for this consultative service was made most famous by Sarah Palin's 2009 Facebook post referencing rationing done by government run insurance programs and popularizing the term "death panel[s]" [40]. The issues of the appropriateness of end of life planning and concerns about the potential for rationing are separate issues; however the two have been conflated and politicized for debate. We are not aware of any evidence that would suggest that compensating providers for taking the time to clarify patients' goals of care has resulted increased rationing. In fact a recent meta-analysis showed that the use of structured communication tools did not change the rate of DNR or the decision to pursue comfort care, but was able to lower the number of hospital days, number of ventilator days, and total cost of care [41].

Fidelity

Ethical care of our patients requires more than simply well-meaning intentions. The tremendous privilege that patients extend to us by inviting us into their most vulnerable moments and trusting us with invasive procedures demands tremendous faithfulness to our promises and implied agreements. The principle of fidelity is broad and encompasses loyalty, fairness, truthfulness, advocacy, and dedication to our patients.

As with the other principles, fidelity cannot be achieved without effective communication. As the leader of the healthcare team, surgeons may find tools like the Omaha system helpful in delineating the commitments made by and the patient needs uncovered by the team. This multidisciplinary tool is incorporated directly into the electronic health record and has been shown to facilitate more open, consistent, and precise communication [42].

The idealized surgeon has been typed as the "triple threat": an outstanding clinician, a teacher, and a research scientist. This completely leaves out the roll of advocate, which is critical to fulfilling our duty of fidelity. Effective advocacy requires a comprehensive knowledge of what goals are important to our patients and what barriers stand in the way of these goals. A recent study by Nadin and colleagues identified that families of patients who had recently died valued the following most highly: the perception of the healthcare workers coming together as a team, honest communication, and demonstrable compassion and support for their family member [43]. Unfortunately, it seems that we are often not successfully identifying or advocating for end of life care that is consistent with what our patients want – fully one in eight Medicare beneficiaries' family members report that the end of life care their love one received was not consistent with their wishes. Pain, dyspnea, and prevention of depression are common concerns; however nearly a third of patients with terminal cancer reported that their financial distress was more severe than their physical, family, or emotional distress [43]. These concerns are rarely identified or addressed beyond referral for discussion with a social worker or case manager.

Although as surgeons we have much of the background necessary to successfully guide end of life care and decision-making, particularly when using structured decision-making tools, it is worth noting the impact of formally involving

a palliative care specialist. A recent study showed that even just a single visit with a palliative care specialist improved physical and psychological symptoms [44]. These providers may have more time to devote to end of life care planning and be more familiar with local resources, allowing them to meet patient and family needs that would be difficult for the surgical team to address. Additionally, surgeons may not be accurate reporters of what experiences in hospice would be like or what could be achieved with optimal symptom control. In a particularly troubling essay, Moss describes how far off the consulting surgeon was when describing the symptom control achievable with medical management of a malignant large bowel obstruction [45].

Concluding Remarks

Providing excellent surgical care for patients at the end of life can present many ethical dilemmas. These range from avoiding unduly influencing preference construction, to decisions to withdraw life support, to attempting to prognosticate outcome in complex situations with incomplete information. At the crux of applying beneficence, non-maleficence, respect for autonomy, justice, and fidelity to these situations is open, frequent communication. Through working to understand who our patients are, discover what they value, and mitigate gaps in medical knowledge, we can provide ethical, high-quality end of life care.

References

1. Gawande A. Being mortal: medicine and what matters in the end. 1st ed. New York: Metropolitan Books: Henry Holt & Company; 2014. 282 p.
2. Pawlik TM, Curley SA. Ethical issues in surgical palliative care: am I killing the patient by "letting him go"? Surg Clin North Am. 2005;85(2):273–86.. vii
3. Sadler E, Hales B, Henry B, Xiong W, Myers J, Wynnychuk L, et al. Factors affecting family satisfaction with inpatient end-of-life care. PLoS One. 2014;9(11):e110860.
4. Hamel RP, Walter JJ. Artificial nutrition and hydration and the permanently unconscious patient: the Catholic debate. Washington, D.C.: Georgetown University Press; 2007. ix, 294 p.

5. Kelly G. The duty to preserve life. Theol Stud. 1950;11:203–20.

6. O'Rourke KD. Artificial nutrition and hydration and the Catholic tradition. The Terri Schiavo case had even members of congress debating the issue. Health Prog. 2007;88(3):50–4.

7. Sulmasy DP. Terri Schiavo and the Roman Catholic tradition of forgoing extraordinary means of care. J Law Med Ethics. 2005;33(2):359–62.

8. McDermott A. Championing mistakes: reclaiming the safe learning environment for family-centered bedside rounds. J Grad Med Educ. 2017;9(2):257.

9. Ingram TC, Kamat P, Coopersmith CM, Vats A. Intensivist perceptions of family-centered rounds and its impact on physician comfort, staff involvement, teaching, and efficiency. J Crit Care. 2014;29(6):915–8.

10. Schenker Y, Dew MA, Reynolds CF, Arnold RM, Tiver GA, Barnato AE. Development of a post-intensive care unit storytelling intervention for surrogates involved in decisions to limit life-sustaining treatment. Palliat Support Care. 2015;13(3):451–63.

11. Silverman E. Sharing and healing through storytelling in medicine. JAMA Intern Med. 2017;177:1409.

12. Barnato AE, Herndon MB, Anthony DL, Gallagher PM, Skinner JS, Bynum JP, et al. Are regional variations in end-of-life care intensity explained by patient preferences?: a study of the US Medicare population. Med Care. 2007;45(5):386–93.

13. Barnato AE. Challenges in understanding and respecting patients' preferences. Health Aff (Millwood). 2017;36(7):1252–7.

14. Cooper Z, Courtwright A, Karlage A, Gawande A, Block S. Pitfalls in communication that lead to nonbeneficial emergency surgery in elderly patients with serious illness: description of the problem and elements of a solution. Ann Surg. 2014;260(6): 949–57.

15. Crippen D. Moral distress in medicine: powerlessness by any other name. J Crit Care. 2016;31(1):271–2.

16. Aultman JM. Ethics of translation: MOLST and electronic advance directives. Am J Bioeth. 2010;10(4):30–2.

17. Sharma RK, Dy SM. Cross-cultural communication and use of the family meeting in palliative care. Am J Hosp Palliat Care. 2011;28(6):437–44.

18. Kon AA, Shepard EK, Sederstrom NO, Swoboda SM, Marshall MF, Birriel B, et al. Defining futile and potentially inappropriate interventions: a policy statement from the Society of Critical Care Medicine Ethics Committee. Crit Care Med. 2016;44(9):1769–74.

19. Swetz KM, Burkle CM, Berge KH, Lanier WL. Ten common questions (and their answers) on medical futility. Mayo Clin Proc. 2014;89(7):943–59.

20. Yadav KN, Gabler NB, Cooney E, Kent S, Kim J, Herbst N, et al. Approximately one in three US adults completes any type of advance directive for end-of-life care. Health Aff (Millwood). 2017;36(7):1244–51.

21. Cooper Z, Mitchell SL, Gorges RJ, Rosenthal RA, Lipsitz SR, Kelley AS. Predictors of mortality up to 1 year after emergency major abdominal surgery in older adults. J Am Geriatr Soc. 2015;63(12):2572–9.

22. Shif Y, Doshi P, Almoosa KF. What CPR means to surrogate decision makers of ICU patients. Resuscitation. 2015;90:73–8.

23. Azoulay E, Pochard F, Chevret S, Adrie C, Annane D, Bleichner G, et al. Half the family members of intensive care unit patients do not want to share in the decision-making process: a study in 78 French intensive care units. Crit Care Med. 2004;32(9):1832–8.

24. Sullivan DR, Liu X, Corwin DS, Verceles AC, McCurdy MT, Pate DA, et al. Learned helplessness among families and surrogate decision-makers of patients admitted to medical, surgical, and trauma ICUs. Chest. 2012;142(6):1440–6.

25. Ornstein KA, Kelley AS, Bollens-Lund E, Wolff JL. A national profile of end-of-life caregiving in the United States. Health Aff (Millwood). 2017;36(7):1184–92.

26. Ladin K, Buttafarro K, Hahn E, Koch-Weser S, Weiner DE. "End-of-life care? I'm not going to worry about that yet." Health literacy gaps and end-of-life planning among elderly dialysis patients. Gerontologist. 58(2):290-299, 2018.

27. Elwyn G, Edwards A, Thompson R. Shared decision-making in health care: achieving evidence-based patient choice. 3rd ed. Oxford, UK/New York: Oxford University Press; 2016. xxi, 309 p.

28. Taylor LJ, Nabozny MJ, Steffens NM, Tucholka JL, Brasel KJ, Johnson SK, et al. A framework to improve surgeon communication in high-stakes surgical decisions: best case/worst case. JAMA Surg. 2017;152(6):531–8.

29. Stacey D, Kryworuchko J, Belkora J, Davison BJ, Durand MA, Eden KB, et al. Coaching and guidance with patient decision aids: a review of theoretical and empirical evidence. BMC Med Inform Decis Mak. 2013;13(Suppl 2):S11.

30. Gillon R. Medical ethics: four principles plus attention to scope. BMJ. 1994;309(6948):184–8.

31. Pecoud A, Cornuz J. Reflexion of the Swiss Academy of Medical Sciences on the professional profile of physicians: some thoughts to remember. Rev Med Suisse. 2008;4(181):2555–6.

32. London DM. Rationing gets official seal of approval from UK Health. Lancet. 2000;355(9197):49.

33. Drummond M, Mason A. Rationing new medicines in the UK. BMJ. 2009;338:a3182.

34. Chin MH. Movement advocacy, personal relationships, and ending health care disparities. J Natl Med Assoc. 2017;109(1):33–5.

35. Chinn PL. Commentary: Lesbian, gay, bisexual, and transgender health: disparities we can change. Nurse Educ. 2013;38(3):94–5.

36. Iezzoni LI. Why increasing numbers of physicians with disability could improve care for patients with disability. AMA J Ethics. 2016;18(10):1041–9.

37. Koss CS, Baker TA. Race differences in advance directive completion. J Aging Health. 2017;29(2):324–42.

38. Barnato AE, Chang CC, Saynina O, Garber AM. Influence of race on inpatient treatment

intensity at the end of life. J Gen Intern Med. 2007;22(3):338–45.

39. Ornstein KA, Aldridge MD, Mair CA, Gorges R, Siu AL, Kelley AS. Spousal characteristics and older adults' hospice use: understanding disparities in end-of-life care. J Palliat Med. 2016;19(5):509–15.

40. Palin S. 2009. Available from: https://www.facebook.com/notes/sarah-palin/statement-on-the-current-health-care-debate/113851103434/.

41. Oczkowski SJ, Chung HO, Hanvey L, Mbuagbaw L, You JJ. Communication tools for end-of-life decision-making in ambulatory care settings: a systematic review and meta-analysis. PLoS One. 2016;11(4):e0150671.

42. Slipka AF, Monsen KA. Toward improving quality of end-of-life care: encoding clinical guidelines and standing orders using the Omaha system. Worldviews Evid-Based Nurs 2017. Gerontologist. 58(2):290-299, 2018

43. Nadin S, Miandad MA, Kelley ML, Marcella J, Heyland DK. Measuring family members' satisfaction with end-of-life care in long-term care: adaptation of the CANHELP lite questionnaire. Biomed Res Int. 2017;2017:4621592.

44. Dalal S, Bruera E. End-of-life care matters: palliative cancer care results in better care and lower costs. Oncologist. 2017;22(4):361–8.

45. Moss DK. Getting it right at the end of life. Health Aff (Millwood). 2017;36(7):1336–9.

Suggested Literature and Resources

Back A, Arnold RM, Tulsky JA. Mastering communication with seriously ill patients: balancing honesty with empathy and hope. Cambridge: Cambridge University Press; 2009.

Gawande A. Being mortal: medicine and what matters in the end. 1st ed. New York: Metropolitan Books: Henry Holt & Company; 2014.

https://www.mypcnow.org/copy-of-core-curriculum

Mosenthal AC, Murphy PA, Barker LK, Lavery R, Retano A, Livingston DH. Changing the culture around end-of-life care in the trauma intensive care unit. J Trauma. 2008;64(6):1587–93.

Taylor LJ, Nabozny MJ, Steffens NM, Tucholka JL, Brasel KJ, Johnson SK, et al. A framework to improve surgeon communication in high-stakes surgical decisions: best case/worst case. JAMA Surg. 2017;152(6):531–8.

Winakur J. Conundrums at the end of life. Health Aff (Millwood). 2017;36(7):1343–4.

Ethics and Surgical Innovation

Maria S. Altieri and Aurora D. Pryor

Introduction

Innovation requires two conditions: a great idea and significant investment. Innovation in the medical field in the past decades has immensely improved patient care. When implementing new technologies and techniques, physicians are confronted with multiple factors influencing the process, such as the drive to provide superior care, the appeal of new technology, and the pressure from health-care systems and competition. Surgical innovation has led to benefits in imaging, techniques, and novel devices, attempting to decrease health-care costs and/or improve patient outcomes [1]. Yet each of these steps has required collaboration between thoughtful leaders and financial backers. In the operating room, surgeons have been able to exercise creativity in regard to procedural modification. However, for surgical devices, the process is much more involved. To bring a new device to the bedside requires prototyping, clinical investigation, manufacturing, and to secure the idea, patent work. Most surgeons lack the experience and the funds to bring a novel device idea to fruition independently. Some partner with internal resources to help meet this need. However, surgeons frequently find support through industry partners.

M. S. Altieri (✉) · A. D. Pryor
Division of Bariatric, Foregut and Advanced GI Surgery, Department of Surgery, Stony Brook Medicine, Stony Brook, NY, USA
e-mail: maria.altieri@stonybrookmedicine.edu

This can lead to ethical conflict on many levels. This chapter will review the ethical challenges of bringing new technologies and techniques into clinical practice in the current era of surgical innovation.

Innovation in Surgery

Introduction of new medications and devices is heavily regulated by the Food and Drug Administration (FDA) in the USA. The process is extremely rigorous and often takes years to bring a new product to the market. Although, the process of new devices does not differ, the area of surgery differs from other areas of medicine, as new techniques and procedures are mostly unregulated [2, 3]. This particular type of innovation starts in the operating room where every patient is different, and no procedure is 100% step by step the same, and some variability in technique is possible and even expected. As repetitive failures and unanswered questions remain, surgeons strive to find solutions. If the surgeon believes that something novel or different may improve outcomes, they do not need approval by the FDA, and without plans to study it even the local institutional review board (IRB), prior to attempting it. Through these attempts, many successes have been discovered and procedures such as laparoscopy and endoscopy have been invented and have revolutionized surgery.

A. R. Ferreres (ed.), *Surgical Ethics*, https://doi.org/10.1007/978-3-030-05964-4_23

Innovation has led to so many advances in surgery, and today even more complex procedures can be performed through the smallest incisions. Because of people who attempted new techniques, minimally invasive surgery has flourished. In the early 1980s, K Semm performed the first laparoscopic appendectomy. Erich Mühe soon after performed the first laparoscopic cholecystectomy on September 12, 1985. Within a couple of years, he performed close to 100 endoscopic cholecystectomies [4]. Initially, his work was not accepted by surgeons; however today, nearly every abdominal surgery is performed laparoscopically, and minimally invasive techniques have spread to other areas such as urology, gynecology, and orthopedics.

Today, laparoscopic cholecystectomy has become the standard of care for treating gallbladder disease, as close to 1.2 million laparoscopic cholecystectomies are being performed each year in the USA alone [5]. But the search for even less invasive techniques has not stopped. More recently, another example of an unregulated new procedure is the introduction of the single-incision laparoscopic cholecystectomy (SILC). The technique was widely being performed prior to its safety being fully assessed. Soon after its implementation, it was noted that the rate of bile duct injuries, which is a dreaded complication following LC, is higher compared to the rate of this complication during LC [6]. The obvious question is whether the timing of its implementation was appropriate and safe.

There are many other examples of unsuccessful or even harmful procedures such as internal mammary artery ligation for the treatment of angina, gastric freezing for treatment of ulcer disease, and jejunoileal bypass for bariatrics. Thus, many believe that implementation of new techniques should be regulated in order to protect the patient and ensure that safety is evaluated before a new technique is widely implemented [7–9]. However regulation itself is not infallible, as evidenced by the Garren-Edwards bubble or laparoscopic adjustable gastric banding for weight loss.

What Constitutes Surgical Innovation in Terms of Procedure/ Technique?

As opposed to devices, where the definition of an innovation is often clear, the definition of innovative procedures is often not as clear cut. As defined by the position statement of the Society of University Surgeons, an innovation is:

> a new or modified surgical procedure that differs from currently accepted local practice, the outcomes of which have not been described, and which may entail risk to the patient. [10]

While the first laparoscopic procedure can clearly be defined as surgical innovation, the problem of oversight of new techniques or procedures today lies in the vague definition of a "novel" or "innovative" procedure versus a modification on an already existing procedure. In an article published in 2002, Reitsma and Moreno hypothesized that surgeons are unaware of regulations, rarely seek IRB approval, and are uncertain of what constitutes innovation and research. The article shows that a minority of surgeons would seek Institutional Review Board (IRB) approval or understand what research requires IRB approval. For those surgeons, who had an IRB approval, a minority would mention the innovative nature of the procedure to their patients [11]. More recently, a study interviewing 18 surgeons on what is innovation, reported that there is no uniform view about the definition of surgical innovation or the difference between innovation and research [12]. Further, Rutan et al. reported knowledge deficits among academic surgeons regarding federal requirements and the role of IRBs in conducting clinical research [13].

In the absence of regulation, the process requires the surgeon to exercise a great deal of professionalism, integrity, and transparency. Some may argue that without oversight, objective decision-making may not be enough. Thus, the need for local IRB and federal oversight is essential to ensure patient safety.

Oversight of Innovation

With regard to innovation of procedures/techniques, several types have been described: minor modification of a standard procedure, major modification of an established technique, or new innovation; in addition there is innovation that has been validated elsewhere, but not performed by the individual surgeon or at the institution [14]. Even minor modifications of a procedure may require oversight [15–17] if the surgeon intends to study the outcomes of the procedure. New innovations that are not intended to be studied, however, do not necessarily require IRB oversight. This does, however, require transparent patient disclosure, as mentioned below. IRB involvement is also helpful for assuring adequate safety measures and consent and encourages surgeons to study the results of their novel procedures to assure they truly improve outcomes.

Oversight of surgical innovation can occur at the local or federal level. The Institutional Review Board (IRB) is an example of local oversight. This committee is comprised of physicians and other institutional personnel and is in charge of reviewing research prior to its initiation involving human subjects and ensuring welfare, rights, and privacy. It is in charge of approval, monitoring, and reviewing any research involving patients.

In the USA, the Code of Federal Regulations (CFR) defines regulations involving human subjects at the national level. There are also specific FDA research regulations for the use of biologics, drugs, and devices. Although these regulations only apply to federally funded research, they are applied by most institutions. Other regulations are provided by national organizations in the form of guidelines or certain committees. Another example of a regulatory system is ClinicalTrials.gov, which is a database created by the US National Library of Medicine, where clinical trials performed in the USA and around the world are registered.

Informed Consent

One of the fundamental ethical challenges for surgical innovation is the appropriate achievement of informed consent. According to the Belmont Report, developed by the US Department of Health and Human Services in 1979, the consent process has three key elements: information, comprehension, and voluntariness [18]. The process of obtaining consent with patients can be challenging, especially when discussing new technology and procedures. Surgeons have an obligation to provide a forthright, balanced, and unbiased presentation of information regarding new technology to the patients, as the goal is for the patient to make an educated and voluntary choice to participate. Information provided should include the innovative nature of the procedure or device, the risks and benefits, the experience of the surgeon with the procedure or device, possible unforeseen risks, the standard procedure or the alternatives to the proposed procedure, and the evidence, or lack of, in terms of success or risks of procedure/device. At the end, the patient should be given the opportunity to answer questions.

In the case of a new procedure or device, there are several challenges. If the surgeon has had some experience with it, she/he can discuss some of the benefits and risks thus encountered. However, sometimes risks and benefits may not be truly known. An example is the risk of common bile duct injuries in laparoscopic cholecystectomy [19]. The lower the incidence of a complication, the more likely is to miss the true incidence until larger scale data is compared. Another challenge is the inherent bias for the term "new" or "innovative" that can affect both the surgeon and the patient [20]. This can prevent a balanced discussion, as the potential benefits may be highlighted and the risks excluded from the conversation.

A systematic review of the literature aimed to identify positions about consent and disclosure regarding surgical innovation [21]. The study reported four major tension points: the use of biasing/biased terminology to characterize innovation, patient vulnerability, the relationship between the surgeon-innovator and the patient, and the practices and associated gaps related to consent and disclosure [21]. Surgeons may also often have a stake in the patient's choice, desiring the patient to undergo the innovative procedure for personal reasons. Any financial or personal bias of the surgeon should be revealed to the patient during the consent process.

Both institutional and national medical associations can provide guidelines and direction in this arena in order to protect both the patient and the surgeons and institutions.

Device Development

Any device development and approval in the USA has to pass through the regulatory pathway which is supervised by the FDA. This path is well established (Fig. 1).

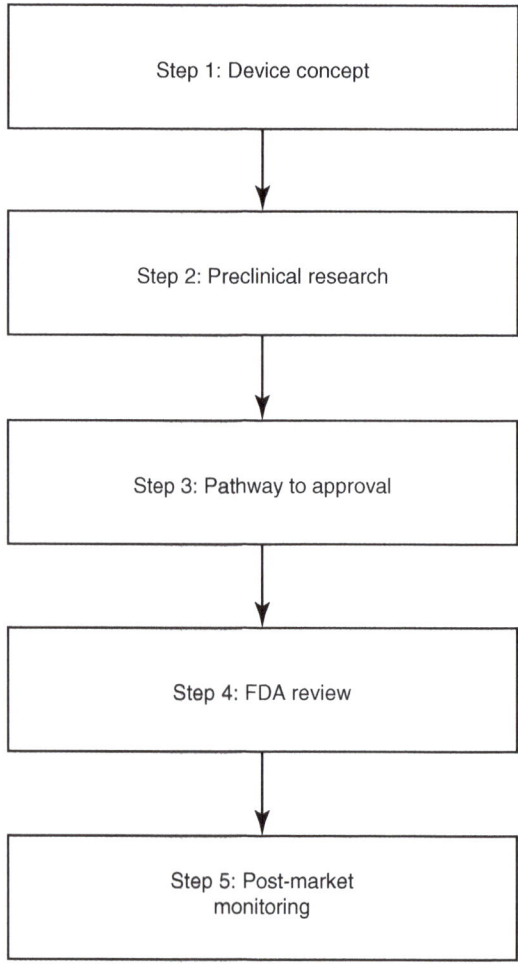

Fig. 1 Path of device development

The first step is device concept which occurs when the surgeon sees an unmet medical need and creates an idea or a concept for a new device. Following that, a "proof of concept" is built that outlines the steps needed to determine is the idea is workable. If it appears to be promising, the device leads to the following stages. There are three classifications of devices depending on the device safety:

- Class 1: General controls
- Class 2: General controls with special controls
- Class 3: General controls and premarket approval

Class 1 devices pose the least amount of risk to the consumers. Examples are surgical instruments. Class 2 devices pose more risks. Class 3 devices can support or sustain life, can have the potential for unreasonable risk of illness or injury, or can be implanted in the body and include examples such as breast implants and pacemakers. Due to the risks, they require premarket approval, which means that the device needs a proof of being safe and effective, often through animal studies and clinical trials.

The second step is building a device prototype, which may not be for human use. The device can undergo testing and refinement in controlled setting. During this testing, the device can be improved in order to reduce risk to the patient.

The third step is the pathway to approval and depends on the risk classification of the device, which is described above. Class 1 is subject to the least regulatory control as most are exempt from premarket approval, compared to Class 3 which requires the most control.

There are two main ways of getting a device to the market through the FDA: premarket notification (510(k)) and premarket approval (PMA). In general the 510(k) pathway is used for a device that is substantially similar to another (predicate) device. The PMA pathway is much more expensive and involved and is used for truly new technologies. Clinical data is necessary to prove that the device has a reasonable safety and effectiveness. Whereas only a small percentage of 510 (k) studies require clinical data, devices going through premarket approval require clinical data through an investigational device exemption (IDE).

An IDE application to the FDA is the first step in getting a device to be investigated in a clinical study. The application must include the belief that the risks to human subjects outweigh the benefits and that the device is scientifically sound and effective. After the application, the sponsor is advised on the trial design, statistical plan, and study endpoints. More information can be found on the FDA website [22].

The two regulatory controls are the premarket notification (aka 510(k)) and premarket approval. The 510(k) path is undertaken for devices that are not exempt from premarket review but do not need premarket approval [23]. In order to qualify for premarket notification, the sponsor needs to prove that the device is equivalent to a marketed device as it has the same intended use and characteristics as a legally marketed device.

The premarket approval (PMA) path is highly regulated and usually evaluates Class 3 devices but can also evaluate Class 1 or 2 devices that are not substantially equivalent to others under the 510(k) process. The process is very involving and requires extensive reports and summaries to ensure that the device has a reasonable safety profile.

Another pathway to market a device is Humanitarian Device Exemption (HDE). Devices falling under this category benefit patients by treating or diagnosing a disease that affects less than 4000 patients and are thus harder to study.

Once submitted, the FDA team will review the data. If needed, the FDA will consult an advisory committee, which consists of groups of experts in order to provide an independent advice. Following approval, the FDA continues to monitor device performance through post-market safety monitoring. The FDA officials can conduct inspections of the device manufacturing facilities through the USA. These inspections can be routine or due to a particular problem. If the FDA notices a problem, they can stop the production of the device. Other ways of monitoring is through reporting programs, such as MedWatch or Medical Product Safety Network (MedSun).

Tracking Outcomes

Tracking outcomes of new devices is coordinated by the FDA (described above). However, tracking outcomes following new technology or procedures is less clear. There is an ethical obligation to early adopters of new technology and procedures to track outcomes and ensure the wellness of the patients. There are a multitude of ways to accomplish this task, including IRB-approved studies, establishing randomized controlled trials, participation in prospective registries and consortia, institutional databases and prospective reviews, and personal databases. This process is not only time-consuming, requiring a great deal of manpower and coordination, but expensive as well. Regardless, the surgeons are duty-bound to properly assess the new technology or procedure to fully elucidate its relevance and effectiveness [24].

Prospective randomized controlled trials (RCT) are the gold standard of evaluating interventions. Initiation of a RCT for a new intervention can prove challenging. One of the challenges is timing of the study. In the case of starting a study too early, the definitive technique may not be fully developed; thus the study will reflect the learning curve and not the true effect of the intervention [25]. Another challenge can be the dependence of the technique or procedure on the operator and sometimes the team [26]. While individual surgeons may be willing to offer the procedure based on their experience, this may not be the general opinion of everyone in a practice, and getting enough participants may prove challenging. The investigators may need to extend the trial beyond a single institution, which may be even more challenging for uniformity of a trial. Designing the trial may be difficult as well, as outcomes need to be clearly defined. Without knowing potential complications, assessing for those may be difficult. Other challenges of RCTs for evaluating surgical innovation have been described [27].

Other forms of tracking new technologies are in form of databases, either national, institutional, or personal. In the new era of improving outcomes, national databases such as NSQIP,

ASMBQIP, and SCIP have proved to play a central role. Originally involving Veteran Affairs hospitals, the ACS NSQIP database now includes more than 400 hospitals nationwide and provides the ability to capture 30-day outcomes. Other statewide databases are also available in certain states and are under the governance of the State Department of Health. Such databases are the New York Statewide planning and research cooperative system (SPARCS) and the California Department of Health Care Services (DHCS) database. However, these databases may not reflect the initiation of a new technique or technology. Examples are procedures such as natural orifice transluminal endoscopic surgery (NOTES), peroral endoscopic myotomy (POEM), and single-incision laparoscopic surgery (SILS). These databases do not accurately capture these procedures. In the case of SILS, most procedures were coded as laparoscopic cholecystectomy, thus inaccurately assessing early complications and long-term outcomes.

As it is imperative to capture adverse events related to medical devices and procedures, further work is needed in designing methods to track outcomes.

Training and Credentialing in New Technologies

Education is essential for safe and effective use of a new procedure, device, or technique. Even if a surgeon-innovator has become proficient, when a procedure has moved beyond the first adopter and is being performed by others, the issue of the learning curve and training others to safety adopt a procedure arises. In surgery, there is a traditional approach of a master-student apprenticeship.

A good example of issues in training and credentialing in surgical technique is laparoscopy. As previously mentioned, while originally not well accepted, it is now a standard of care for many surgical procedures. Laparoscopy differs compared to open surgery in loss of three dimensions and sense of touch. While today, residents have the opportunity to train in laparoscopic procedure, when initially introduced, surgeons had

to learn this new technique in a safe and effective manner. This leads to development of laparoscopic simulators and trainers. In addition, mentored cases or visiting apprenticeships have been used to help obtain surgical skills. Mentoring is a form of training, where a more experienced surgeon can observe or scrub in order to supervise.

Credentialing refers to the next step following training. Credentialing involved the verification of a surgeon's licensure, education, training, experience, and ultimately competence to perform a technique or procedure. It should be the result of a standardized, competency-based peer review evaluation. At an institutional level, it usually involves the review of length of training and number of procedures performed. This usually has been the task of a multidisciplinary committee, bringing together clinical, academic, administrative, financial, and other providers who can assess new technology/procedure implementation and grant the approval. More recently, "training competence" has emerged as a form of credentialing. An example is the Fundamentals of Laparoscopic Surgery (FLS), which was developed by the American College of Surgeons and the Society of American Gastrointestinal and Endoscopic Surgeons (SAGES). FLS curriculum is used for certifying trainees in laparoscopic skills, not only at an institutional level but also at the national level. The success of FLS has encouraged the development of programs like the Fundamentals of Endoscopic Surgery (FES) and Fundamentals of Robotic Surgery (FRS) [24].

Working with Industry

Another potential controversial topic is the surgeon-industry relationship. Sometimes these relationships can lead to significant financial gain, and therefor bias, for surgeons. The concern regarding these relationships is, regardless of the scale of the rewards, a perceived notion that people generally tend to feel obliged to return favors, no matter how small the gifts may be. The general public is specifically concerned and critical of physicians who own stocks in companies where the financial gain is even more direct [28,

29]. While of these concerns exist, the collaboration between surgeons and industry has been necessary for the progress of innovation. A survey of 822 surgeons reported that most surgeons believe that industry-surgery relationship is positive and necessary for surgical innovation to occur [30]. The collaboration involves merging the surgeons' clinical expertise and skills with the knowledge and resources of technical experts, engineers, and businessmen. It is essential for surgical innovation.

There are many reasons for this relationship to exist. The process of getting a device to the market is lengthy and very involving, as described above under the Device Development section. Industry can provide the resources for the development and marketing of many devices or technology. Physicians, who invent, usually are not involved in the manufacturing or marketing of the product [31]. Thus, industry can provide help in this area. In the start of a new technology or technique, industry can help with maintenance of databases for new innovations, which often requires funding and expertise. In addition, industry can help with research and education, as industry can provide research opportunities, grants, and courses involving the new technology or technique.

Because of this relationship, the term conflict of interest (COI) was coined, which refers to a situation in which the physician's judgment or decision-making can be influenced due to a relationship. Strategies to address potential COIs is the enhancement of disclosures by physicians, known as transparency [32].

Disclosures

Disclosure is the act of revealing all relevant information that may influence the decision of the patient. As mentioned above, the enhancement of disclosures by physicians is known as transparency. This can refer to the involvement of the physician with industry. Under the Patient Protection and Affordable Care Act, all physicians should exercise transparency and companies have to publicly release details of payments made to physicians. Other disclosures, in context of new devices or procedures, include personal experience of the surgeon, outcomes, complications, and how the device or procedure/technique compare to current standard of treatment.

A study published in 2013 examined what information patients and surgeons consider essential to disclose in an informed consent [33]. The study consisted of a survey of 85 surgeons and 383 patients. The authors concluded that important information that should be disclosed includes the innovative nature of the surgery, potentially unknown risks and benefits, and the experience of the surgeon with the surgery. The majority of the patients (80%) indicated that one of the most important variables in their decision was whether it would be the surgeon's first time doing the procedure [33].

Conclusions

- Innovation in the surgical field has immensely improved patient care.
- In order to bring new devices and procedures collaboration between physicians, scientists, engineers, and industry is key.
- The surgeon must have the ethical principles to ensure that the patients' well-being is not affected when bringing new technology or techniques, while striving to improve outcomes.

References

1. Riskin DJ, Longaker MT, Gertner M, Krummel TM. Innovation in surgery: a historical perspective. Ann Surg. 2006;244(5):686–93.
2. Spodick DH. Numerators without denominators. There is no FDA for the surgeon. JAMA. 1975;232(1): 35–6.
3. Love JW. Drugs and operations. Some important differences. JAMA. 1975;232(1):37–8.
4. Litynskic GS. Erich Mühe and the rejection of laparoscopic cholecystectomy (1985): a surgeon ahead of his time. JSLS. 1998;2(4):341–6.
5. Truven Health Analytics (Thomson/Solucient), USA procedure volumes 2014 National Data. Ann Arbor.
6. Strasburg SM. Single incision laparoscopic cholecystectomy and the introduction of innovative surgical procedures. Ann Surg. 2012;256(1):2.

7. McKneally MF, Daar AS. Introducing new technologies: protecting subjects of surgical innovation and research. World J Surg. 2003;27:930–7.
8. Steinbrook R. Improving protection for research subjects. N Engl J Med. 2004;346:1425–8.
9. Roy DL, Black PM, McPeek B. Ethical principles in research. In: Principles and practice of research: strategies for surgical investigation. New York: Springer-Verlag; 1991.
10. Biffl WL, Spain DA, Reitsma AM, Society of University Surgeons Surgical Innovations Project Team, et al. Responsible development and application of surgical innovations: a position statement of the Society of University Surgeons. J Am Coll Surg. 2008;206(6):1204–9.
11. Reitsma AM, Moreno JD. Ethical regulations for innovative surgery: the last frontier? J Am Coll Surg. 2002;194(6):792–801.
12. Rogers WA, Lotz M, Hutchison K, Pourmoslemi A, Eyers A. Identifying surgical innovation: a qualitative study of surgeons' views. Ann Surg. 2014;259(2):273–8.
13. Rutan RL, Deitch EA, Waymack JP. Academic surgeons' knowledge of Food and Drug Administration regulations for clinical trials. Arch Surg. 1997;132:94–8.
14. Broekman ML, Carrière ME, Bredenoord AL. Surgical innovation: the ethical agenda: a systematic review. Medicine (Baltimore). 2016;95(25):e3790.
15. Sundaram V, Vermana G, Bhayani SB. Institutional review board approval and innovation in urology: current practice and safety issues. BJU Int. 2014;113(2):343–7.
16. Lieberman I, Hemdon J, Hahn J, Fins JJ, Rezai A. Surgical innovation and ethical dilemmas: a panel discussion. Cleve Clin J Med. 2008;75(Suppl 6):S13–21.
17. Tan VK, Chow PK. An approach to the ethical evaluation of innovative surgical procedures. Ann Acad Med Singap. 2011;40(1):26–9.
18. Research, The National Commission for the Protection of Human Subjects of Biomedical and Behavioral. *The Belmont report: ethical principles and guidelines for the protection of human subjects of research*, E. Department of Health, and Welfare, editor. U.S. Department of Health and Human Services; 1979. www.hhs.gov.
19. Bernard HR, Hartman TW. Complications after laparoscopic cholecystectomy. Am J Surg. 1993;165(4):533–5.
20. Frader JE, Caniano DA. Research and innovation in surgery. In: McCullough LD, Jones JW, Brody BA, editors. Surgical ethics. New York: Oxford University Press; 1998. p. 217–41.
21. Bracken-Roche D, Bell E, Karpowicz L, Racine E. Disclosure, consent, and the exercise of patient autonomy in surgical innovation: a systematic content analysis of the conceptual literature. Account Res. 2014;21(6):331–52.
22. Device advice: investigational device exemption (IDE). June 2014. http://www.fda.gov/MedicalDevices/DeviceRegulationandGuidance/HowtoMarketYourDevice/InvestigationalDeviceExemptionIDE/.
23. PremarketNotification (510k). Aug 2014. http://www.fda.gov/?MedicalDevices/DeviceRegulationandGuindance/HowtoMarketYourDevice/PremarketSubmissions/PremarketNotification510k.default.htm.
24. Strong VE, Forde KA, MacFadyen BV, Mellinger JD, Crookes PF, Sillin LF, Shadduck PP. Ethical considerations regarding the implementation of new technologie s and techniques in surgery. Surg Endosc. 2014;28(8):2272–6.
25. Yang SH, Zhang YC, Yang KH, et al. An evidence-based medicine review of lymphadenectomy extent for gastric cancer. Am J Surg. 2009;197(2):246–51.
26. Barkun JS, Aronson JK, Feldman LS, et al. Evaluation and stages of surgical innovations. Lancet. 2009;374(9695):1089–96.
27. Ergina PL, Cook JA, Blazeby JM, et al. Challenges in evaluating surgical innovation. Lancet. 2009; 374(9695):1097–104.
28. Orlowski L, Wateska JP. The effects of pharmaceutical firm enticements on physician prescribing patterns. There's no such thing as a free lunch. Chest. 1992;102(1):270–3.
29. Perry JE, Cox D, Cox AD. Trust and transparency: patient perceptions of physicians' financial relationships with pharmaceutical companies. J Law Med Ethics. 2014;42(4):475–91.
30. Altieri MS, Yang J, Wang L, Yin D, Talamini M, Pryor AD. Surgeons' perceptions on industry relations: a survey of 822 surgeons. Surgery. 2017;162(1): 164–73.
31. Chaterji AK, Fabrizio KR, Mitchell W, Schulman KA. Physician-industry cooperation in the medical device industry. Health Aff (Millwood). 2008;27(6):1532–43.
32. Katz D, Caplan AL, Merz JF. All gifts large and small: toward an understanding of the ethics of pharmaceutical industry gift-giving. Am J Bioeth. 2003;3: 39–46.
33. Lee Char SJ, Hills NK, Lo B, Kirkwood KS. Informed consent for innovative surgery: a survey of patients and surgeons. Surgery. 2013;153(4):473–80.

Ethics and Breast Cancer

Amtul R. Carmichael and Kerstin Sandelin

Key Points

- Informed consent for breast surgery is the autonomous authorisation by the patient to undergo a breast procedure, ranging from lumpectomy to bilateral mastectomy and reconstruction.
- Patient's autonomy during the process of informed consent for breast surgery can only be truly respected, if appropriate and adequate information is given to them in a manner, which is understandable by the patient.
- Enabling, empowering and educating a woman with breast cancer to make the right choice between breast-conserving surgery and mastectomy are in concordance with the fundamental principle of bioethics, that is, respect for autonomy.
- For responsible healthcare professionals, it is important to gain clear and unambiguous knowledge of handling genetic information in order to protect the patients, their families and the public from undesirable social and ethical consequences.
- While the underutilisation of screening mammography can be attributed to socio-economic and cultural and geographic barriers, ethical principles need to be taken into account and highlighted especially information regarding the potential harm in false-positive tests and overdiagnosis.
- Few decision aids exist and health professionals need training in the communication process about DCIS [1]. Patient participation can be enhanced through printed materials, patient navigators, consumers, group aids, use of electronic media, etc.

A. R. Carmichael (✉)
University Hospitals of Derby and Burton NHS
Foundation Trust, Queens Hospital, Burton-on-Trent,
UK

University of Aston, Birmingham, UK
e-mail: amtulcarmichael@nhs.net

K. Sandelin
Department of Molecular Medicine and Surgery,
Karolinska Insitutet, Stockholm, Sweden

Information and Consent: Disclosure

Informed consent for breast surgery has many ethical considerations ranging from the obligations of healthcare professionals to disclose adequate information (regarding risks and benefits such as the risk of local recurrence, cosmetic outcomes and survival) to the capacity of the patient to understand the given information in order to make informed choices. In other words, informed consent for breast surgery is the autonomous

© Springer Nature Switzerland AG 2019
A. R. Ferreres (ed.), *Surgical Ethics*, https://doi.org/10.1007/978-3-030-05964-4_24

authorisation by the patient to undergo a breast procedure, ranging from lumpectomy to bilateral mastectomy and reconstruction. The process of informed consent for the breast surgery is bound by the same ethical principles that govern the practice of medicine and provide a framework for ethical practices. Nowhere else than in the practice of breast surgery, the principles respect for autonomy, non-maleficence, beneficence and justice become more important. The judgement to apply these ethical principles is a subjective process. Therefore, for breast surgery, a judgement based on the collective wisdom of the multidisciplinary teams is deemed to ensure that these principles have been applied in an objective manner. Though decisions regarding how to treat women with breast cancer are made in a multidisciplinary setting, however, it is the task of an individual surgeon to convey the decision to the patient and seek her consent for the proposed treatment. It is the responsibility of the surgeon to apply an ethical judgement while practising the process of informed consent. Injudicious use of this can lead to disastrous consequences for women. The Kennedy report questioned issues of informed consent in the context of management of surgical margins, local recurrence rates and cosmesis after breast cancer treatment [2]. There are three main elements of the process of informed consent relating to breast surgery (preconditions, information giving and proper consent), consisting of two elements of preconditions (competence and voluntariness), three elements of information giving (disclosure, recommendation and understanding) and two elements of consent (decision and authorisation) [3].

To consent for breast surgery, the first precondition is competence from the side of the individual, in the sense that they have the capacity to understand and decide about the information that has been provided. Women after receiving the diagnosis of breast cancer may be nervous, usually affected or distracted, which may negatively impact on their understanding. Inadequate understanding will most probably lead to a decision which may not be necessarily in the best interests of women [4, 5]. If a patient lacks the capacity of understanding because of mental health issues,

then a case can be made for surgery without her consent in extremely rare circumstances [6]. The second precondition for consenting for breast surgery is that the individual takes part in the process voluntarily. Clinical experience and scientific evidence show a wide variation in the understanding of patients regarding diagnosis, treatment options, possible benefits, risks, complications and prognoses [7]. A consent for breast surgery can be regarded as truly voluntary, if the patients are given information and choices in a manner which instils confidence in them. The whole process of giving information and enabling right choices requires fostering an environment where women regard themselves as the legitimate source of the authority and perceive themselves to be able and authorised to speak for themselves [8].

The second element of informed consent for breast surgery is the information (disclosure, recommendation and understanding) element. It entails disclosure of material information and recommendation of a treatment plan for breast cancer, in a way which is comprehensible and easily understood by the patient. Healthcare professionals need to disclose a core set of information to enable the patient to refuse or consent to a proposed intervention or treatment of breast cancer. For example, when consenting women for mastectomy, they need to understand that mastectomy may not remove all breast tissue, and in any such events, infrequent as it may be, re-excision may be required [9]. The consent for mastectomy and type and timing of reconstruction is a vast subject. The outcome and type of breast reconstruction not only has an impact on body image; it also affects women's psychosocial and sexual well-being [10]. The evidence from the published literature, a clear and full explanations of all potential complications such as breast implant-associated anaplastic large cell lymphoma (ALCL), surgeon's own results supported by clinical photographs and access to past patients, who have had breast reconstruction for support and discussion, are absolutely mandatory to ensure effectiveness of the informed consent process in breast reconstruction. This information can help patients to better understand the expected outcomes and make an informed choice about the type and timing of reconstruction. These discussions need to be appropriate, adequate and fully documented [11]. The third element of the informed consent for breast surgery is consent (decision and authorisation) itself, which in the case of breast surgery is extremely sensitive. The overriding principle in consenting women for breast surgery is respect for their autonomy. A patient's autonomy during the process of informed consent for breast surgery can only be truly respected, if appropriate, and adequate information is given to them in a manner, which is understandable by the patient. After understanding the risks, benefits, alternatives and complications of a particular treatment or intervention, a patient decides in favour of the treatment and authorises this chosen plan.

Breast Conservation vs Mastectomy

Giving women a true choice between breast-conserving surgery and mastectomy for the treatment of early breast cancer is one of the most important aspects in the management of breast cancer. Enabling, empowering, and educating a woman with breast cancer to make the right choice between breast-conserving surgery or mastectomy are in concordance with the fundamental principle of bioethics, that is, respect for autonomy. It is vitally important that women have a clear understanding of the survival and recurrence rate associated with these surgical procedures. There is evidence to suggest that most women have inadequate knowledge with which to make informed decisions about breast cancer surgical treatment [12]. This is in contradiction with the basic principles of ethical practice. The respect for autonomy can only be of any value, if women are empowered to decide from a position of knowledge. Several decision aids have been used to facilitate true knowledge. Decision aids have shown to be beneficial to enhance shared decision-making, decreasing decisional conflict and increasing knowledge and satisfaction [13].

It must be understood that decision-making between breast-conserving surgery and mastectomy is more complex than simply making rational choices between two treatments [14]. Despite all the research on this topic, there is no evidence to suggest that breast-conserving surgery offers overall psychological benefits. After breast-conserving surgery, concerns for body image are replaced by fear of recurrence as a source of psychological morbidity. It is interesting to note that mastectomy rates have remained the same despite women being offered the choice of treatment. This could be due to the self-determination and independent decision-making, reflecting that women have become independent and active decision-makers. There is also evidence to suggest that some women like to take a more passive or collaborative role in decision procedures and prefer to accept the recommendations of clinicians [15]. Where women choose to accept the recommendations of clinicians, it is absolutely mandatory for the healthcare professionals to ascertain that they have disclosed adequate information to the patient; also the healthcare professionals need to probe and ensure that patient is accepting this decision from a position of knowledge and understanding [16]. The healthcare professionals obtaining the consent need to exercise ethical judgement. This ethical judgement will depend upon the character, moral standing, sense of responsibility and accountability of the healthcare professional.

The diagnosis of cancer leads to intense anxiety, mainly for the reason of survival, but other factors also play a part. It is important to recognise these factors to facilitate a woman's autonomy to make a treatment decision. There are several reasons identified why women choose breast-conserving surgery or mastectomy, when they have had a true choice, but our understanding about the reasons for this choice is far from clear [17–19]. When faced with the choice, women need time and space to practise a degree of self-reflection, exploring their sense of self and identity [20]. The support of specialist nurses, family members and friends provides a safe environment for a woman to reflect her sense of identity and decide on her preferences. While thinking about the impact of surgery on their identity and self, women may take into consideration other underlying reasons. While exercising her right for autonomy, views formed by a woman during her experiential encounters have an impact on the choice of surgical treatment. For example, the evidence suggests that if women observe a negative experience after breast-conserving surgery in a close family member or friend or witness a positive mastectomy experience, they are more likely to choose mastectomy as the part of treatment for early breast cancer [21]. The evidence suggests that for some women, experiential interpretation rather than statistical probabilities are the deciding factors, when making a choice for the type of surgery for the treatment of early breast cancer [19, 22]. Women making such choices are influenced by the perceived risk of each treatment options, more than the objective information and explanation given by the healthcare professionals [23].

Genetic Testing and Implications: Prophylactic Surgery

The immense scientific breakthrough in the field of genetics in the last decade has led to several ethical challenges relating to the individuals, their families and the society as a whole and highlighting the importance of the informed consent process. It is vital for a practising breast healthcare professional to be able to sensitively and thoughtfully navigate through the process and implication of genetic testing and its consequent prophylactic surgery. A clear understanding of the contemporary ethical and social issues relating to genetic testing for breast cancer is necessary to develop a practical approach for counselling, testing and treating patients with a genetic disposition to breast cancer [24]. It is well recognised that there are significant gaps in the genetic knowledge of the public and the professionals [25, 26]. The respect for the autonomy of the patient remains the overarching principle while addressing the ethical issues relating to genetic testing for breast cancer and offering risk-reducing surgery. The implications of

genetic testing are based on the principle of 'primacy of patient' and ensure that the patient remains the primary concern of the healthcare professionals in terms of respect for patient's autonomy, individuality, welfare and freedom [27]. The respect for the autonomy of the patient demands that healthcare professionals respect and promote the capacity of an individual to make a sound decision. The fundamental principle that is ingrained in the informed consent process also applies to the genetic testing in the sense that the healthcare professional should facilitate patients to decide for themselves without personally influencing them, termed as nondirective genetic counselling [28].

The implications of the results of genetic testing are not limited exclusively to the patients themselves, because family members share genes; and the detection of a germline mutation of BRCA genes may have deep implications for the individuals genetically related to the patient. It is a moral obligation and legal duty to warn at-risk relatives, if there is a potential for serious and avoidable harm by the disease. The duty of healthcare professional is to encourage patients to communicate with the relatives to convey the results of abnormal genetic testing. If attempts to encourage patients to communicate with their families fail, and the condition is considered to be serious and treatable, and harm of disclosure is deemed to be less than the harm of nondisclosure, then it may be appropriate to breach the confidentiality of the patient [29]. This can be justified on the ethical principle that nondisclosure undermines the decision-making autonomy of at-risk relatives [30]. This, a major ethical challenge of decision-making while handling genetic information, contests the ethical principle of patient's autonomy and forces the recognition of familiar and global implications [24]. For responsible healthcare professionals, it is important to gain clear and unambiguous knowledge of handling genetic information in order to protect the patients, their families and public from undesirable social and ethical consequences [31]. Increasing availability of refined genetic information has created new moral choices and challenges. There are several ethical considerations when deciding to undergo prophylactic mastectomy and reconstruction by patients diagnosed with a germline mutation of BRCA genes. Women can perceive this genetic knowledge as a means to enhance control of their personal health and well-being, thus seeking interventions to minimise the risk of developing breast cancer [32]. This increase in demand poses the challenge of balancing the ethical principles of autonomy and beneficence against non-maleficence and justice.

According to Jacobson, the first prophylactic mastectomy reported was carried out in 1917 for a rather paradoxical reason: '… the fear of having the breast mutilated keeps patients away and allows a tumour to run a progressive course' [33]. The breast is a special organ, and it is deeply connected with a sense of femininity, sensuality, sexuality, adulthood and motherhood. Therefore, it is argued that irreversible mutilation of this organ deserves caution and attention [34]. The ethical responsibility of the healthcare professional is to explain to patients the risks associated with the risk-reducing mastectomy and reconstruction and convey the expected protection from cancer in a meaningful and understandable manner along with its potential side effects. It is an ethical responsibility of each clinician in this particular field to enable the woman to understand the benefits and the personal costs involved in such an emotive procedure. In this circumstance, the final word belongs to the woman, and the legal, economic or psychological and cultural factors have much lighter-weighting in this ethical consideration.

Access to Treatment in Resource Constrained Environments: Compliance with Guidelines

Ethnic inequities, disparities and timeliness to treatment and its prognostic significance on breast cancer mortality have been studied in several populations worldwide. Being cancer the leading cause of death among non-communicable diseases prompted the World Health Organization in 2017 to present key drivers that cover the structure, process and outcome of cancer including palliative care and rehabilitation [35–38].

The Breast Health Global Initiative duly collaborates with national and international health organisations. Evidence-based guidelines stratifying recourses into basic, limited, enhanced and maximal levels were set in the early 2000 and further implemented. Basic resources are defined as having the core facilities for any cancer, limited includes availability of both diagnostic and the least expensive treatment options and enhanced includes third tier services that can improve outcome and quality for patients. Maximum level resources include those included in guidelines without resource constraints. However, they must depend on existence and functionality [37].

Guidelines give a structured framework for management in being evidence-based and having quality indicators to ensure equal care. Some countries have modified existing guidelines that better correspond to the epidemiologic and social structure of the countries [39]. A recent Cochrane review focusing on the effect of interventions for raising awareness found a paucity of high-quality studies addressing the issue [40]. The review concludes that the combination of written information and face-to-face communication improved awareness. Mammography screening as a rapid cost-efficient tool now used for 40 years has been an instigator for early detection and awareness. There is disagreement among experts about its value and whether screening yields more harm than benefit. A difference in screening behaviour was noted among Haitian women residing in Florida compared to other women living in the same region. No specific risk factor was found albeit late presentation and poorer outcomes for the Haitian women prevailed. Previous studies encountered communication barriers among the residents [41]. While the underutilisation of screening mammography can be attributed to socioeconomic and cultural and geographic barriers, ethical principles need to be taken into account and highlighted especially information regarding the potential harm in false-positive tests and overdiagnosis. Moreover, there is a discrepancy between the public perception of the screening benefits and the real benefits for the individual. Autonomy according to personal values is not being met in the screening programs [42]. To improve personal decision-making process, a framework tool was developed for ethical decision-making. The tool can serve as an adjunct for discussion of important issues in screening mammography where values are incorporated concurrently with evidence [43].

Ductal Carcinoma In Situ

Ductal carcinoma in situ (DCIS) often diagnosed in asymptomatic women through mammography screening of cases represents a complex heterogeneous pathologic entity in its management. The risk of invasive recurrence without robust risk factors leads to radical surgery and radiotherapy although overall prognosis for treated DCIS is very good and mortality rates for DCIS are very low. A scoping review found that women are not knowledgeable about the disease nor about its prognosis and that physicians had difficulties in communicating facts about DCIS to patients whereby dissatisfaction with information occurred. Few decision aids exist and health professionals need training in the communication process about DCIS [1]. Patient participation can be enhanced through printed materials, patient navigators, consumers, group aids, use of electronic media, etc. Recently active surveillance protocols are being offered to women with low-grade DCIS which raises concerns in the informed decision process and the ethical concepts of no harm and autonomy that apply [44, 45].

Final Remarks

- Breast surgery should be guided by stringent ethical principles to prevent paternalistic approaches, inequities and unfairness in access to therapeutic conservative options.
- Compliance with an adequate and thorough informed consent process is mandatory.
- In restrictive resources settings, agreement with clinical guidelines and pathways should be promoted.

- Tailored approaches and precision medicine in breast malignancies request an ethical approach from all the members of multidisciplinary teams.

References

1. Kim C, Liang L, Wright FC, Hong NJL, Groot G, Helyer L, Meiers P, Quan ML, Urquhart R, Warburton R, Gagliardi AR. Interventions are needed to support patient-provider decision-making for DCIS: a scoping review. Breast Cancer Res Treat. 2018;168(3):579–92. https://doi.org/10.1007/s10549-017-4613-x.
2. Kennedy Report. Review of the response of Heart of England NHS Foundation Trust to concerns about Mr I Paterson's Surgical Practice: lessons to be learned and recommendations.
3. Beauchamp TL, Childress JF. Respect for autonomy from the book principle of biomedical ethics. 7th edition. New York, Oxford: Oxford University press.
4. Mendick N, Young B, Holcombe C, Salmon P. The ethics of responsibility and ownership in decision-making about treatment for breast cancer. Triangulation of consultation with patient and surgeon's perspective. Soc Sci Med. 2010;70(12):1904–11.
5. Caldon L, Walters SJ, Reed MW. Changing trends in decision-making preferences of women with early breast cancer. The. Br J Surg. 2008;95(3):312–8.
6. Dyers C. Woman who rejected breast cancer diagnosis may undergo surgery without her consent. BMJ. 2017;j5358:359.
7. Bernhardt BA, et al. Educating patients about cystic fibrosis carrier screening in a primary care setting. Archives of family medicine five; 1996. p. 336–40.
8. McKenzie C. Relational autonomy, normative authority and perfectionism. J Soc Philos. 2008;39:512–33.
9. Bundred NJ, Thomas J, Dixon JMJ. Whither surgical quality assurance of breast cancer surgery (surgical margins and local recurrence) after Paterson. Breast Cancer Res Treat. 2017;165:473.
10. Pusic AL, Matros E, Fine N, Buchel E, Gordillo GM, Hamill JB, Kim HM, Qi J, Albornoz C, Klassen AF, Wilkins EG. Patient-Reported Outcomes 1 Year After Immediate Breast Reconstruction: Results of the Mastectomy Reconstruction Outcomes Consortium Study. J Clin Oncol. 2017;35(22):2499–506.
11. Association of Breast Surgery at BASO. Surgical guidelines for the management of breast cancer. Eur J Surg Oncol EJSO. 2009;1:1–22.
12. Fagerlin A, Lakhani I, Lantz PM, Janz NK, Morrow M, Schwartz K, Deapen D, Salem B, Liu L, Katz SJ. An informed decision? Breast cancer patients and their knowledge about treatment. Patient Educ Couns. 2006;64(1–3):303–12.
13. Zdenkowski N, Butow P, Tesson S, Boyle FA. Systematic review of decision aids for patients making a decision about treatment for early breast cancer. Breast. 2016;26:31–45.

14. Twomey M. Autonomy and reason: treatment choice in breast cancer. Eval Clin Pract. 2012;18(5): 1045–50.

15. Swainston K, Campbell C, van Wersch A, Durning P. Treatment decision-making in breast cancer: a longitudinal exploration of women's experience. Lit Health Psychol. 2012;17(1):155–70.

16. Beauchamp TL, Childress JF. Principles of biomedical ethics. New York: Oxford University press; 2009.

17. Temple WJ, Russell ML, Parsons LL, et al. Conservation surgery for breast cancer as the preferred choice: a prospective analysis. J Clin Oncol. 2006;24:3367–73.

18. Covelli AM, Baxter NN, Fitch MI, McCready DR, Wright FC. "Taking control of cancer": understanding women's choice for mastectomy. Ann Surg Oncol. 2015;22:383–91.

19. Molenaar S, Oort F, Sprangers M, et al. Predictors of patients' choices for breast-conserving therapy or mastectomy: a prospective study. Br J Cancer. 2004;90:2123–30.

20. Keller J. Autonomy, relationality and feminist ethics. Hypatia. 1997;12:152–64.

21. Jeffrey G, et al. Understanding women's choice of mastectomy versus breast conserving therapy in early-stage breast cancer. Clin Med Insights Oncol. 2017;11:1179554917691266. *PMC*. Web. 19 Aug. 2018.

22. McVea KL, Minier WC, Johnson Palensky JE. Low-income women with early stage breast cancer: physician and patient decision-making styles. Psychooncology. 2001;10:137–46.

23. Zikmund-Fisher BJ, Fagerlin A, Ubel PA. Risky feelings: why a 6% risk of cancer does not always feel like 6%. Patient Educ Couns. 2010;81: S87–93.

24. Qullin JM, Lyckholm LJ. A principle-based approach to ethical issues in predictive genetic testing for breast cancer. Breast Dis. 2007;27(2006): 137–48.

25. Lanie AD, Jayaratne JP. Exploring the public understanding of basic genetic concept. J Gene Couns. 2004;13:305–20.

26. McInerney J. Education in a genomic world. J Med Philos. 2002;27:369–90.

27. National Society of genetic counsellors. NSGC: code of ethics.

28. Lo B. Overview of the doctor-patient relationship. In: Resolving ethical dilemmas: a guide for clinicians. 2nd ed. Philadelphia: Lippincott Williams & Williams; 2000. p. 195.

29. ASHG statement, professional disclosure of familiar genetic information. the American Society of human genetics social issue subcommittee unfamiliar disclosure. Am J Hum Genet. 1998;62: 474–83.

30. Wilson BJ, Forrest K, van Teijlingen ER, McKee L, Haites N, Matthews E, Simpson SA. Family communication about genetic risk: the little that is known. Community Genet. 2004;7:15–24.

31. Surbone A. Social and ethical implications of BRCA testing. Ann Oncol. 2011;22(Suppl 1):I 60–6.

32. Surbone A. Balance between science and mortality. Ann Oncol. 2004;15(Suppl 1):i60–4.

33. Jacobson N. The socially constructed breast: breast implants and the medical construction of need. Am J Public Health. 1998;88:1254–61.

34. Eisinger F. Prophylactic mastectomy: ethical issues. Br Med Bull. 2007;81(82):7–19.

35. Seneviratne S, Campbell I, Scott N, Lawrenson R. A cohort study of ethnic differences in use of adjuvant chemotherapy and radiation therapy for breast cancer in New Zealand. BMC Health Serv Res. 2017;17:64. https://doi.org/10.1186/s12913-017-2027-4PMCID.

36. Raphael MJ, Biagi JJ, Kong W, Mates M, Booth CM, Mackillop WJ. The relationship between time to initiation of adjuvant chemotherapy and survival in breast cancer: a systematic review and meta-analysis. Breast Cancer Res Treat. 2016;160(1):17–28.

37. Al-Sukhun S, de Lima Lopes G Jr, Gospodarowicz M, Ginsburg O, Yu PP. Global Health Initiatives of the International Oncology Community. Am Soc Clin Oncol Educ Book. 2017;37:395–402. https://doi.org/10.14694/EDBK_100008.

38. Distelhorst SR, Cleary JF, Ganz PA, Bese N, Camacho-Rodriguez R, Cardoso F, Ddungu H, Gralow JR, Yip CH, Anderson BO. Optimisation of the continuum of supportive and palliative care for patients with breast cancer in low-income and middle-income countries: executive summary of the Breast Health Global Initiative, 2014. Breast Health Global Initiative Global Summit on Supportive Care and Quality of Life Consensus Panel Members. Lancet Oncol. 2015;16(3):e137–47.

39. O'Mahony M, Comber H, Fitzgerald T, Corrigan MA, Fitzgerald E, Grunfeld EA, Flynn MG, Hegarty J. Interventions for raising breast cancer awareness in women. Cochrane Database Syst Rev. 2017;2:CD011396.

40. Abulkhair O, Saghir N, Sedky L, Saadedin A, Elzahwary H, Siddiqui N, Al Saleh M, Geara F, Birido N, Al-Eissa N, Al Sukhun S, Abdulkareem H, Ayoub MM, Deirawan F, Fayaz S, Kandil A, Khatib S, El-Mistiri M, Salem D, Sayd el SH, Jaloudi M, Jahanzeb M, Gradishar WI. Modification and implementation of NCCN guidelines on breast cancer in the Middle East and North Africa region. MENA Breast Cancer Regional Guidelines Committee. J Natl Compr Cancer Netw. 2010;8(Suppl 3): S8–S15.

41. Kobetz E, Mendoza AD, Barton B, Menard J, Allen G, Pierre L, Diem J, McCoy V, McCoy C. Mammography use among Haitian women in Miami, Florida: an opportunity for intervention. J Immigr Minor Health. 2010;12(3):418–21.

42. Parker L, Carter S, Williams J, Pickles K, Barratt A. Avoiding harm and supporting autonomy are under-prioritised in cancer-screening policies and practices. Eur J Cancer. 2017;85:1–5. https://doi.org/10.1016/j.ejca.2017.07.056.

43. Parker L. Including values in evidence-based policy making for breast screening: An empirically grounded tool to assist expert decision makers. Health Policy. 2017;121(7):793–9. https://doi.org/10.1016/j.healthpol.2017.03.002.

44. Francis A, Thomas J, Fallowfield L, et al. Addressing overtreatment of screen detected DCIS; the LORIS trial. Eur J Cancer. 2015;51(16):2296–30.

45. Elshof LE, Tryfonidis K, Slaets L, et al. Feasibility of a prospective, randomised, open-label, international multicentre, phase III, non-inferiority trial to assess the safety of active surveillance for low risk ductal carcinoma in situ - The LORD study. Eur J Cancer. 2015;51(12):1497–510.

Ethical Issues in Pediatric Liver Transplantation

Imventarza Oscar Cesar and Rojas Luis Daniel

Key Points
- Pediatric liver transplantation.
- Ethical issues in pediatric liver transplantation.
- Who should receive an organ and who should pay the high cost of the procedure?
- Living donor and/or cadaveric liver transplantation.

Introduction

The first experimental trials of organ transplants were started in 1902 by Alexis Carrel.

Until the mid-1950s, the definition of death was cardiorespiratory arrest. The technical measures of resuscitation showed the possibility of keeping pulmonary oxygenation, the heart beating, and blood pressure in patients with cerebral electrical silence. The first criteria of brain death were developed in 1968 by the Ad Hoc Committee of the Harvard Medical School [1].

I. O. Cesar (✉)
Hospital Argerich- Hospital Garrahan,
Buenos Aires, Argentina

R. L. Daniel
EAIT (Transplant Institute of Buenos Aires City),
Buenos Aires, Argentina

Since 1980, the transplantation of solid organs has become a regular practice. As well as the indications expand, the waiting lists grow exponentially almost disproportionate. Donors remain stable with a small increase without expectations of meeting the demand needs for enough organs for transplantation [2]. This situation generated ethical issues and conflicts within the medical community as well as in the society as a whole. The questions that frequently arise include the following:

- Does everyone have the same right to access an organ?
- Who should be an organ recipient, the one in the most severe clinical condition or the one with the best chance of long-term survival?
- Who should cover and pay for the high costs of the procedure and the subsequent medication requested for life?
- Is the use of the living donor approach the solution to the shortage of some organs such as the kidney, the liver, the lung, and eventually the intestine [3, 4]
- Throughout the chapter, approaches and suggestions will be offered to provide answers and solutions to these questions from the perspective of bioethics.

The organ and tissue transplantation started as a concept to treat several terminal diseases. This type of treatments began in the early 1954 with corneal transplantation. In the middle of

© Springer Nature Switzerland AG 2019
A. R. Ferreres (ed.), *Surgical Ethics*, https://doi.org/10.1007/978-3-030-05964-4_25

the same century, the first attempts to perform transplants with irrigated organs in humans began. The first successful kidney transplantation was performed by Joseph Edward Murray in 1954 on Ronald and Richard Herrick, who were identical twins, at Brigham and Women's Hospital Boston, MA [5].

The damages caused by the ischemia and reperfusion process and rejection as an immunological phenomenon are the most difficult obstacles to overcome from the technical point of view. Some attempts were made to try to solve those problems with post-cardiac arrest donors or with identical siblings' live donors. In 1968, the Committee of the Harvard Medical School integrated by ten physicians, a lawyer, a theologian, and a historian laid out the first criteria for the determination of death based on a total and permanent encephalic damage, hence developing the concept of brain death [1]. Thus, the foundations for the diagnosis of death under neurological criteria were laid, in this way it is possible to diagnose death without the need to wait for cardiac arrest and therefore to harvest irrigated and oxygenated organs with greater chance of good function after transplantation. Subsequently, preservation solutions were developed to improve the quality of harvested organs and increased preservations times after harvesting. Finally, the last problem that appeared was the rejection control, omnipresent ghost after the transplant. At the beginning of the 1980s, the drug cyclosporine A appeared as a new immunosuppressive agent that changed the evolution of patients in a radical way [6]. From this moment on, the growth and development of transplantation did not stop growing, and the results were increasingly promising. At the same time, there was a need to create organ procurement agencies. At that time the National Transplant Organization (ONT) model – wordly known as the "Spanish model" – began to increase the donor pool in Spain and has proved to be the most efficient so far [7, 11].

Pediatric Liver Transplant

The first liver transplant was performed by Dr. Thomas Starzl in a pediatric patient in Denver, Colorado, in March 1963 [8]. This procedure was initially considered experimental, and the first results were not encouraging, but Dr. Starzl continued working in this procedure. Liver transplantation was considered as a standard practice and finally approved by the US National Institute of Health in 1983 and became a regular practice around the world [9].

Although pediatric liver transplantation is a procedure recognized as a standard procedure of care with very good long-term results, it still presents challenges which request an ethical approach and solutions [2]. The four principles of medical bioethics (beneficence, non-maleficence, autonomy, and justice) must be observed for the greatest benefit to the patient and in order to avoid harm to him or her or third parties and always respecting the autonomous decision of the patient and/or his family.

To solve the issue of donor shortage, different strategies have been implemented to adequate supply of organs to the increasing demand. There are two options: cadaveric and living donors.

Splitting the liver is the best choice procedure in the first group. There are two types of living donors: the living related donors (LRD) and the non-related living donor (NRLD) [10]. Regarding the non-related type, it is rejected by many physicians and ethicists, except in very exceptional situations, since it promotes situations that are unclear and could be associated with organs' sale and trafficking. Despite this risk, some utilitarian advocates have tried to introduce it as a further strategy [12]. The benefit of the cadaveric donor is to avoid the harm to a healthy person, but, on the other hand, the living donor avoids the prolonged time in the waiting list.

Living related liver donors are universally accepted even more in pediatric patients, since the results for the recipient are comparable with standard techniques. However, from the perspective of bioethics, this approach deserves further considerations and discussion. The donor does not profit in any sense and could suffer harm due to the partial liver resection from another point of view, we should also consider the satisfaction and welfare state that their parents, uncles or grandparents who suffer great anguish to see the child suffer and he or she has been recoverd after a living related donor surgery [12]. This situation alters family dynamics as well as the health of donors and their environment, according to the World Health Organization definition of health as an optimal biopsychosocial balance. On the contrary, a full benefit is achieved for the pediatric recipient who receives an optimal organ with a minimum of cold ischemia time and avoiding prolonged waiting list times, which is known to be longer and more deleterious to the general state of the potential recipient.

The informed consent process (IC) is a very important tool because many times, the donors suffer pressures, and they feel guilty or fear being discriminated against, among other things for not being donors. These circumstances make them vulnerable, and their autonomy may be compromised to a greater or lesser degree. Therefore, the IC should not be a mere procedure but a process where autonomy and full freedom of decision are guaranteed without pressure or manipulation. As far as possible, the donor must be approached by a multidisciplinary team that will guarantee the absolute transparency and safety of the process. In addition, it must be discussed if the benefit the recipient will obtain and the probability of success of the procedure justify the potential risk the donor will undertake, given that if the transplant fails or the recipient dies, a feeling of guilt or futility may be developed by the donor due to the loving act of donation.

In Argentina as well as in many other countries, the legal regulations establish the age of 16 years as the limit for a patient to accept or neglect from a given treatment. Under that age, parental consent is requested. These principles not only emerge from bioethics but also from the Convention on the Rights of Children established since 1989 by the General Assembly of the United Nations, signed by the majority of the countries represented at the UN, and introduce the concept of childhood as a subject of rights and promoting the Comprehensive Protection of Children.

Regarding NRLD, the position is a little bit different from the bioethical perspective. Being an individual without a close relationship or affinity, the possibility of manipulation or some form of retribution may affect the act in an imperfect way. Although some authors favor the NRLD strategy, it is generally rejected and condemned by most bioethical trends, such as the personalist one. Despite the permanent scientific advances in all fields of transplantation, the shortage of organs is an unresolved problem and therefore the greatest pitfall of this practice at present times. When increasing the applicability, the immediate consequence is a greater demand of organs and a significant increase of the waiting lists since the demand for organs is not satisfied.

Anyway, although pediatric liver transplantation should be considered a standard practice with very good results, its indication must be carefully evaluated to comply not only with the precepts of good clinical practice but also with basic bioethical precepts.

The waiting list is a fundamental aspect to consider, regulations in many countries determine that donors under 18 years old are first awarded to the pediatric recipients or pediatric recipients in the waiting list are given extra scores in order to prioritize this population. This is justified because the list of adult patients far exceeds the list of pediatric ones, and children would be harmed by prolonging their placement in the waiting list. Children under 3 years with terminal liver failure suffer a cognitive decline with quick evolution, as shown by Robertson [13]. These patients usually present hepatopulmonary syndrome, pulmonary arterial hypertension, or hepatic osteopathy, among other complications, added to the nutritional status and the cirrhosis; this is the reason why it is imperative to perform the transplantation as early as possible to avoid the irreversible damage in the central nervous system.

Although the waiting list can prioritize the pediatric patient, it is not as effective in practice as in many cases the wait list time is also extended, so different strategies have been implemented such as the split liver (SL) technique that allows transplanting children up to 25 kg with the liver left lateral segment (LLS) of a cadaveric donor and in turn implant the remaining liver in an adult or adolescent. As a last measure, in cases of extreme need, this technique may be applied to a living related donor, with the due ethical considerations raised above. Many countries have adopted the model used by the US United Network Organ Sharing (UNOS) called PELD-MELD (pediatric end-stage liver disease-model end-stage liver disease). Through a mathematical formula, this tool predicts in a relative accurate way the risk of life after 3 months, but at the same time, there are numerous exception pathways requested to achieve the upgrade of the score since this model does not always reflect the real severity of the patient.

Since not all authors agree to prioritize the most severe patient or the one with the highest risk of death, but the patient with the greater chances of survival after transplantation, a survival-benefit model was recently proposed to answer requests to increase efficiency and reduce futile transplants, as proposed by Schaubel and Keller [14]. This proposal was evaluated by UNOS/OPTN (United Network Organ Sharing/Organ Procurement and Transplantation Network), but they considered there was no need for change. This model can be considered as utilitarian by personalistic ethics; in this way there are two almost antinomic approaches: the implant should be assigned to the sickest or the one who has the best possibilities to survive. This antinomy may be overcome if the assessment is finely balanced to generate fair decisions and allow an optimization in the awarding of the donated livers.

Transplantation is recognized as a standard procedure, but it can sometimes become an extraordinary and disproportionate practice if the risks and benefits of its implementation are not correctly assessed, so as not to incur in therapeutic futility, which is contrary to human dignity. The surgeon must put aside his or her almightiness and recognize that patients with critical and terminal illnesses cannot always be saved, and they should be just accompanied through the death process in the best and most dignified way. On the other hand, using a scarce resource such as a cadaveric organ or exposing a healthy person to the risk of complications when the chances of

success are very low is also a position opposite to ethics and sound medical practice, also removing a single possibility of survival to those who would have had the opportunity of a favorable evolution.

The ethical issues in pediatric liver transplantation not only go through the questions of proportionality and the chances of success of the transplantation but also that the accessibility to treatment is unrestricted as long as there are no formal contraindications. This accessibility not only refers to the possibility of obtaining an organ but to sustain medical monitoring and immunosuppressive medication over time, thus fulfilling the principles of beneficence, equity and justice. Unfortunately, many parents migrate with their sick children from countries where transplantation is not performed or is inaccessible for them to countries where they can do it, if all the pitfalls may be overcome such as the inclusion of an alien patient in the local waiting list, or forcing the use of a related living donor. If the transplant takes place, several problems will arise: with whom and where the medical follow-up will continue as well as the coverage for the high-cost medication in the long-term postoperative follow-up. This is a problem that occurs in some countries with high poverty rates throughout the world [15].

Once presented the main ethical issues and implications in pediatric liver transplantation, its resolution should be framed through the principles described by Beauchamp and Childress: autonomy, beneficence, non-maleficence, and justice [16].

The living related donor scenario is crossed by the four principles. From the point of view of autonomy, the individual possesses the full right to be a donor and may exercise it; as a safety measure, the donor must and should be informed in a truthful, complete, understandable fashion and without manipulation regarding the potential risks of a donor and the chances of success. The beneficence principle is explained by the satisfaction and well-being caused by the fact of having helped a loved one. On the other hand, non-maleficence must be considered because the damage to the donor must never be greater than the benefit provided to the recipient; in order to safeguard this principle, a complete and thorough psychophysical evaluation of both actors is necessary, the donor and the recipient, to avoid in this way a futile transplant or an excessive risk for the donor. Finally, the principle of justice involves the need for equitable treatment for donor and recipient, as well as the profit for the society as a whole.

Another point to address is the one that refers to the conflict between the allocation of the organ to the sickest patient or to the one who is more likely to achieve a better survival, with specific reference to the pediatric recipients, who are usually in a disadvantageous position. The first principle that emerges is that of justice: an assessment and evaluation of each patient are required to avoid inequities; failure to choose the correct recipient harms the patient in the only chance he/she can survive. In these situations, the application of the principle of unrestricted beneficence can invalidate the principle of non-maleficence, since the medical possibility is not always ethical.

The number of adult recipients at least triples the pediatric recipients in the waiting list for liver recipients. This situation generates inequities and would not comply with the principle of justice. This explains why several strategies have been devised to solve this issue, for example, (1) multiplying the PELD score × 3, (2) every pediatric donor organs are granted in the first instance to pediatric recipients, (3) the split liver is mandatory in all optimal donors, and (4) the LLS is granted to a pediatric recipient first.

Finally, it must be recognized that all over the world, there are vulnerable children due to their social, socioeconomic, politic, ethnic, and/or racial status and not because of their medical condition, who will probably never access expensive and complex treatments such as liver transplantation. Global ethics should provide an answer and a solution to this dilemma.

Key Points

- The biggest problem that faces organ transplantation is the shortage of donors.
- Pediatric liver transplantation has a numerical inferiority in the waiting list in relation to adults.
- In spite of multiple strategies, this problem is not solved without showing positive impact in the waiting list.
- Ethical problems are related to LD and the selection of the donor, equity in the distribution, and equipoise between the sickest and best match.
- Transplantation is a high-cost practice and needs lifelong support. Who pays for this vulnerable population? Argentina law protects these patients.

Conclusions

1. Liver transplantation is the last resort available to solve pediatric patients with terminal liver disease, having this procedure been validated since 1983 by the NIH.
2. The biggest issue is the lack of organs, which is a constraint problem for all world health systems.
3. The shortage of donors has led the medical community to develop different techniques to achieve the best use of the cadaveric organs and develop the living donor technique.
4. This problem raises not only medical challenges but also bioethical conflicts such as the allocation of living donors' organs, patients' prioritization, competition between pediatric and adult patient, as well as the accessibility and the high cost of chronic treatment.
5. The sound advice from a bioethics committee should be part of the evaluation of the patient's candidates for liver transplant in questionable situations, even more if the patient is a child.

References

1. Ad Hoc Committee of the Harvard Medical School to examine the definition of brain death. A definition of irreversible coma. JAMA. 1968;205:337–401. Eric J. Keller, Paul Y.
2. Keller EJ, Kwo PY, Helft PR. Ethical considerations surrounding survival benefit–based liver allocation. Liver Transpl. 2014;20:140–6.
3. Wells WJ, Barr ML. The ethics of living donor lung transplantation. Thorac Surg Clin. 2005;15(4):519–25.
4. Mendoza Sánchez F, Sereno Trabaldo S, Montemayor Cantú G, et al. Trasplante intestinal de donador vivo relacionado: Primer reporte en Latinoamérica. Rev Invest Clin. 2011;63(Suppl 1):96–100.
5. Welman T, Villani V, Shanmugarajah K, et al. From kidney transplants to vascularized composite allografts: the role of the plastic surgeon in transplantation vascularized composite. Allotransplantation. 2015;2(4)
6. Borel JF, Kis ZL. The discovery and development of cyclosporine (Sandimmune). Transplant Proc. 1991;23:1867–74.
7. Matesanz R (Editor). El modelo español de coordinación y transplantes. 2ª Edición. 2008. GRUPO Aula Médica, Madrid, 2008.
8. Starzl TE, Putnam CW, editors. Experience in hepatic transplantation. Philadelphia: WB Saunders Company; 1969. p. 131–5.
9. Liver Transplantation. NIH Consensus Statement. 1983;4(7):1–15. US Department of Health and Human Services. Hepatology. 1984;4(1 Suppl):107s–110s.
10. Singer PA, Siegler M, Whintington PF, et al. Ethics of Liver transplantation with living donors. N Engl J Med. 1989;321:620–2.
11. Organizació Catalana de Trasplantaments, Servei Català de la Salut, Societat Catalana de Transplantament. Etica de la donación en vivo (Editorial). Butlleti Trasplantaments. 2007;36:1.
12. Chugh KS, Jha V. Treatment modality options: Problems and outcomes of living unrelated donor transplants in the developing countries. Kidney Int. 2000;57:S131–5.
13. Robertson CM, Dinu IA, Joffe AR, et al. Neurocognitive outcomes at kindergarten entry after liver transplantation at 3yr of age. Pediatr Transplant. 2013;17:621–30.
14. Schaubel DE, Guidinger MK, Biggins SW, et al. Survival benefit based deceased-donor liver allocation. Am J Transplant. 2009;9:970–81.
15. www.who.int/about/mission/en/. Accessed 17 Mar 2018.
16. Beauchamp TL, Childress JF. Principles of biomedical ethics. Oxford: Oxford University Press; 1989.

Suggested Bibliography

Beauchamp TL, Childress JF. Principles of biomedical ethics. New York: Oxford University Press; 1979.

Brown H. La nueva filosofía de la ciencia. Madrid: Tecnos; 1983.

Popper K. La lógica de la investigación científica. 13° ed. Madrid: Tecnos.

Sgreccia E. Manual de Bioética, vol. I. Madrid, BAC; 2009.

Valapour D. Living donor transplantation: the perfect balance of public oversight and medical responsibility. J Clin Ethics. 2007 Spring;18(1):18–23.

Ethical Issues in Cardiothoracic Surgery

Richard I. Whyte and Douglas E. Wood

Introduction

If one defines ethics as the study of conduct and moral judgment, it is not hard to understand Pellegrino and Relman's statement that "medicine is, in essence, a moral enterprise" [1]. We, as physicians, strive to do the "right" thing for our patients, yet, as anyone who has practiced medicine has realized, controversies emerge on a daily basis and, on longer time frames, techniques and practices constantly evolve making it difficult to always know what is "right." Given the fact that what we do often has negative as well as beneficial effects on our patients, many controversial areas in medicine can be seen to have an ethical component. One can easily extrapolate this to the field of surgery where we are enjoined to use our surgical skills and knowledge for the benefit of our patients (beneficence) and with a minimum of associated harm (non-maleficence). Similarly, we are expected to do this after we have obtained informed consent (Respect for Persons) and to treat all patients with respect and dignity (Justice). These four ethical principles, as defined by Beauchamp and Childress, form a useful, but not exclusive, framework for examining ethical issues pertaining to the field of surgery [2]. The specialty of cardiothoracic surgery shares these moral principles and related ethical challenges with other branches of surgery yet must deal with them in contexts unique to the field. The goal of this chapter is to identify some of the ethical challenges that, while perhaps not unique to cardiothoracic surgery, are currently topics of current controversy.

Cardiothoracic surgery is a field of high technical complexity and one which is associated with substantial benefit to patients—albeit at the cost of significant risks. The field is diverse and encompasses general thoracic surgery, adult cardiac surgery, and congenital heart surgery and includes subjects that traverse all three areas, critical care and transplantation being two such examples. As with any high-risk area within surgery, certain topics have clear ethical dimensions that have been extensively covered in this and other works [2–4]. While the context and underlying disease are different depending on whether one is dealing with an adult cardiac surgical patient, a general thoracic surgical patient, or congenital heart patient, the concepts of balancing beneficence, the prolongation of suffering, and honoring the wishes (or presumed wishes) of the patient regarding prolonged ICU care and predicted outcomes are concepts that have been extensively written about [5–7].

R. I. Whyte (✉)
Harvard Medical School, Boston, MA, USA
e-mail: rwhyte@bidmc.harvard.edu

D. E. Wood
University of Washington, Seattle, WA, USA

© Springer Nature Switzerland AG 2019
A. R. Ferreres (ed.), *Surgical Ethics*, https://doi.org/10.1007/978-3-030-05964-4_26

Ethical Frameworks

In many clinical scenarios in which the surgeon is given a choice of action, the morally correct one is obvious. Although a duty-based ethical framework (respect for autonomy, beneficence, non-maleficence, and justice) may not be overtly considered, experience and common morality make it simple to pick the right moral choice. In other situations, the available solutions to a problem face one of the duties or virtues to another. For example, in a situation where withdrawal of care is being considered, one may have to prioritize autonomy over beneficence. In other situations, such as allocation of organs for transplantation, justice for a group of patients may prevail on beneficence or autonomy. Other situations arise where the duties-based theories of morality are not helpful and one may have to utilize an alternative theory such as consequentialism. Consequentialist theory asserts that it is the outcome more than the intention that determines what is right. One variant of this is utilitarianism in which the morally superior choice is that which creates the greatest good for the greatest number of people [8]. While utilitarianism has its drawbacks—notably difficulties in quantifying "good" and requiring one to be able to predict future effects of one's actions—a clear example of the practical application of this theory is the lung allocation score used for allocation of lungs for transplantation among patients on a waiting list. In this methodology, the expected gain of life is balanced against the loss of life while awaiting an organ and lungs are allocated to maximize life-years gained—a quantifiable and reasonably predictable "good."

Limitations on Care, Withdrawal of Life-Sustaining Care, and Futile Care

As referred to above, there are times when one duty conflicts with others and the surgeon has to prioritize these duties. One such example that has come up in the context of cardiothoracic surgery has arisen when a surgeon agrees to take on a high-risk operation only if the patient (and family) agrees not to withdraw support within a certain window following the surgery. While no surgeon wants to waste his or her time performing a futile operation, it is frustrating to take on a long and complex operation, for the patient then to have a predictable but potentially reversible complication and then to have the patient's family withdraw support only a few days into an expectedly long postoperative course. While a surgeon may want a window of time to get the patient through the procedure, is it appropriate to request, or even require, such a window of time as a contingency of taking on the case? Is such a request coercive? Or, if the family originally agrees to the stipulation, can they change their minds later?

Although there may be certain circumstances where attempting to place limitations on the patient's ability to withdraw care may be appropriate, one of the primary duties of the surgeon is to act on behalf of the patient—respecting his or her choices and acting accordingly. While a surgeon may request a certain therapeutic window and he may go to significant lengths to justify his position, ultimately it is the patient's choice, and the surgeon has the options of either agreeing or not performing the operation and allowing the patient to go elsewhere. Obviously, for this latter choice to be appropriate, there must be an alternative surgeon who would agree to the patient's stipulations, and the situation cannot be an emergency—a situation where forcing certain stipulations would clearly be coercive. Nonetheless, in an elective situation, the patient and the surgeon enter into a voluntary agreement and neither the surgeon nor the patient is forced to accept the other's stipulations.

What, then, happens if the patient agrees to a surgeon's requirement for a window of care and then later changes his mind, or the family wishes to withdraw care sooner than planned? The two are different situations with the first being easier to address: assuming that the patient is competent to make decisions, if the patient changes his/her mind and opts to withdraw care, this is his/her prerogative under the principle of autonomy. While frustrating to the surgeon, it is rarely appropriate to force care upon an individual who refuses or declines it. It is more complicated if the family wants the surgeon to change course. While the family may have the legal ability to make healthcare decisions when an individual is unable to do so, the surrogate decision-maker generally does so under the doctrine of substituted judgment, i.e., deciding what patient would decide if he/she could make the decision himself/ herself [9]. In general, if the patient is unable to provide consent, the caregiver (surgeon) needs to abide by the decision of the family (or designated healthcare proxy). The surgeon may argue that the patient agreed to a course of treatment and there is no reason to change plans, but unless there is a reason to suspect treacherous activity on the part of the family or surrogate decision-maker, the surgeon generally must respect those decisions made on behalf of the patient. While the surgeon can attempt to convince the patient or healthcare proxy to "be more reasonable," "give the patient some more time," etc., ultimately, it is not the surgeon's choice when to withdraw care—it is his/her job to work on behalf of the patient and respect the patient's decision.

A similar set of ethical concerns arises when a surgeon is asked to do an operation that he/she thinks is likely futile. Some situations are relatively straightforward: a patient requests a lung resection in an advanced stage of disease and for which there are other or, in fact, better alternatives. In this case, the surgeon is not forced to offer an inappropriate procedure and can justifiably turn the patient down. Cannons of professionalism do not include the statement that "the customer is always right" and, as professionals, we are obligated to use our best clinical judgment on behalf of the patient. Other situations are less clear. For example, consider a situation in which a patient has an aortic dissection and an unclear neurological status. Outcomes of aortic repairs in the setting of severe neurological damage are extremely poor, and it may well be appropriate not to offer an operation in such a setting [10]. But there are times when the neurological status is unclear, sometimes due to the administration of sedatives or other impairing medication. In such settings, it is probably best to err on the side of beneficence and proceed with surgery.

Another situation is even more unclear: that of recurrent endocarditis in a drug-addicted patient who has repeatedly gone back to intravenous drug use and who may have no intention of changing his/her habits. Surgery in this setting will likely not be beneficial and probably has a high chance of causing further harm. Arguably, not operating also has a high probability of harm, and surgery is likely the only, even potentially, effective course of treatment. Is it ethical to turn the patient down for surgery under these circumstances?

Empirically, surgeons have gone both ways on this: some may feel obligated to offer an operation on the grounds that failure to do so is tantamount to a death sentence; others feel justified in not offering an operation that is both unlikely to be helpful and fails to address the core issue of intravenous drug addiction [11]. In a setting of fixed resources, one could make an argument on the basis of justice—that failure to offer an operation is justified in that enormous amounts of resources will go to an individual with an almost certainly poor outcome and, as a result, not be available to others. In the situation where there is no limitation of resources, one could still argue that some sort of patient buy-in—a willingness to, or interest in, giving up intravenous drug use, for example— could be required by the surgeon prior to entering into an agreement to undertake the high-risk procedure. As with the earlier case though, if there is no other surgeon available, the refusal to take on a life-saving operation stands on more tenuous ethical grounds. One could argue that surgery is not in the best interest of the patient, i.e., that the outcome would be the same with or without surgery or even that the outcome of debilitating neurological impairment may be even worse from the patient's perspective than death and that failure to accept this risk changes the matter from an issue of beneficence/non-maleficence to one of patient autonomy. Finally, one could assert that the patient's repeated use of intravenous drugs puts the onus of responsibility on the patient, not the surgeon. Such an argument may be reasonable in the rare situation where the patient was told, on a prior operation, that "this is the last time we are going to operate on you."

Should a surgeon refuse to operate on a patient, it is reasonable to expect him/her to assist the patient in finding another surgeon who may not feel similarly encumbered in offering what is likely to be an operation with a relatively low long-term benefit.

Surgical Outcomes Databases

In both of the above cases, the neurologically impaired patient with an aortic dissection and the drug addict with endocarditis, the end result of the operation is likely to be death of the patient. While obviously not what the surgeon or the patient had hoped for at the outset, this is, unfortunately, a potential outcome of any high-risk area of medicine or surgery. Over the past two decades, quantitative assessment of surgical outcomes has gained increasing importance in not only determining where patients go for care and how value in healthcare is determined but also reimbursement—at least at the hospital level. While self-improvement has long been part of the professional ethos, measurement of surgical outcomes, the development of outcomes databases (National Surgical Quality Improvement Program (NSQIP), the Society of Thoracic Surgeons Cardiac Database, and others) as well as improved quantitative methods for data analysis have been associated with an increasing focus on quality of care [12, 13]. While offering a firm basis for quality improvement, the use of such databases can lead to ethical controversies.

The STS (Society of Thoracic Surgery) Cardiac Surgery Database began in 1989 as an initiative designed to improve quality and safety in adult cardiac surgery. As of January, 2018, the initiative has expanded to include adult cardiac surgery, congenital heart surgery, general thoracic surgery, and, most recently, mechanical circulatory assistance (typically ventricular assist device) outcomes. The database has records of over six million patients, is audited, and has become an invaluable resource for both quality improvement and research.

While detailed outcomes results of a specific program are typically only made available to that program, certain elements are made publicly available, with appropriate qualification, on a voluntary basis. The degree of specificity of this publicly available information varies with the different specialty databases, but the aim is to benefit patients by creating transparency in outcomes using audited, credible, and risk-adjusted data.

Early attempts at public reporting of surgical outcomes did not adjust for surgical risk [14]. As a result, surgeons who took on cases with higher-risk profiles and had higher expected (and actual) mortality rates looked worse in the public eye. Risk-averse surgeons could appear to have better than expected results simply on the basis of patient selection. A result of such a system would be to encourage aggressive case selection by turning down more complex, high-risk cases and transferring them to other, less risk-averse institutions. The ultimate effect could be to deny potential benefits of surgery to all high-risk patients—a clear affront to the principle of justice. More recent efforts at public reporting have involved risk adjustment that, while not perfect, substantially decreases the incentive of cherry picking—at least from the standpoint of public reporting of outcomes.

Different databases report results back to the participating programs in different formats. The American College of Surgeons National Surgical Quality Improvement Program (ACS NSQIP), for example, reports mortality and a number of domains of morbidity as a risk-adjusted observed to expected (O-E) ratio and a decile ranking in comparison to all other participating institutions. Some specific indexes of morbidity and mortality are made publicly available for this. The STS Cardiac Database has created robust statistical models of several operations and reports outcomes as both observed to expected (O-E) ratios as well as raw occurrence frequencies. Results of an individual institution's outcomes are presented in comparison to outcomes of all participating institutions as well as institutions of similar size and case composition—defined as "like" institutions. Additional statistical manipulation and creations of "star ratings" were developed to complement the increasingly common availability of publicly available outcome data as reported data in *hospitalcompare.gov*, *US News and World Report*, and *Consumer Reports*. "Star ratings" combine morbidity and mortality, as well as other quality indices, with 2-star institutions providing risk-adjusted outcomes within two standard deviations of the mean, and 1 and 3-star ratings being associated with results that are statistically significantly lower or higher than the mean. Institutions can choose to have their star ratings made public or may choose to use them only internally.

Since outcomes databases, such as STS and NSQIP, were established as quality improvement tools, rules governing use and dissemination of the data are incorporated into "data use agreements" (contracts, essentially) that generally limit use of the data for institutional marketing purposes. Although the surgeons themselves are technically responsible for use of their institution's data, it is not difficult to imagine use of star ratings and publicly available O-E ratios to be used to directly compare outcomes between institutions. While STS rules specifically prohibit or limit such uses, once data gets into the public realm, it is increasingly difficult, if not impossible, to control data use and dissemination.

While certainly better than non-risk-adjusted models, risk-adjusted outcomes take both outcomes and case mix into account. As such, simple comparison of two O-E ratios of two institutions' results can be misleading, i.e., appearing to favor one institution over another, yet may be due solely to differences in case mix and not to differences in quality. For example, imagine a situation where one institution that does a large number of noncomplex procedures and no complex ones yet has a similar or even better O-E ratio than an institution that does a significant number of complex cases. A statistically naïve patient with a complex condition could easily be misled into thinking that the institution with a slightly better O-E ratio but focuses on less complex cases also does well with more complex ones.

A related, although somewhat niche, issue is the whole concept that one operation can be used for several underlying diagnoses and outcomes databases may look only at the outcome of a specific operation—not an outcome of an operation for a specific underlying diagnosis. For example, pulmonary lobectomy, while most commonly performed for cancer, tends to have higher complication rates when done for infectious causes. Since diagnosis may not be taken into account in a risk model, an institution that has a patient bias toward infectious lung disease may appear to have a higher than expected morbidity and mortality when, in fact, the worse outcomes are due solely to the underlying diagnosis and not poor performance on the part of the institution or surgeon. Another example, from the area of congenital heart surgery, would be to present risk-adjusted outcomes for a Fontan procedure, a procedure which is performed for the correction of hypoplastic left heart syndrome and tricuspid atresia, as well as other congenital anomalies. If (a) outcomes are clearly worse for the one of these diagnoses, (b) the risk model does not take this into account, and (c) two institutions have marked differences in case mix, there may appear to be significant differences in risk-adjusted outcomes even though these may be due solely to case mix differences and not to actual quality. While one may simply attribute such an occurrence to shortcomings of the risk adjustment models, direct comparisons of such data can be misleading—particularly to unsophisticated audiences.

The last issue related to databases is that not all institutions participate in them. While the STS Adult Cardiac Database has near 100% participation, in part this may be because there are mandatory regulatory reporting requirements of many of the database elements in many states. Parenthetically, such regulatory requirements do not apply to congenital cardiac surgery or general thoracic surgery, and perhaps as a result, participation in the STS Congenital and General Thoracic databases has, historically, not been as robust [15]. If participation is selective, one then wonders whether the public can truly rely on these databases to provide true comparative data. None of the above, however, should be construed to diminish the remarkable value of the STS Database as a tool for institutional quality improvement.

Innovation

As with many areas of surgery, cardiothoracic surgery has witnessed significant technological innovation over the past several years. Cardiac surgery has seen percutaneous coronary intervention change both the volume and clinical presentation of coronary artery bypass grafting (CABG) [16]. Patients are now older, have more comorbidities, and have had more previous interventions. In the area of valve surgery, transcatheter aortic valve replacement (TAVR) has resulted in a dramatic drop in what are now termed "surgical" aortic valve replacements (SAVR's), and the technology of minimally invasive mitral valve surgery is developing at a rapid pace. Cardiac surgeons must work collaboratively with interventional cardiologists, and the phenomenon of a "heart team" has developed. The aim is for patients to profit from the care of a skilled team of physicians and associated practitioners—all with improved outcomes and, hopefully, lower costs. Duplication, or overlapping, of services can be minimized, financial incentives can be based on overall outcomes and not individual productivity, and patient outcomes can be prioritized. Such a system, however, requires significant restructuring of the traditional fee-for-service and volume-based payment/reimbursement strategies. New strategies for inter-specialty collaboration, which will require trust and fair distribution of gains and liabilities, must be developed.

Innovation is not limited to cardiac surgery however. General thoracic surgery has seen a dramatic shift from large open cases to minimally invasive or VATS (video-assisted thoracic surgery) techniques. In fact, recent STS Database reports suggest that up to 60% of lobectomies (at STS Database participating institutions) are now done using minimally invasive techniques [17]. The increasing use of surgical robotics has complicated this progression from open to less invasive surgery, and it is unclear whether its use is also associated with patient benefits.

The use of the robot is associated with a number of other ethical issues such as how to teach residents robotic techniques, how does a surgeon obtain "informed" consent for the initial procedures with which he has little experience, and how does a surgeon ethically replace a procedure with which he/she has a great deal of experience with one where he/she is on the early stage of the surgical "learning curve." Finally, why should the patient pay the price of longer surgery, potentially unknown risks, and little in the way of proven benefit? Is the patient truly informed of these issues when he or she is asked to provide consent for the procedure—or is the consent really a matter of "entrustment" as McKneally has described [18]?

In the area of cardiothoracic critical care, the increasing use of extracorporeal membrane oxygenation (ECMO) has generated a number of ethical and practical issues [19]. Some of the reasons for the increased use of this technology is attributable to general improvements in outcomes of ECMO in critically ill adults in general but may also be attributable to increasing ease of initiation. Early models of ECMO required surgical cannulation of large vessels (internal jugular vein and carotid artery), assembly of complex devices incorporating separate pumps and oxygenators, and essentially meant having a patient on bypass in the ICU. Technological innovations have resulted in single cannulas (for blood return and infusion), intervention on only the venous side of the circulatory system (veno-venous ECMO vs venoarterial ECMO), more efficient oxygenators, lower requirements for anticoagulation, and small, portable, less complex devices. The increased use of ECMO technology has resulted in several ethical issues: (1) those related to who should go on ECMO, (2) the costs and resource utilization required by the technology, and (3) discontinuing ECMO [20].

Ethical issues related to the initiation of ECMO include who should go on ECMO, who should make this determination, and how can informed consent be obtained. While ECMO has been termed a "bridge to survival," if cardiopulmonary failure is irreversible, ECMO becomes a "bridge to nowhere." Since ECMO is often initiated urgently, if not emergently, how then is one obtain informed consent and present a balanced view of risks, benefits, and alternatives to a patient, or family of a patient, in extremis? Likely posed as "this is the only way to save the patient's life," critical issues related to outcome, such as withdrawal from ECMO, and risks of stroke and bleeding are probably covered in a cursory fashion—if at all.

In terms of maintaining patients on ECMO, the substantial resources that are utilized request special consideration. Because any single institution must allocate its resources efficiently and equitably, it is incumbent on institutions that have ECMO services to define clear and strict criteria for who and how many patients can be put on ECMO. Such criteria clearly involve balancing the goals of beneficence and justice. For example, if a hospital with the capability of running two patients on ECMO at any given time already has two such patients, should high-risk cardiac surgical or catheter laboratory (cath lab) cases— which may eventually require ECMO—be delayed until a time when additional ECMO capability is available. In terms of justice, is it appropriate to tie up several intensive care unit (ICU) beds for patients on ECMO and end up with a situation where resources to accomplish this cannibalize from other services that would then be under-resourced and not capable of appropriately caring for their normal caseload [19].

The ethical issues of withdrawing ECMO, while often qualitatively similar to withdrawal of other life-sustaining treatments (e.g., dialysis, mechanical ventilation, and enteral feedings), may, however, be unusual in that the patient may be awake, alert, and able to participate in the decision process. It is well documented that complications related to ECMO increase with duration of the treatment, and it is also well documented that prognosis after discontinuing ECMO depends on adequate recovery of cardiopulmonary function. Emotionally and ethically charged situations may arise when a patient who has been placed on ECMO shows no underlying cardiopulmonary improvement over a reasonable time frame and could continue on ECMO for a prolonged period of time with no reasonable chance of recovery. It would seem that discontinuing ECMO would be qualitatively similar to discontinuing tube feeds, dialysis, or mechanical

ventilation—balancing beneficence and non-maleficence: are we prolonging life with the technology, or are we prolonging death? Many difficult questions arise: At what point is non-recovery guaranteed? Is high-risk transplantation an option? And how does one deal with the situation where the patient is awake and fully aware that discontinuing ECMO will likely result in a rapid demise? Is this the same as discontinuing mechanical ventilation in a patient with multisystem organ failure and delirium? While some care withdrawal issues are unique to ECMO, Courtwright has pointed out that the majority of ethics committee consultations arising in the setting of ECMO withdrawal resemble traditional concerns about withdrawal of life-sustaining treatments [21].

Ethics in Cardiothoracic Surgical Education

Ethical issues related to resident education cross all subspecialties within cardiothoracic surgery—although they exist primarily in academic institutions and less so in community or non-teaching environments. While this is no different from other branches of surgery, resident education must be balanced with duty to the patient. In the teaching environment, faculty surgeons have clear responsibilities to both the resident and to the public at large—for if teaching is ineffective, the future of other patients, and perhaps the specialty as a whole, is at risk. On the other hand, no patient is likely to be willing to undergo a substandard operation for the benefit of some future hypothetical patient—particularly when the price paid of a substandard operation in cardiac surgery is high. Arguably, this problem is greatest in the pediatric cardiac surgical population where the operations are the most challenging, the cost of error is the greatest, and the training is the most protracted [22].

How then do cardiothoracic surgeons balance these competing missions? And how much are patients aware of these trade-offs and the roles of trainees. While again, this is not qualitatively different in any other area of high-acuity/high-risk surgery, cardiothoracic surgical residency training programs in the United States are having to adapt from 2-year programs—where entering residents who have already completed residencies in general surgery are reasonably well prepared technically and intellectually to do cardiothoracic surgery—to integrated 6-year programs, where junior residents may be only days out of medical school. In the former model, cardiothoracic surgical faculty may assume some substantial baseline knowledge and technical skill level; in the new training model, such technical expertise cannot be assumed, and the balance of teaching, oversight, and independence is more complex.

Another ethical area related to the teaching of residents, as well as the use of non-physicians in the operating room, concerns the degree to which patients are aware of the roles of these individuals. While there is little in the literature indicating how much patients understand about the role of residents and physician extenders, Kent demonstrated that patients have both strong and highly varied opinions on how much independence trainees should be permitted in the operating room [23]. Furthermore, while consent forms often make note of assistants in surgery, the specific role of these assistants is generally not apparent. In 2017, the Federal Government of the United States considered the requirement that patients be informed of the role of all individuals involved in the operation—including their names and specific roles. Fortunately, this was aborted but largely because it was argued to be highly impractical, not because of what was better or worse for the patient. To be more specific, when consent is obtained well in advance of the case, the patient can better participate in the decision process. On the other hand, at that time, it is often impossible to know which resident or advanced practice provider (nurse practitioner or physicians' assistant) would be available for at the operation. If the consent is obtained immediately before the procedure, the ancillary staff may be known, but there are great, potentially coercive, pressures on the patient to sign the consent form. While it would be possible to get a second, or amended, consent with the added information, changing teams at such a late date

would arguably be impractical at best and potentially harmful at worst.

Another controversial area that is related to education and that has engendered controversy recently is that of the role of broadcasting live surgery in professional educational conferences. More specifically, the controversy is whether live surgery should be permitted as an educational tool: the pros being that live surgery has historically been used to train colleagues and that there is a value to observing both real-time decision-making and the results, either good or bad, of such decisions. The cons include (1) the potential for the surgeon being distracted by an audience, (2) the possibility that the presence of a live audience may influence the conduct of an operation and direct it toward an optimal educational outcome instead of optimal patient care, and (3) the possibility that audience enthusiasm is based more on a voyeuristic basis—the surgical equivalent waiting for a crash in a NASCAR race—rather than an educational one. There is also the concern that while audiences in professional educational sessions are reasonably well regulated, people in the audience may have nonprofessional relationships with the live patient and may witness either an untoward event or nonprofessional behavior on the part of the audiences. Individuals and organizations that oppose live surgical broadcasts assert that edited videos would provide as much information—generally in a tighter, more efficient time frame—than a live broadcast that the surgeon can devote his full attention to education, even pausing the "operation" to address a technical matter, and that there is no possibility of the surgeon diverting his full attention from the patient at the time of surgery. Needless to say, whether the operation is recorded or broadcast live, the patient must be apprised of the educational nature of the modifications to the standard procedures and be given the opportunity, without coercion, to agree or disagree.

Professionalism

The final area to address involves the topic of professionalism; in particular, the ethical issues related to the increasing incidence of cardiothoracic surgeons being employed by hospitals and large medical organizations rather than the previously more common model, at least in the United States, of being in small group practices. Both hypothetical and real cases have anecdotally come up where surgeons feel pressured to comply with corporate goals rather than what they may feel is best for their patient [24]. An example might be where a young surgeon is hired by a hospital to develop a specific area of cardiothoracic surgery—aortic surgery, for example. The surgeon is well trained but may not be highly experienced. In the hypothetical case, the surgeon feels that his/her institution and its nursing or anesthetic team are not yet prepared to handle the complexities of the case without more preparation and training, yet he/she is being pressured by hospital administration to take on the case for which he/she was hired. Should he/she bow to the pressure and do the case, or should he/she transfer the patient to a different institution and risk the displeasure, and potentially adverse actions, of his/her employer.

While one may simply dismiss this as poor judgment and shortsightedness on the part of the hospital administrators, it illustrates the potential conflict between fiduciary duties to two entities. The surgeon has clear duties to his/her patient—to provide excellent care and to act in the patient's best interests. The surgeon also has duties to his/her employer. The tenets of professionalism, at least as described by Friedson, tend to put priority on duties toward the patient—at least as far as clinical performance and judgment are concerned [25]. More specifically, Friedson has pointed out that professionals must have some freedom to exercise discretionary judgment—even if it may not be in the obvious best interests of the employer. In a professional setting, as opposed to a conventional vendor/purchaser relationship, the customer is not always correct. Friedson went on to assert four other characteristics of a successful relationship between employer, client (patient), and professional: (1) adequate resources to do a job well, (2) a formal organizational structure that features some sort of "carve-out" for professionals to maintain some discretion in activities, (3) a recognition of the specialized knowledge of

the professional, and (4) an element of performance measurement that is recognized within the peer professional community.

In some areas within cardiothoracic surgery: academic institutions, closed model health maintenance organizations, multispecialty clinics, and even the military as examples, the employed surgeon has long been the common model. Surgeons have learned to balance duties to two masters, yet it is not surprising that surgeons, or small groups of surgeons, who have historically been self-employed or in independent practice may struggle with the transition to one where there are clear delineations of responsibility (a hierarchical organizational structure), particularly if a physician is not at the helm.

Another area of professionalism that has come under increasing discussion lately has been the role of the surgeon in public policy and in the allocation of scarce resources. The resource in question may be organs for transplantation, money to support ventricular assist devices, or even something as commonplace as ICU bed allocation. Surgeons clearly have a duty to advocate for their patient but should society's needs at large come under consideration, and if so, how does the surgeon balance the needs of his patient with those of society, the hospital, and/or other physicians' patients? One example would be the case of a patient with severe heart failure and who may benefit from a left ventricular assist device (LVAD). While the obvious answer as to whether the individual should get an LVAD is "yes," the specific situation may be that the patient has such severe comorbidities that his/her life expectancy is extremely poor even with an LVAD. What if the individual does not have insurance that will cover the several hundred thousand dollars of anticipated expenses and the hospital organization has to cover it on a fixed budget? What other vital programs must be sacrificed for the expected, but not even guaranteed, minimal benefit of a single patient? Should the patient's surgeon be involved in these decisions? Should they be decided by a committee (potentially for diffusion of responsibility)? And can a surgeon, or any physician for that matter, compartmentalize his/her thoughts and motivations arguing on one hand for access to an LVAD for his/her patient but against the use of high-cost technology for likely end-of-life treatments for patients in general?

End-of-Life Care

Issue of end-of-life care, are ones that essentially all cardiothoracic surgeons will deal with at some point in their careers. While discussed earlier in the context of futile care or when to say "no" to surgery, the concept of withdrawing care, and particularly terminal sedation, raises emotional levels and issues such as self-doubt and recrimination and raises concerns for lawsuits and even violence directed toward the surgeon. Fortunately, violence directed to physicians is rare in the United States, and there are generally well-recognized and accepted ways to deal with bad outcomes vis-à-vis involvement of risk management personnel within a healthcare organization. Yet the moral, emotional, and practical matters related to withdrawal of care can be challenging.

Issues that have come up through the lay press related to persistent vegetative states, prolonged coma, and even brain death, challenge the conventional notions of respect for autonomy, beneficence and non-maleficence, and even neurophysiology [26]. Is prolonging the life of someone in a persistent vegetative state beneficent or maleficent? How can one be sure of a person's wishes when they are in a coma? Could their wishes have changed? Is not instituting care morally the same as withdrawing care? And finally, in cases of terminal extubation, is the provision of narcotics hastening death or diminishing suffering? To the latter point, two theories have addressed the matter [27]. The first, the principle of double effect, postulates that the *intention* of the sedation is what is important, not the actual action. To expound, if one sedates with the goal of alleviating suffering, it is not unethical even if the outcome is death of the patient. On the other hand, if the goal is to hasten death, the action is unethical. The other theory, known as

the moral equivalence hypothesis, claims that if allowing a patient to die is ethical (or unethical), then physician-assisted suicide, active euthanasia, or any other means that hastens death are morally equivalent and, hence, equally ethical (or unethical). While there are well-crafted and justified arguments for both hypotheses, from the practical and political standpoints, the concept of actively assisting death (physician-assisted suicide) has gained legal justification in only a few jurisdictions in the United States. Having said that, the risk of accepting the moral equivalence hypothesis at a policy level is that if the decision comes down that actively assisting death (physician-assisted suicide and euthanasia) is morally wrong, withdrawal and non-initiation of care, both of which are currently well accepted, would be equally unacceptable. Should this be the case, our ICUs would likely be far more crowded, and there would be many more patients being maintained on ventilators and tube feeds with a minimal quality of life. Ultimately, without significant increases in healthcare resources, our ability to care for patients with far better prognoses may be compromised.

Conclusions

- Cardiothoracic surgery shares many of the ethical challenges associated with most other branches of surgery.
- There are specific issues facing the cardiothoracic surgical community at this point in time, and it is likely that consensus will be reached on these issues and that policies may render some controversies mute, but that other ethical issues and controversies will eventually take their places.
- This chapter has attempted to describe some of the current ethical controversies facing the specialty and to offer a framework of both duty-based ethics and consequentialist ethics to help the reader analyze these controversies and come to his own conclusions regarding their resolution.

References

1. Pellegrino ED, Relman AS. Professional medical associations: ethical and practical guidelines. JAMA. 1999;282(10):984–6. https://doi.org/10.1001/jama.282.10.984.

2. Beauchamp TL, Childress JF. Principles of biomedical ethics. 7th ed. New York: Oxford University Press; 2013. p. 13.

3. Ferreres AR, Angelos P, Singer EA, editors. Ethical issues in surgical care. Chicago: American College of Surgeons; 2017.

4. Sade RM, editor. The ethics of surgery. New York: Oxford University Press; 2015.

5. Wicclair MR, White DB. Surgeons, intensivists, and discretion to refuse requested treatments. Hast Cent Rep. 2014;44(5):33–42. https://doi.org/10.1002/hast.356.

6. Chaet DH. The AMA *Code of Medical Ethics*' opinions on patient decision-making capacity and competence and surrogate decision making. AMA J Ethics. 2016;18(6):601–3. https://doi.org/10.1001/journalofethics.2016.18.6.coet1-1606.

7. Meisel A, Snyder L, Quill T. Seven legal barriers to end-of-life care: myths, realities, and grains of truth. JAMA. 2000;284(19):2495–501.

8. Brock D. Chapter 9: Utilitarianism. In: Regan T, VanDeVeer D, editors. And justice for all: new introductory essays in ethics & public policy. Totowa: Rowman and Littlefield; 1982.

9. Beauchamps TL, Childress JF. Principles of biomedical ethics. 7th ed. New York: Oxford University Press; 2013. p. 226.

10. Eusanio M, Patel HJ, Nienaber CA, et al. Patients with type a acute aortic dissection presenting with major brain injury: should we operate on them? J Thorac Cardiovasc Surg. 2013;145(3 Suppl):S213–21.e1. https://doi.org/10.1016/j.jtcvs.2012.11.054.

11. DiMaio JM, Salerno TA, Bernstein R, Araujo K, Ricci M, Sade RM. Ethical obligation of surgeons to noncompliant patients: can a surgeon refuse to operate on an intravenous drug-abusing patient with recurrent aortic valve prosthesis infection? Ann Thorac Surg. 2009;88:1–8.

12. https://www.facs.org/quality-programs/acs-nsqip.

13. https://www.sts.org/registries-research-center/sts-national-database/sts-adult-cardiac-surgery-database.

14. Chassin MR, Hannan EL, DeBuono BA. Benefits and hazards of reporting medical outcomes publicly. N Engl J Med. 1996;334(6):394–8. PMID 8538714.

15. Jacobs JP, Shahian DM, Prager RL, et al. The society of thoracic surgeons national database 2016 annual report. Ann Thorac Surg. 2016;102(6):1790–7. https://doi.org/10.1016/j.athoracsur.2016.10.015. PMID: 27847042.

16. Cornwell LD, Omer S, Rosengart T, et al. Changes over time in risk profiles of patients who undergo coronary artery bypass graft surgery: the veterans affairs surgical quality improvement program (VASQIP). JAMA Surg. 2015;150(4):308–15. https://doi.org/10.1001/jamasurg.2014.1700.

17. Seder CW, Wright CD, Chang AC, Han JM, McDonald D, Kozower BD. The society of thoracic surgeons general thoracic surgery database update on outcomes and quality. Ann Thorac Surg. 2016;101(5):1646–54. https://doi.org/10.1016/j.athoracsur.2016.02.099. Epub 2016 Mar 31. PMID: 27041451.

18. McKneally MF, Martin DK. An entrustment model of consent for surgical treatment of life-threatening illness: perspective of patients requiring esophagectomy. J Thorac Cardiovasc Surg. 2000;120(2):264–9.

19. Abrams DC, Prager K, Blinderman CD, Burkart KM, Brodie D. Ethical dilemmas encountered with the use of extracorporeal membrane oxygenation in adults. Chest. 2014;145(4):876–82. https://doi.org/10.1378/chest.13-1138. PMID: 24687709.

20. Makdisi T, Makdisi G. Extra corporeal membrane oxygenation support: ethical dilemmas. Ann Transl Med. 2017;5(5):112. https://doi.org/10.21037/atm.2017.01.38. PMID: 28361077.

21. Courtwright AM, Robinson EM, Feins K, Carr-Loveland J, Donahue V, Roy N, McCannon J. Ethics Committee Consultation and Extracorporeal Membrane Oxygenation. Ann Am Thorac Soc. 2016;13(9):1553–8. https://doi.org/10.1513/AnnalsATS.201511-757OC. PMID: 27299991.

22. Ohye RG, Jaggers JJ, Sade RM. Must surgeons in training programs allow residents to operate on their patients to satisfy board requirements? Ann Thorac Surg. 2016;101(1):18–23. https://doi.org/10.1016/j.athoracsur.2015.08.049.

23. Kent M, Whyte R, Fleishman A, Tomich D, Forrow L, Rodrigue J. Public perceptions of overlapping surgery. J Am Coll Surg. 2017;224(5):771–778.e4. https://doi.org/10.1016/j.jamcollsurg.2017.01.059. Epub 2017 Feb 11.

24. Fenton K, Ellis J, Sade RM. Should a thoracic surgeon transfer a complicated case to a competing medical center against the hospital's order? Ann Thorac Surg. 2015;100(2):389–93. https://doi.org/10.1016/j.athoracsur.2015.04.053.

25. Freidson E. Chapter 12: Nourishing professionalism. In: Professionalism reborn: theory, prophecy, and policy. Chicago: University of Chicago Press; 1994.

26. Aviv R. What does it mean to die? New Yorker Magazine. 2018, February 5, Issue. https://www.newyorker.com/magazine/2018/02/05/what-does-it-mean-to-die.

27. Brody H. Physician-assisted suicide in the courts: moral equivalence, double effect, and clinical practice. Minn Law Rev. 1998;82:939–1695.

Ethics of Surgical Intervention in Jehovah's Witness Patients

Edward E. Cho and D. Rohan Jeyarajah

Key Points
- The principle tenant in Jehovah's Witness patients in relation to health care is their religious prohibition to accepting blood products. However, while most Jehovah's Witness patients firmly reject actual blood products, some individuals may be lenient on blood analogues or isolated coagulants. Therefore, it is important to respect the autonomy of the patient by having a detailed and comprehensive informed consent regarding each of those agents.
- Most Jehovah's Witness patients will have a liaison that can help with a checklist of products that are permissible for each individual patient. Each patient must be treated on a case-by-case basis.
- During the informed consent process, it is imperative that the patient understands the higher risk of death. After careful discussion, they need to clearly express their desire that they would rather die rather than receiving life-saving transfusions.

- The surgeon must look at all alternatives and weigh other treatments balancing the efficacy of the treatment versus the risk of death.
- Multidisciplinary care and preoperative planning with all necessary departments are crucial to optimizing the patient's preparation prior to surgery.
- Meticulous surgical technique to minimize blood loss and having protocols in place in case massive bleeding is encountered intraoperatively are important aspects in treating for any patients, especially Jehovah's Witness patients.
- Early vigilance, recognition, and intervention in the postoperative period will minimize blood loss and safely guide the Jehovah's Witness patients through recovery.
- Perioperative techniques to minimize blood loss and transfusions should be employed in all patients, not just Jehovah's Witness patients.
- Access to health care and any surgical interventions should not be denied to any patients solely on the grounds of their religious beliefs.

E. E. Cho (✉) · D. R. Jeyarajah
Department of Surgery, Methodist Richardson
Medical Center, Richardson, TX, USA
e-mail: edwardc@tsgsurgical.com

© Springer Nature Switzerland AG 2019
A. R. Ferreres (ed.), *Surgical Ethics*, https://doi.org/10.1007/978-3-030-05964-4_27

Introduction

The right of every person to either approve or reject medical and surgical therapies is well established in ethics and law. Jehovah's Witness patients represent a well-known group of individuals that pose a major challenge in their surgical care. The most defining tenant for Jehovah's Witness patients in the setting of health care is their strict prohibition against receiving blood. This limitation poses a higher risk of complications from profound anemia, should bleeding occur for these patients. This is especially the case in high-risk surgeries, with increased possibility of morbidity and mortality. In an era where a surgeon is defined and judged on publicly reported quality metrics, it seems illogical and irresponsible that a patient can be exposed to the risk of death from exsanguination that can be prevented with transfusion. However, this is the challenge that the modern surgeon faces when tasked with providing surgical care to a Jehovah's Witness patient. Their belief in rejecting blood products and other medical resources also opens up ethical and moral implications that the surgeon has to respect and comply with.

The contract between the surgeon and the Jehovah's Witness patient includes two parties – one of these is the surgeon him-/herself. It is important that the additional stress of taking care of Jehovah's Witness patients on the psychological state of the surgeon be recognized. In fact, the Jehovah's Witness community tends to be very understanding of the responsibility that the surgeon is taking on. However, does the medical community feel the same way? Or is the surgeon judged by the same standards as if he or she had the luxury of using blood products? These are ethical questions raised for further discussion.

This chapter provides a succinct overview of the history and beliefs of Jehovah's Witnesses followed by the discussion on ethical and legal ramifications of their beliefs that may affect surgical practice and the various contingencies and options that the authors utilize, which are not only applicable to Jehovah's Witnesses but to all patients to prevent and minimize complications.

Historical Background and Transfusion Beliefs of Jehovah's Witnesses

The Jehovah's Witness religion was initially instituted under the name of the Watchtower Bible and Tract Society founded in 1879 by Charles Taze Russell, a Western Pennsylvania businessman [1]. The Society was restructured under the direction of a society of international Bible students in 1931, and the name was changed to Jehovah's Witnesses. The religion is primarily based on the prophecy of Armageddon or "the end of the world" as described from the Bible. Teachings from Jehovah's witnesses specify that as "true" Christians, Jehovah's witnesses will be saved at the time of Armageddon and the second coming of Christ and will be ushered into heaven and eternal life. Today there are over six million Jehovah's Witnesses in 235 countries and territories. Nearly one million of them are in the United States. Their numbers are increasing, particularly in Central and South America, Italy, Japan, and Eastern Europe.

As a matter of firm religious belief, Jehovah's Witnesses are prohibited by their governing body for utilizing blood products and blood-like substances. The Watchtower Bible and Tract Society instituted this policy of refusal of transfusions in 1945. This prohibition is based on at least three citations from the bible:

- "But you shall not eat flesh with its life, that is, its blood." (Genesis 9:4. English Standard Version)
- "There I say to the Israelites, None of you may eat blood, nor may any foreigner residing among you eat blood." (Leviticus 17:12. English Standard Version)
- "…that you abstain from what has been sacrificed to idols, and from blood, and from what has been strangled, and from sexual immorality. If you keep yourselves from these, you will do well…" (Acts 15:29 English Standard Version)

The reason for this policy is based on the belief that "blood, irrespective of the manner of consumption, serves as a nutrient," and acceptance would be defying divine precepts. Based on this policy, the refusal of transfusions of whole blood (including preoperative autologous donation) and primary blood components – red cells, platelets, white cells, and unfractionated plasma – remains nonnegotiable for nearly all Jehovah's Witnesses. However, acceptance of blood product alternatives and/or components such as albumin, all clotting factors, all immunoglobulins, interferons, and interleukins is up to individual patients (Transfusion Handbook 2014). This directive further complicates surgical care of Jehovah's Witness since now it is up to each Jehovah's Witness patient to determine what blood product alternatives and/or components they will and will not accept.

Moral Framework

The "Four Principles" of medical ethics were introduced by American philosophers Tom Beauchamp and James Childress in the 1970s [2]. These principles of beneficence, autonomy, nonmaleficence, and justice provide a moral framework in which to discuss the ethical implications for providing medical care to any patients. The authors will use these principles in discussing the ethical implications of caring for a Jehovah's Witness patient.

Beneficence

Beneficence refers to the commitment by the medical professional to benefit patients by acting in their best interest. This means having comprehensive knowledge of the patient's wishes and beliefs. In the case of the surgeon and Jehovah's Witness, this would involve perioperative planning in such a way to minimize harm. In this sense, there is a close association of beneficence to autonomy and nonmaleficence which will be discussed below. The onus is on the surgeon to conduct him-/herself in a responsible and professional manner with full disclosure of the disease process that the patient has, the appropriate steps

to work up the problem, the ideal surgical plan with frank discussion of realistic chances of a cure and/or control of symptoms, and a realistic discussion of the expected postoperative recovery with disclosure of chances of possible complications. For the best interest of the patient, the surgeon has a duty to continuously develop his/her knowledge base and technical skills through professional development, exercise the utmost competence during surgical care, and display the ability to exercise sound judgment. As the sole advocate for the patient's life, it falls on the surgeons to maximize the conditions surrounding themselves and the environment in which the patient will receive their care to minimize potential risks of hospital-borne infections, poor nutrition, deconditioning, and other potentially preventable complications to the patients. If the surgeon has personal issues occurring that prevents him or her from maximally performing for the patient, it is the ethical duty of that surgeon to disclose that to the patient and allow the patient the choice of being cared for by a different surgeon. If the surgeon feels that they have inherent bias that would not allow for the principle of beneficence, they should recluse themselves from taking care of the Jehovah's Witness patient.

Even if the surgeon is at full functional capacity, before taking on a case involving a Jehovah's Witness patient, the surgeon has to be willing to take on the risk themselves. Self-reflection and honesty with oneself are critical elements in this process. It is imperative that the surgeon asks himself/herself if he/she is willing to accept the higher chance of death in the surgery involved. It is the authors' experience that the devout Jehovah's Witnesses will ask the surgeon if he/she is "okay with proceeding?" This question usually informs the surgeon that the patient has a clear insightful understanding that there is a contract between the patient and the surgeon regarding the proposed procedure which is risker than normal.

Not all surgeons are willing to take on this additional risk. In the event of an outcome that could have been altered by the addition of blood elements, the surgeon has to be very secure in his/her decision to operate on the patient. There are many factors that impact this decision. Some critical elements are:

1. The likelihood of death without surgery. In the authors' opinion, there has to be a high likelihood of death due to the patient's disease process in order to take on the risk of surgery in a Jehovah's Witness patient.
2. The likelihood of death from bleeding with the surgery. This is of great importance in many surgical fields, such as cardiac, vascular, and hepatopancreaticobiliary, as they are all at high risk for bleeding [3].
3. The relationship between the surgeon and the patient. There has to be an excellent rapport between the two. This may mean more detailed and frequent meetings to discuss perioperative complications and care with the patient. Documentation is paramount and having the liaison (see below) present may be helpful.
4. Importance of comorbidities. The impact of comorbidities that may be especially affected by anemia or inability to correct blood coagulation may be of greater importance in the Jehovah's Witness patient. For example, in a patient with metastatic tumor to the liver in the presence of chronic liver disease (CLD), a surgeon may agree to a minor resection in a well-compensated CLD patient knowing that red blood cells, platelets, and fresh frozen plasma (FFP) are available should there be an issue. However, such a procedure may be too morbid in a Jehovah's Witness patient, and the surgeon's decision may be altered. The authors feel that the risk of death from comorbidities at 1 year must be less than the risk of death from the process requiring surgery. For example, if the patient has a resectable hepatocellular carcinoma (HCC) in a non-cirrhotic liver and has coronary disease that is well compensated, we would ask the cardiologist to give us the risk of death from heart disease at 1 year. If this is less than the risk of death from unresected HCC, it would be our practice to consider surgery in that patient.
5. The option of other modalities that may require less blood products. The surgeon must consider other options that might be as effective for the condition being treated. Taking the example above, such consideration is critical in a patient with a metastatic liver tumor with CLD where ablation may be the second best option in the surgeon's mind compared to resection. However, in a Jehovah's Witness patient, ablation might rise to the top of the list in order to provide a safer option for the patient with significantly less potential for bleeding. It is also the responsibility of the surgeon, with his/her comprehensive knowledge of the disease process, to protect the patient from harmful treatment options. For a surgical oncologist, for example, it is unreasonable to expect that chemotherapy is a realistic option in a patient that would require agents that would substantially cause marrow suppression and high risk of blood component transfusion [4].
6. Are there options to decrease bleeding ahead of surgery? The use of adjuncts to assist in blood loss intraoperatively should be investigated. The surgeon must not feel that the use of these measures makes them any "less" of a surgeon. An example of a surgeon adjunct would be the use of transarterial chemoembolization (TACE) of a liver lesion prior to surgery. While this seems attractive at first glance to decrease the risk of bleeding, there is a trade-off in that there is an increased inflammatory response to TACE that can make the dissection more technically challenging. Such potential pros and cons of adjuncts must be weighed by the surgeon prior to surgery.

Autonomy

The literal meaning of autonomy is "self-rule," and it refers to the right of an individual to make a choice based on his/her belief and value. In the context of surgical care, this means obtaining an informed consent of all aspects of perioperative care, not just the actual surgical intervention. A patient has the legal right to decide to forego treatments that are clinically necessary if the patient is deemed to be competent to make that decision. It is important for the reader to understand that all patients exercise this choice to some degree – we, as physicians, are just more aware of autonomy in the Jehovah's Witness patient. For example, a patient choosing to forego a recommended colonoscopy is exercising their right of autonomy. We do not recognize this as such, as the consequences are felt to be minimal in this specific case. For any procedure, it is important to have a thorough discussion with the patient and obtain an informed consent. The authors often follow a specialized informed consent form for Jehovah's Witness patients with emphasis on discussion of complications that would normally require transfusion of blood products. Such a checklist is crucial in comprehensively reviewing with the patient all the available options for optimizing, correcting, and repleting the patient's hemoglobin level perioperatively (Table 1). Since

Table 1 Informed consent checklist for Jehovah's Witness

Informed consent tailored for Jehovah's Witness
Checklist
1. Check to see if patient has advanced directive. Review of all relevant documentation
2. Explanation of preoperative planning
(a) Discuss all preoperative tests and imaging
(b) Consultation to relevant specialties and follow-up on all documentations and/or tests run by those consulting physicians
(c) Discussion of all medications to optimize patient's condition
(i) Obtain patient's permission for use after explanation of these medical interventions
(d) Explanation of follow-up visit schedule prior to surgery
3. Explanation of procedure
(a) Especially highlight any points where risk of hemorrhage is high
4. Explanation of all risks
(a) Discuss potential for significant and/or fatal hemorrhage
(i) Confirm that patient will not consent to blood products (packed RBC, WBCs, FFP, cryoprecipitate, platelets)
(ii) Determine whether patient consents to synthetic colloid solution (albumin, hetastarch, dextran, gelatin), hemoglobin-based substitutes (perfluorocarbons) and recombinant proteins (erythropoietin, activated factor VII)
(iii) Preoperative strategy – Iron sulfate, folic acid, vitamin B12, erythropoietin, granulocyte colony-stimulating factor, hyperbaric oxygen therapy
(iv) Intraoperative strategy – Hemostatic agents (Gelfoam, Surgicel, Evarrest, etc.), injectable agents (desmopressin, ε-aminocaproic acid, tranexamic acid, vitamin K), acute normovolemic hemodilution, intraoperative blood salvage (cell saver)
(v) Postoperative strategy – Same as above
(b) Discuss potential for acute kidney injury (if relevant) and the use of dialysis
(i) Closed circuit usually employed with no blood prime used, no blood storage
(c) Discuss potential for thromboembolic event (if relevant)
(i) IVC filter? (if relevant)
(ii) Discuss possible use of anticoagulation if indicated unless patient has a higher risk of hemorrhage
(d) Discuss potential for other events (if relevant) that may increase chance of hemorrhage
5. Explanation of potential benefits
(a) Discuss outcome for patient if surgical procedure is completed
6. Explanation of alternative treatment
(a) Discuss outcome for patient if surgical procedure is not completed
(b) Discuss other interventions and their outcomes compared to surgery
(c) Weigh the risk of death due to uncontrolled hemorrhage during surgical intervention versus risk of morbidity/mortality if procedure not performed and discuss with patient
7. Discuss with patient his/her wishes if fatal massive hemorrhage is encountered. Is the patient willing to die rather than receiving life-saving transfusion?
8. Explanation of postoperative care
(a) Discuss expected routine postoperative course
(b) Discuss all possible complications again
(i) Discuss plans on how we will monitor for these complications
(ii) Discuss interventional plans and obtain patient's approval
(c) Discussion of all medications to optimize patient's condition
(i) Obtain patient's permission for use after explanation of these medical interventions
(d) Explanation of follow-up visit after hospital discharge
9. Give patient and family ample opportunity to ask any questions/concerns

each Jehovah's Witness patient may differ on what hematopoietic alternatives he/she may consent to, it is imperative for the clinician to explain what each medication or solution is comprised of so that the patient can make an informed decision on what he/she will allow to be infused into their body.

Before speaking with the patient, all relevant documents are reviewed, and special attention is paid to the patient's advanced directive if there is one. Preoperative steps are explained in detail, as well as medications that may be used to improve patient's hemoglobin and clotting levels. Then the procedure is explained in detail, highlighting the surgical steps where bleeding may be an issue. Each type of blood products is reviewed with the patient, and the authors take note of whether the patient would approve of products like fresh frozen plasma, cryoprecipitate, and/or platelets. Various colloid solutions are all reviewed to see whether the patient would permit infusion. Various hematopoietic medications (iron, folic acid, vitamin B12, erythropoietin) as well as anticoagulation medications are reviewed and are approved or disapproved by the patient. Intraoperative hemostatic devices and agents are reviewed, with clear disclosure that some of these agents contain human or bovine fractions of blood. All other complications are discussed in detail with the patient. Potential benefits, alternative treatment options other than the proposed surgical procedure, and outcomes if the procedure is not performed are all reviewed with the patient. Perhaps the most important portion of the consent is to convey to the patient that there may be a real risk of death and that the patient would prefer death rather than consenting to a life-saving transfusion [5]. Then our consent is signed by the patient, the physician, and a staff witness.

The surgeon must have good insight to discern whether the patient's understanding, and agreement, of the consent was clouded by emotional factors. Such emotions such as fear, anxiety, embarrassment, pressure from family, spiritual guides, etc. or stress from such things as finances, etc. can all negatively influence the patient's decision. If such factors do exist, counseling should be provided by appropriate personnel prior to obtaining informed consent. Persuasion, manipulation, and coercion are various influential forces that can also mar an informed consent. Persuasion can be a negative if it incites an emotional reaction that drives a patient's decision. Manipulation occurs when a physician presents the relevant information in a biased way, misrepresenting or even withholding information and is an ethical violation. Coercion, the use of force or threats, is the ultimate underminer of autonomy [6]. The surgeon must be cognizant of the possibility of coercion by other family members or friends. If this is detected, we recommend interviewing the patient alone and asking them to designate a power of medical attorney that they choose. This person should be included in all discussions and be tasked with communicating with the family. There are circumstances where the family appears to be coercing the patient into refusing blood when the patient him-/herself is fine with this.

In general, it is best if the entire family is included in all discussions. There is little conflict when the patient and the family are all Jehovah's Witnesses. However, when some or all members of the family are not Jehovah's Witnesses, real conflict can arise. The issue becomes who has the right to decide to allow blood products if the patient is in extremis. It is vital that the surgeon and the team have a clear discussion with all involved and make it clear that the patient's wishes will be honored should there be an issue of profound anemia that could lead to death. Indeed, the authors have experience where the family wanted transfusion when the Jehovah's Witness patient did not. The family called for the ethics team to get involved. This can be a tough situation that can create friction between the treating physician and the family. Our practice is to have the patient work with the Jehovah's Witness liaison regarding a "checklist" of products that the patient will accept. This is a qualified officer whose main aim is to ensure that the patient can make an informed decision with manipulation or coercion. The liaison will generally present a checklist that will be filled out with the patient.

Legal and ethical standards regarding the autonomy of Jehovah Witness minors (patients under the age of 18) can be confusing to the medical community. It is important to note that although the patient's parents may be devout Jehovah's Witnesses, the minor might not be. US federal statute gives physicians the authority to provide emergency medical care to minors including blood transfusions without the consent of the parents or without a court order, provided that the physician determines that there is an immediate need for treatment and a second physician concurs. All surgeons should be encouraged to find out their respective state's laws regarding treatment of minors in other medical circumstances. In most cases, emancipated minors can consent to their own procedures. Non-emancipated minors are generally granted right to seek treatment in specific medical situations (i.e., pregnancy, psychiatric disturbance, substance abuse, treatment of sexually transmitted diseases).

Nonmaleficence

The principle of nonmaleficence refers to the moral and ethical obligation to not cause any intentional net harm to the patient. This principle is often considered in conjunction with the principle of beneficence. Nonmaleficence is rarely an overt issue with a treating surgeon, as it is unusual for a caring physician to intentionally harm a patient. The real question is one for the true inner soul of the surgeon: do they believe, at some level, that they are harming the patient by withholding blood products? This is an important self-realization process that the surgeon must go through to ensure that they can answer this question to the negative.

In the context of Jehovah's Witness patients, the principle of nonmaleficence stresses the importance of preoperative planning, optimizing the patient's condition for the upcoming surgery, and having contingencies in place intraoperatively and postoperatively in case there are complications, especially bleeding complications, to surgery.

The perioperative management of Jehovah's Witnesses requires a multidisciplinary strategy compatible with their religious beliefs. It is imperative for the surgical team to meet preoperatively with other specialities such as the anesthesiologist, hematologist, cardiologists, pulmonologists, and other medical disciplines to discuss preoperative optimization, intraoperative strategies, and postoperative blood conservation and bleeding surveillance plans. The hematologist, especially in the setting of blood dyscrasias, can be an important resource. Cardiac risk is relevant because relative ischemia can be made worse with hemodilution and decreased oxygen carrying capacity. Similarly, impaired pulmonary function can lead to potential challenges for the patient if there is decreased oxygen carrying capacity with blood loss and anemia.

Preoperative optimization of a Jehovah's Witness patient must start weeks to months prior to surgery if possible. The authors start with basic blood work as baseline measurement. Many of these non-transfusion strategies take days to weeks to see the effect, and thus early detection to optimize the patients is crucial. The authors routinely use iron, folic acid, vitamin B_{12}, and/or erythropoietin to replenish the patient's blood storage. Hematology is involved early to help optimize management. If further therapy is needed and the patient is agreeable, granulocyte colony-stimulating factor or other hematopoietic agents such as erythropoietin can be considered.

If the patient is on anticoagulation, interventions to reverse the anticoagulant affects are initiated. This aspect can be especially challenging, as there are scenarios with non-Jehovah's Witnesses where the surgeon will accept a less than perfect coagulation profile knowing that they can use blood components to correct these abnormalities. With the Jehovah's Witness patients, the surgeon has to take a calculated risk in stopping the anticoagulants. Usually the risk of clotting is a greater concern than the risk of bleeding, as long as there is the option to transfuse. In the Jehovah's Witness patients where there is no such option, the surgeon may have to accept a higher risk of a clotting phenomenon in order to minimize the risk of bleeding. Patients can be taken off their anticoagulation medication at appropriate times preoperatively for the effects to wear off. Appropriate services such as cardiology are contacted beforehand so that we can safely take the patient off their medications. If the patient has current or history of thromboembolic disease, appropriate workup can be initiated, and preventative measurements such as inferior vena cava filters can be used to minimize future thromboembolic events. All of these interventions must be carefully discussed and agreed upon with the patient prior to activation.

Regarding intraoperative strategies, surgical planning with the anesthesiologist and the OR staff is crucial. The authors routinely meet and discuss care regarding our patients prior to the operative day, aiming for minimal blood draws during procedures and focusing on intraoperative monitoring devices to assess the patient's condition. Appropriate lines such as arterial and central venous lines are planned to be placed with minimal blood loss for monitoring purposes. Foley is placed to trend urine output as a measure of resuscitation. Permissive hypotension is employed in the operating room to minimize blood loss. This is especially the case during liver resections where the aim is low central venous pressure (CVP) anesthesia to minimize the bleeding from hepatic veins. The authors also routinely meet with our OR circulators and staff prior to the operation to make sure all medications and equipment are ready in the OR prior to starting the case.

Meticulous attention to hemostasis and minimizing technical blood loss during procedures is crucial. Detail-oriented surgical technique is employed while striving for hemostasis throughout the planned procedure. Each surgical procedure employs techniques to minimize blood loss. The surgeon must be familiar with blood-saving maneuvers and techniques in case complications arise during surgery. There are also a number of coagulating energy devices and hemostatic agents are available in the market, which may be used if the patients are informed and agreeable to them. Advanced energy devices such as the Harmonic (Ethicon™), Ligasure (Covidien™), etc. can be used for tissue transection. The authors

recommend that each surgeon use devices that they have the most experience with and is the most comfortable. It is also important for the surgeon to have an in-depth knowledge of all the resources available and ready in case any bleeding is encountered during the operation.

Carefully surgical planning and proper imaging prior to surgery often gives us a roadmap to follow and allows us to anticipate any variations in blood vessel distribution, such as the often-encountered replaced right hepatic artery coming off of the superior mesenteric artery during a Whipple procedure. Any appropriate imaging modalities such as CT and/or MRI should be done leading up to the operation, with the images loaded up and viewable in the operating room on the day of surgery. The authors often employ intraoperative ultrasound as an adjunct in liver and pancreatic surgeries, identifying critical structures such as major blood vessels. Anticipating these structures prior to encounter will ensure that those vessels will not be accidentally clipped or ligated prior to proximally and distally control.

There are a number of blood-saving and blood-salvaging techniques that are described. The surgeon must be familiar with these techniques and must have held a discussion with the patient regarding the usage of such techniques prior to surgery. Some Jehovah's Witness may agree on employing some of these techniques. Acute normovolemic hemodilution (ANH) is an autologous blood collection and volume management technique that may have a role in managing Jehovah's Witness patients intraoperatively [7]. The rationale for this technique is that if the hematocrit level is lowered before any blood loss, lower concentration of red blood cells will be lost if there is any hemorrhage. The patient's blood is removed at the time of surgery before any acute blood loss occurs and acellular fluid, either crystalloid or colloid, is used to maintain circulating intravascular volume. It is important to note that some Jehovah's Witness may refuse colloid infusion in which case the only option for ANH would be crystalloid replacement. Normally the blood that has been removed is in continuous circuit with the patient via an outflow and inflow tubing connected from the patient to the blood collection bag. Given that the blood is in continuous circuit, some Jehovah's Witness patients may decide that this technique does not conflict with their faith.

The intraoperative cell salvage (ICS) is another possibility for volume management in Jehovah's Witness patients. The ICS machine, commonly called a "cell saver," separates, washes, and concentrates collected red blood cells (RBCs) [8]. Just like ANH, the blood that has been removed can also be in continuous circuit with the patient via an outflow and inflow tubing connected from the patient to the cell saver machine. Again, given that the blood is in continuous circuit, some Jehovah's Witness patients may decide that this technique does not conflict with their faith [9].

Once the patient is guided safely through the surgery, steps can be taken to optimize the safest postoperative course. Multiple studies have shown that surgical patients can tolerate an acute drop in hemoglobin, although levels less than 5 g/dL have been associated with increased mortality. The author's overall postoperative approach in managing Jehovah's witness patients are to:

1. Minimize bleeding and blood loss.
2. Optimize physiological tolerance of anemia.
3. Encourage hematopoiesis.

Overall theme in dealing with acute anemia in our postoperative Jehovah's Witness patients is early vigilance and intervention. Early recognition of any bleeding episodes and intervention is crucial in minimizing blood loss. Experienced clinical judgment is crucial to determine whether the patient needs to return to the operating room or whether the bleeding will stop on its own. Jehovah's Witness patients will require a lower threshold for surgical intervention for blood loss compared to those that will accept blood and factors to halt bleeding.

Jehovah's Witness patients are routinely placed in the ICU setting in the early postoperative period for hemodynamic monitoring. The routine use of measures such as heart rate, blood pressure, CVP, and urine output (as long as the patient does not have ESRD) as markers of resuscitation is highly recommended. Antihypertensive medications such as beta-blockers or calcium channel blockers can be employed to keep the blood pressure under control. Crystalloid solutions are used to replete intravascular volume if extreme hypotension and/or tachycardia ensues. If the patient consents to colloids, solutions such as albumin and hetastarch are options for resuscitation and volume repletion. There are downsides to overusing these solutions, such as hemodilution. Balancing the use of these solutions for adequate resuscitation is crucial. Blood substitutes such as hemoglobin-based oxygen carriers (Hemapure) and perfluorocarbon emulsions are under development. Although protocols are in place in select centers for use of Hemapure (HBOC-201 – bovine hemoglobin), this is not FDA approved.

Blood conservation techniques should be extended to the postoperative period. Multiple past studies have shown significant blood losses in the medical/surgical ICUs with prolonged daily phlebotomies. The authors advocate for minimizing daily phlebotomies. Again, the surgeon must use his/her clinical judgment to avoid needless blood draws and only order labs for specific indications. Routine use of blood draws without specific indications may harm the Jehovah's Witness patients and thus violate the principle of maleficence. In addition, the use of pediatric tubes and/or ISTAT devices can minimize blood losses due to blood draws.

Medications are used judiciously in Jehovah's Witness patients. The authors minimize antiplatelet medications (i.e., aspirin, nonsteroidal anti-inflammatory drugs) immediately after surgery. Patients are encouraged to get out of bed and ambulate starting a few hours after coming out of surgery, and anti-embolic stockings and/or sequential compression devices are employed while the patient is in bed for DVT prophylaxis. Chemical DVT prophylaxis is started after ensuring that the patient does not have any ongoing postoperative bleed. There may be some trepidation from the surgeon to start any anticoagulation in these patients. The risk of thromboembolism must be weighed against the risk of bleeding by the surgeon. When the risk of complications from thromboembolism becomes higher than the risk of bleed, blood thinners can be started at the optimal time. It is important to emphasize that Jehovah's Witness patients must have access to every medication that routinely is used on a non-Jehovah's Witness patient. Fear of a bleeding complication must not impair the surgeon from using any medications as long as the benefit of that medication is greater than the event of a postoperative bleed.

Any symptomatic decreases in hemoglobin levels are treated with combination of iron, folic acid, vitamin B12, and/or erythropoietin (see Preoperative Strategies section). Any coagulopathies are aggressively treated. Elevation in INR can be treated with vitamin K injections and any platelet dysfunction secondary to uremia can be treated with desmopressin. If the patient is coagulopathic and bleeding, 4-factor prothrombin complex (Kcentra™) or factor VII can be given, as long as those agents are approved by the patient during informed consent. Other injectable agents such as ε-aminocaproic acid and tranexamic acid are also options although very few studies exist regarding their use in Jehovah's Witness patients in the setting of postoperative bleeding after complex GI surgeries.

Meticulous planning, excellent surgical technique, and early vigilance are the keys to minimizing blood loss and complications in Jehovah's Witness patients. Surgeons are ethically bound under the nonmaleficence clause to provide abundant expertise and resources to ensure that the patient has this level of care.

Justice

Justice refers to the physician's obligation to equally disperse health-care resources to all individuals regardless of religion, sex, creed, ethnicity, or other differences. In the contexts of Jehovah's Witness patients, it is legally and morally wrong for the surgeon to deny any surgical intervention solely due to the religious beliefs of that patient. Even if the patient has specific clause in their religious tenet that forbids them from receiving blood products, withholding care to these individuals is against the ethical that the surgeon must abide by. If the surgeon has the technical ability to perform the procedure in question with minimal blood loss, has the resources around to provide adequate perioperative care and has fully informed the patient on the risks of that procedure, and has gained the approval and trust of that patient, the surgeon has the moral obligation to perform that procedure in the safest manner possible. This tenant is specifically challenged when a surgeon is faced with performing a procedure that has minimal chance of blood loss in a Jehovah's Witness patient: for example, if the patient requires inguinal hernia repair and the surgeon is an experienced groin hernia surgeon. In this circumstance, the surgeon has to ask themselves if they are withholding care simply because of the Jehovah's Witness status of the patient and whether this is a violation of the Justice clause.

It must be emphatically noted that this obligation is different from a surgeon who honestly confesses due to legitimate reason(s) (lack of expertise in that field, lack of operative experience, lack of resources to adequately provide safe perioperative care, etc.) that it is not safe for the patient to receive surgical care with that particular surgeon. Such declaration shows high moral fiber and maturity on the part of the surgeon to admit his/her deficiency as an act of beneficence and nonmaleficence for the patient. In such a case, a frank discussion with the patient that encompasses but does not trespass the limits of the surgeon's expertise should be held. Then the surgeon should provide honest admission of his/her limitations and a plan of referring or transferring the patient to a center with the expertise and the resources for the patient to be properly taken care of.

Concluding Remarks

- The relationship between the surgeon and their patient is like no other relationship.
- The decision to perform surgery on a patient that puts them at higher risk for complications solely based on religious belief is a challenge.
- The surgeon must examine the impact on themselves and on the patient when making the difficult decision to proceed with surgery.

- The surgeon must be both skilled technically and medically and must be knowledgeable about all aspects of perioperative blood conservation when treating the Jehovah's Witness patient.
- The surgeon must decide honestly if their decision to not treat a Jehovah's Witness patient is based on inherent bias rather than hard data. Only he or she can answer that question.

References

1. Lawson T, Ralph C. Perioperative Jehovah's Witnesses: a review. Br J Anesth. 2015:1–12.
2. Adedeji S, Sokol Dk, Palser T, et al. Ethics of surgical complications. World J Surg. 2009. https://doi.org/10.1007/s00268-008-9907-z.
3. Konstantinidis IT, Allen PJ, D'Angelica MI, et al. Pancreas and liver resections in Jehovah's witness patients: feasible and safe. J Am Coll Surg. 2013;217:1101–7.
4. Araujo RL, Pantanali CA, Haddad L, et al. Does autologous blood transfusion during liver transplantation for hepatocellular carcinoma increase risk of recurrence? World J Gastrointest Surg. 2016;8(2):161–8.
5. Waters JH. Intraoperative blood recovery. ASAIO J. 2013;59(1):11–7.
6. Naunheim KS, Bridges CR, Sade RM. Should a Jehovah's witness patient who faces imminent exsanguination be transfused? Ann Thorac Surg. 2011;92:1559–64.
7. Monk TG. Acute normovolemic hemodilution. Anes Clin N Am. 1995;23:271–81.
8. Transfusion handbook. 5th ed. 2014. http://www.transfusionguidelines.org.uk/transfusion-handbook.
9. Kumar N, Lam R, Zaw AS, et al. Flow cytometric evaluation of the safety of intraoperative salvaged blood filtered with leucocyte depletion filter in spine tumour surgery. Ann Surg Oncol. 2014;21(13):4330–5.

Ethical Issues in Bariatric Surgery

Antonio J. Torres, Oscar Cano-Valderrama, and Inmaculada Domínguez-Serrano

Introduction

Bariatric surgery refers to the surgical procedures that are performed to achieve long-term weight lost in obese patients. The number of patients that underwent bariatric surgery increased from 40,000 in 1997 [1] up to 579,517 in 2014 [2]. This increase can be explained by two factors. First of all, prevalence of obesity has been increasing [3], and nowadays it is considered an epidemic. Secondly, bariatric surgery has demonstrated to be a safe and effective way to treat obese patients [4].

This increase in the number of obese patients and patients submitted to bariatric surgery has provoked the appearance of new ethical concerns. For example, Schneider et al. reported the case of a patient that required three ethics consultations in a short period of time [5]. He was a patient with morbid obesity, obesity hypoventilation syndrome, and numerous ICU admissions. The first ethics consultation was requested about forcing bariatric surgery against his will, the second one was about the nursing staff requesting no longer attempt to mobilize him, and the last one was because the patient refused to be discharged. This case report exemplifies the ethical challenges that can appear during the treatment of patients with morbid obesity.

The aim of this chapter is to discuss the main ethical concerns that are related to obesity and bariatric surgery.

Medicalization of Obesity

The first issue that we must analyze is medicalization of obesity. We are nowadays living in a society that gives a high relevance to physical appearance; and the current aesthetic canon is related to thinness. Therefore, everything that is related to overweight and obesity is seen as ugly and wrong. Nevertheless, this aesthetic canon has been changing for centuries. During the Middle Ages, when food was a scarce resource, obesity was considered a sign of wealth and prosperity.

The main concern about medicalization of obesity is that obesity can be treated as a disease just because it is against our aesthetic canon [6, 7]. In a society that judges you because of your physical appearance, obesity is considered undesirable; therefore obesity is medicalizated and considered a disease to remove it from the society.

Medicalization of obesity is also related to financial interests. Surgeons, endocrinologist, nutritionists, fitness centers, pharmaceutical companies, sport shops, food companies, and many other lobbies have a huge economic interest in obesity. They are earning a lot of money

A. J. Torres · O. Cano-Valderrama (✉)
I. Domínguez-Serrano
Department of Surgery, Hospital Clínico San Carlos, Madrid, Spain

© Springer Nature Switzerland AG 2019
A. R. Ferreres (ed.), *Surgical Ethics*, https://doi.org/10.1007/978-3-030-05964-4_28

treating obese patients, so they could be pressing to overstate the problem of obesity just to improve their financial status [8].

These concerns about medicalization of obesity can be reasonable for overweight and low-grade obesity. Nevertheless, when talking about morbid obesity and bariatric surgery, they make no sense. Morbid obesity has been associated with a higher mortality, development of severe comorbidities such as diabetes or high blood pressure, and a lower quality of life [9–11], and bariatric surgery has also demonstrated an improvement in mortality, comorbidities resolution, and quality of life [4, 11, 12]. So, there is little doubt that morbid obesity is a real problem that must be faced.

When talking about overweight and low-grade obesity, medicalization can be a real problem. Nevertheless, overweight is an important risk factor for obesity. So, it seems reasonable to treat obesity in an early stage, before it has provoked severe complications.

Discrimination and Prejudices

As we have already said obesity is not well seen by our society, so obese patients are usually discriminated [13]. Puhl et al. reviewed some of the areas in which discrimination has been seen [14]:

- *Employment settings*: obese patients have a lower salary and a higher rate of employment termination. Also, obtaining a job is more difficult for patients with obesity. There are several reasons for this discrimination. The clearest reason is the importance of physical appearance. Obesity is a problem for all those working with the public, for example, a flight attendant or receptionist. However, obese patients also have problems with employments that are not related to working with the public because obesity is usually associated with negative features such as laziness and low self-control.
- *Education*: although some cases of notorious discrimination have been published, such as not admitting someone to a college due to

obesity, the most important problem is stigmatization. Children and teenagers with obesity are usually stigmatized during high school and college, and they can develop severe problems, for example, depression and/or eating behavior disorder.

- *Insurance and healthcare cost*: obesity is an important risk factor for many disorders. This is the reason why obese patients have problems to get a health insurance. This lack of health coverage provokes a vicious circle because obese patients cannot access to obesity treatment; therefore they remain obese.
- *Jury selection*: some authors believe that obese people could be discriminated during jury selection. Negative attributions have been applied to obese persons (e.g., lazy), so exclusion of jurors could be possible. More research is needed to study this point.
- *Public accommodation*: obese people can experience problems in public settings (e.g., trains, buses, restaurants, theaters, airplanes) because of inadequate seat size.
- *Housing*: an interesting paper published in 1977 by Karris [15] demonstrated that obese patients had more problem to rent a house than nonobese people.
- *Adoption*: adoption could be another point of discrimination although there is not research to be sure about this issue.

Discrimination is not only provoked by obesity; other factors must also be taken into account. For example, some campaigns against obesity stigmatize these patients [16]; nevertheless these campaigns are not effective, and they only help to increase their discrimination.

Gender is another important point when analyzing discrimination in obese patients. Female patients are considered to have a higher discrimination rate; in fact they are considered to be "fat girls," while male patients are only "big guys" [17]. This discrimination is also noted in the way that bariatric surgery is perceived. In a study by Newhook et al., female patients who underwent bariatric surgery referred not being supported by their husband, while most male patients recognized that they were undergoing surgery because of their wife support [17]. This difference proves that obesity is usually more relevant for females than for males.

Other important issue is prejudices by health professionals. Several studies have demonstrated that health professionals associated obesity with poor hygiene, noncompliance, hostility, dishonesty, lacked self-control, and laziness [18–20].

Finally bariatric surgery can be another factor that provokes discrimination. Obesity is more prevalent in low-income populations; nevertheless bariatric surgery is an expensive treatment. So, patients with a good economic status are able to undergo bariatric surgery and get away from the discrimination associated with obesity, while patients with less economic resources cannot afford this surgery, and the differences with wealthy people increase [7, 13, 21].

Information and Bariatric Surgery

One of the main ethical concerns about bariatric surgery is information and informed consent [7, 13, 20–22]. Madan et al. demonstrated that most patients were not able to remember the possible complications after bariatric surgery [23]. So, can we talk about informed consent in these patients?

During the last decades, patient's autonomy has gained more and more importance when talking about decisions in the medical area. Nevertheless, autonomy is based in knowledge; therefore information must be given to patients, and they have to understand this information in order to give a valid informed consent. Taking into account the paper by Madan [23], preoperative information should be improved. Some measures that could be put into practice would be:

- *Written information* that allows the patient to study it.
- *Repeated outpatient appointments*: it is difficult that patients understand all the information in 1 day. Therefore, patients should be seen at least twice before the surgery is decided. In this way the patient can think about the surgery and ask the questions during the next appointment. Some patients will even require more appointments before we are sure that they have understood all the information.
- *Online information*. In the digital age, online information is very important. One of the main problems is that sometimes this information is wrong or at least it is biased. So, it would be interesting that medical societies and hospitals gave high-quality information about bariatric surgery.
- *Assuring that the patient has understood the procedure*: it is difficult to be sure that the patient has understood the information that has been given. An interesting option is asking the patient to write what he/she knows about the surgery. In this way, we can perceive the information that the patient has not yet understood.
- *Preoperative meeting with patients who had already undergone bariatric surgery*: a meeting with patients who have already undergone surgery could be interesting as they can explain the postoperative course and the changes in their life after the procedure.

An interesting point about information and informed consent in bariatric surgery is who should choose the procedure that is going to be performed. There is not a perfect procedure that is indicated for all the patients, so patients must be evaluated to choose the most appropriate procedure. Aforetime medical paternalism advocated that physicians should choose because patients were unable to understand the procedure and its consequences. Nevertheless, autonomy has emerged as a key ethical value. Therefore, for years it has been thought that patient must choose the technique. He must be informed, as it has been previously discussed, and then he/she can select the procedure that it is going to be performed. The main problem with this model is that it is difficult that the patient can understand all the information that must be taken into account to make a decision. Sometimes different surgeons don't agree about the best procedure for one patient, how could a patient choose under these circumstances? Probably, there is an intermediate option, the "soft paternalism." Information must be given to the patient, and then he/she must be guided during the decision process. Evaluating the success is a difficult task after bariatric surgery (factors that can be taken into account are, e.g., weight lost, food tolerance, comorbidities remission, quality of life, bowel movements, etc.), so the patient should set his/her aim for the procedure. Then, patient and doctor can analyze the advantages and disadvantages of each procedure and discuss which procedure will probably get the outcomes that the patient is looking for.

We have described three different models to choose the procedure that is going to be performed, which one is better? Probably it depends on the circumstances. In Spain, patients are used to be guided by their physician; therefore a decision based on autonomy only will be difficult. Meanwhile in the United States of America, patients are used to use their autonomy, and paternalism wouldn't be an available option. Another point that should be taken into account is patient's features. A highly educated patient interested in choosing the technique cannot be treated in the same way that an illiterate patient who ask for your advice.

Access to Bariatric Surgery

Access to bariatric surgery is also an important ethical concern. Bariatric surgery needs a lot of resources, where are we going to get them? Are we going to shunt these resources from other treatments, or are we going to improve the health budget by imposing a tax to fast food restaurants? And finally, should these resources be used to fund bariatric surgery, or should we use them for preventive measures?

It is very difficult to answer these questions. Moreover, each country has a different health system and must adapt these financial issues to their circumstances.

Regardless of the circumstances, some authors advocate that bariatric surgery should be given high priority [24], regardless which moral perspective we consider (greatest needed, utility, or personal responsibility). The arguments to give

bariatric surgery high priority depending on the moral perspective used are:

- *Greatest needed*: the number of obese patients is increasing, and they have a poor quality of life and important health problems. Therefore morbid obesity is an important problem that must be faced.
- *Utility*: bariatric surgery is an effective treatment to treat morbid obesity [25], and it has proved to be cost-effective [26]. Moreover, there are indirect benefits such as motivate friends and colleagues to adopt a healthier lifestyle or improve their productivity. So, investing funds in these procedures is a good option.
- *Personal responsibility*: some authors argue that morbid obesity is a self-inflicted disorder because patients decide to eat more than they need; therefore, they should pay for the treatment. However, there are two points that must be taken into account.

First of all, other self-inflicted disorders such as smoking-related diseases, drug addiction, and sport or crash accidents are funded by the health system; why should obesity be different?

In the second place, obesity can't be considered a self-inflicted disorder. It is true that obese patients eat more than they need, but this behavior is based on multiple factors and not only in gluttony. Many studies have demonstrated that obesity is based in biological alterations [27] such as microbiome, hormone, or inflammatory changes. This biological factor is clearly seen in some patients; for example, patients with Prader-Willi syndrome or twins with obesity. Moreover, there is an important social influence. Some factors such as fast food, sweetened beverages, or video-games have had a key influence in the increase of obesity prevalence. Many companies with financial interest in these products have promoted unhealthy lifestyles and probably should be forced to fund obesity treatment.

To sum up, many factors are related to obesity; therefore obese patients can't be blamed and punished for developing this disorder.

Psychosocial Assessment and Preoperative Weight Lost

Psychosocial assessment and preoperative weight lost are usually considered compulsory prior to bariatric surgery [20, 28]. However, these interventions have not proved to improve postoperative outcomes after bariatric surgery. Therefore, there is an ethical concern about requiring them before bariatric surgery.

Marek et al. have seen that presurgical psychological evaluation can predict weight loss after bariatric surgery; nevertheless it is not clear that these evaluations really change the management of the patients [20]. Taking into account that during psychosocial assessment private information is collected and that this information is usually available to all the health professionals, we must think if it is really needed.

The main reason to get a psychosocial assessment is to detect those patients with a psychopathology that could provoke problems after the surgery, such as anorexia nervosa, or bulimia. Therefore, psychological evaluation is important to ensure that patients with a severe psychopathology don't undergo bariatric surgery, but this information should be maintained confidential. Psychoeducational group interventions could also be interesting to improve patient's psychological state before the surgery, without risking their private information [29].

Compulsory preoperative weight loss is another preoperative condition that is usually requested although there is not a clear evidence that it improves the postoperative outcomes [28]. Preoperative weight lost is requested to prove that the patient is committed with the surgery and that he/she will follow the postoperative recommendations. Also preoperative weight lost can decrease liver volume, making surgery easier. However, if we deny bariatric surgery to patients who doesn't achieve preoperative weight lost, we are condemning them because bariatric surgery is the only effective treatment for morbid obesity. Therefore, this requirement must be applied carefully and patients who don't achieve enough weight lost shouldn't be blamed and they should be submitted to a more intensive nutritional and psychoeducational program.

Centers for Bariatric Surgery and Medical Education

Bariatric surgery is considered a difficult surgery that requires a highly trained surgeon. Moreover, bariatric surgery needs a multidisciplinary team that includes surgeons, endocrinologist, nutritional therapists, physical therapists, anesthesiologists, etc. Therefore, bariatric surgery can only be performed in some faculties. Some scientific societies such as IFSO (International Federation for the Surgery of Obesity and Metabolic Disorders) or ASMBS (American Society for Metabolic and Bariatric Surgery) have developed accreditation programs to ensure that bariatric surgery is performed in centers that are fully equipped. However, taking into account that morbid obesity is an epidemic disorder, restricting the number of centers that can perform this surgery is not a good option. Probably, patients and centers should be stratified; the more complex patients should be submitted to high-volume centers, while low-risk patients can be managed in smaller departments. Also, medical education programs must be developed to ensure that there are enough trained professionals to treat these patients [22].

Medical education is another ethical concern. Governments and scientific societies should guide medical education; because if medical education is promoted by pharmaceutical companies, it could be biased.

All the bariatric surgeons should be able to perform at least two or three different procedures. In this way they can choose the best technique for each patient. A surgeon that always performs the same procedure will probably choose an inadequate technique for same patients. An example of this problem could be a surgeon that only performs gastric bypass. Gastric bypass is a procedure that can be performed in most obese patients; nevertheless a different procedure is needed for some patients, for example, patients with an inflammatory bowel disease. These patients should be submitted to a surgeon who can perform the appropriate procedure.

Finally, revisional bariatric surgery is another important point. Revisional bariatric surgery is more complex, and its outcomes are not as good as expected for primary bariatric surgery. Some payers, insurance providers, and hospitals have tried to avoid revisional bariatric surgery; however revisional bariatric surgery is a moral obligation [30]. Patients who have undergone such a risky treatment can't be abandoned when it fails. One of the reasons to deny this surgery is the perception that bariatric surgery fails because the patient doesn't follow the postoperative recommendations. However, it has already been discussed that patients cannot be blamed for their obesity, so this argument cannot be used to deny revisional bariatric surgery. The cause of the failure must be fully studied, and then an appropriate treatment must be proposed, probably in a high-volume center.

New Procedures in Bariatric Surgery

With the development of bariatric surgery, new procedures have appeared to improve its outcomes. However, these procedures must be carefully tested before they are used [31–33]. Sometimes new techniques are used before there is enough evidence about its effectiveness and safety. There are many reasons to use new procedures that have not been appropriately studied; for example, promote a technique that you have developed, financial pressure by a pharmaceutical company, or trying to use the up-to-date techniques. However, these pressures should be avoided, when using a developing procedure informed consent from the patient and ethic committee approbation should be obtained before the surgery.

Bariatric Surgery in Children and Teenagers

Children and teenagers are one of the main ethical challenges in bariatric surgery. Morbid obesity in children is an important health problem because they can develop comorbidities and long-term complications caused by obesity. However, bariatric surgery can affect children's growth and is associated with severe complications and sequels; therefore submitting a child or teenager to bariatric surgery is a difficult decision.

First of all, we have to analyze if bariatric surgery is effective for children and teenagers. Durkin et al. have studied the evidence for pediatric/adolescent bariatric surgery, and they have found that adolescent bariatric surgery outcomes are comparable to adults, with similar sustainable weight loss, comorbidities resolutions, and complication rates [34]. Therefore, those patients with morbid obesity who fail after conservative treatment can be submitted to surgery to avoid further complications related to obesity. The main doubt is when the surgery should be performed. In this population educational therapies are important because patients are developing and they can change their lifestyle, avoiding the surgical treatment. Therefore, aggressive conservative treatment must be tried before performing the surgery.

Another important issue is preventive measures [35]. Obesity in children and teenagers is an epidemic problem, so preventive measures should be implemented in order to avoid obesity and bariatric surgery. Promoting healthy lifestyles, avoiding fast food and sweetened beverages, and promoting physical exercise are some of the measures that could help to decrease obesity prevalence.

Medicalization of obesity is also an interesting point in children. This issue has been previously discussed, but it is more important in children and teenagers. Taking into account the prevalence of obesity and the possible stigmatization of obese children, only the more severe cases must be treated as a disease. However, patients with mild obesity must be submitted to educational programs and dietary counsel avoiding an excessive medicalization of their state and the impression that beauty and appearance are too important.

Finally, informed consent is also an ethical challenge. It is difficult that children and teenagers can understand the problems that can be associated with obesity, the possible complications of bariatric surgery, and the long-term sequels of these procedures. Therefore, informed consent is a problem [35]. Unless the patient is very mature, parents will have a key role during information and informed consent. Some studies have seen that parents who believe that obesity is a "biological" problem are willing to accept the surgery, while parents who believe that obesity is an "educational and self-inflicted problem" are willing to deny bariatric surgery [36]. It is also important that children and teenagers understand that bariatric surgery is only the beginning of the treatment, but they also have to change their lifestyle. If they cannot understand this, the procedure will probably fail, and it shouldn't be performed [36].

Medical Tourism and Bariatric Surgery

Medical tourism is a big problem for all kind of surgeries, including bariatric surgery. Several moral challenges are related to medical tourism for bariatric surgery [37, 38].

First of all complications can appear after bariatric surgery. In these cases the patient is usually back in his/her country and complications must be treated by surgeons who didn't take part in the procedure and who usually don't have all the details about the technique performed and the early postoperative course. There is also a problem about who must pay these expenses.

Another ethical concern is that medical tourism is a proof of inequity in the patient's country. Medical tourism means that access to bariatric surgery is limited by financial issues and only wealthy people can afford this treatment. Taking into account that morbid obesity is associated with low incomes, medical tourism for bariatric surgery increases the financial difference between wealthy people who can afford the surgery or are not obese and patients who have to invest the few savings that they have to undergone surgery abroad.

Finally, medical tourism is also unfair for the people who live in the country where the surgery is performed. Medical tourism means that there are health facilities that they usually can't use because they can't afford them. This inequity is more meaningful if these facilities are funded with public budgets.

Body Contouring Surgery After Bariatric Surgery

Body contouring surgery has proved to improve quality of life after bariatric surgery [39]. However, only about 6% of the patients who undergo bariatric surgery are submitted to body contouring surgery [40]. This problem is due to a financial issue. Body contouring surgery is usually considered an aesthetic procedure; therefore it is denied by insurance companies. The ethical concern is, can we deny a surgery that will improve quality of life after bariatric surgery?

There is not a simple answer to this question. On the one hand, body contouring surgery is effective, and it would be desirable that all patients who undergo bariatric surgery could later undergo body contouring surgery. On the other hand, body contouring surgery is expensive. So, in a setting of a limited budget, these procedures, with a high aesthetic component, can't be paid for all the patients.

To sum up, body contouring surgery would be desirable for all the patients who undergo bariatric surgery, but insurance companies and public health systems can't afford it. Therefore patients should be carefully selected before body contouring surgery, but denying it for all the patients is not a moral option.

Should We Deny Surgery for Benign Disorders to Patients with Obesity?

Surgery for benign disorders in obese patients is a controversial issue. On the one hand, some authors think that outcomes after surgery for benign disorders are worse in obese patients, so these procedures should be denied to them until they lose enough weight [41]. Another reason to deny the surgery is that if the patient loses some weight he/she will probably improve, so the surgery could be unnecessary. On the other hand, other authors defend that obese patients shouldn't be discriminated. Obesity is another feature of the patient, so if the surgery is indicated, it should be performed [42].

Once again, horse sense is the answer. If outcomes after the surgery are much worse in obese patients and there is a chance for the patient to lose weight, the surgery should be delayed. For example, bariatric surgery could be performed before knee replacement. However, if the patient has failed to lose weight and the disorder is invalidating, the surgery should be performed, even if the patient has to accept that outcomes are worse than for nonobese patients. The only moral option that will never be admissible is giving up the patient without a treatment program. We can treat the obesity or the disorder, but something must be done.

Conclusion

Bariatric surgery is a field with many ethical concerns. These moral challenges are related to morbid obesity, access to surgery, patient selection, and informed consent. We can summarize these concerns according to the four principles of bioethics:

- *Autonomy*: information and informed consent for bariatric surgery is one of the main concerns. It is difficult that the patient understands the problems associated with obesity, the risk of the surgery, and the possible complications and long-term sequels. These problems are even bigger for children and teenagers.
- *Justice*: access to bariatric surgery is also a moral challenge. Obesity is more prevalent in low-income populations, but bariatric surgery is an expensive treatment; therefore improving access to bariatric surgery is important. Preventive programs are also essential.
- *Beneficence*: we must always look for the best available treatment. All the financial and personal interests must be forgotten when deciding the best treatment for the patient.
- *Non-maleficence*: medical treatments and preventive measures can provoke discrimination, stigmatization, or excessive medicalization of obese patients.

Bibliography

1. Scopinaro N. The IFSO and obesity surgery throughout the world. International Federation for the Surgery of Obesity. Obes Surg. 1998;8:3–8.
2. Angrisani L, Santonicola A, Iovino P, et al. Bariatric surgery and endoluminal procedures: IFSO worldwide survey 2014. Obes Surg. 2017;27:2279–89.
3. Peralta M, Ramos M, Lipert A, Martins J, Marques A. Prevalence and trends of overweight and obesity in older adults from 10 European countries from 2005 to 2013. Scand J Public Health. 2018. https://doi.org/10.1177/1403494818764810.
4. Sjostrom L. Bariatric surgery and reduction in morbidity and mortality: experiences from the SOS study. Int J Obes. 2008;32(Suppl 7):S93–7.
5. Schneider PL, Li Z. Ethical challenges in the care of the inpatient with morbid obesity. Narrat Inq Bioeth. 2016;6:143–52.
6. de Vries J. The obesity epidemic: medical and ethical considerations. Sci Eng Ethics. 2007;13:55–67.
7. Brandon AR, Puzziferri N, Sadler JZ. Stuck in the middle: what should a good society do? Am J Bioeth. 2010;10:18–20.
8. Macgregor AM, Macgregor CC. Economic theory and physician behavior in bariatric surgery. Obes Surg. 2000;10:4–6.
9. Long MT, Fox CS. The Framingham Heart Study – 67 years of discovery in metabolic disease. Nat Rev Endocrinol. 2016;12:177–83.
10. Hayes M, Baxter H, Muller-Nordhorn J, Hohls JK, Muckelbauer R. The longitudinal association between weight change and health-related quality of life in adults and children: a systematic review. Obes Rev. 2017;18:1398–411.
11. Mazer LM, Azagury DE, Morton JM. Quality of life after bariatric surgery. Curr Obes Rep. 2017;6:204–10.
12. Schauer PR, Bhatt DL, Kirwan JP, et al. Bariatric surgery versus intensive medical therapy for diabetes - 5-year outcomes. N Engl J Med. 2017;376:641–51.
13. Saarni SI, Anttila H, Saarni SE, et al. Ethical issues of obesity surgery – a health technology assessment. Obes Surg. 2011;21:1469–76.
14. Puhl R, Brownell KD. Bias, discrimination, and obesity. Obes Res. 2001;9:788–805.
15. Karris L. Prejudice against obese renters. J Social Pshycol. 1977;101:2.
16. Vartanian LR, Smyth JM. Primum non nocere: obesity stigma and public health. J Bioeth Inq. 2013;10:49–57.
17. Newhook JT, Gregory D, Twells L. 'Fat girls' and 'big guys': gendered meanings of weight loss surgery. Sociol Health Illn. 2015;37:653–67.
18. Klein D, Najman J, Kohrman AF, Munro C. Patient characteristics that elicit negative responses from family physicians. J Fam Pract. 1982;14:881–8.
19. Price JH, Desmond SM, Krol RA, Snyder FF, O'Connell JK. Family practice physicians' beliefs,

attitudes, and practices regarding obesity. Am J Prev Med. 1987;3:339–45.

20. Rouleau CR, Rash JA, Mothersill KJ. Ethical issues in the psychosocial assessment of bariatric surgery candidates. J Health Psychol. 2016;21:1457–71.

21. Hofmann B. Stuck in the middle: the many moral challenges with bariatric surgery. Am J Bioeth. 2010;10:3–11.

22. Puia A, Puia IC, Cristea PG. Ethical considerations in bariatric surgery in a developing country. Clujul Med. 2017;90:268–72.

23. Madan AK, Tichansky DS, Taddeucci RJ. Postoperative laparoscopic bariatric surgery patients do not remember potential complications. Obes Surg. 2007;17:885–8.

24. Persson K. Why Bariatric surgery should be given high priority: an argument from law and morality. Health Care Anal. 2014;22:305–24.

25. Buchwald H, Avidor Y, Braunwald E, et al. Bariatric surgery: a systematic review and meta-analysis. JAMA. 2004;292:1724–37.

26. Terranova L, Busetto L, Vestri A, Zappa MA. Bariatric surgery: cost-effectiveness and budget impact. Obes Surg. 2012;22:646–53.

27. Upadhyay J, Farr O, Perakakis N, Ghaly W, Mantzoros C. Obesity as a disease. Med Clin North Am. 2018;102:13–33.

28. Glenn NM, Raine KD, Spence JC. Mandatory weight loss during the wait for bariatric surgery. Qual Health Res. 2015;25:51–61.

29. Wild B, Hunnemeyer K, Sauer H, et al. Sustained effects of a psychoeducational group intervention following bariatric surgery: follow-up of the randomized controlled BaSE study. Surg Obes Relat Dis. 2017;13:1612–8.

30. Buchwald H. Revisional metabolic/bariatric surgery: a moral obligation. Obes Surg. 2015;25:547–9.

31. Dixon JB, Logue J, Komesaroff PA. Promises and ethical pitfalls of surgical innovation: the case of bariatric surgery. Obes Surg. 2013;23:1698–702.

32. De Ville K. Bariatric surgery, ethical obligation, and the life cycle of medical innovation. Am J Bioeth. 2010;10:22–4.

33. Shikora SA. A call for maintaining ethical behavior in bariatric surgery. Obes Surg. 2012;22:849–50.

34. Durkin N, Desai AP. What is the evidence for paediatric/adolescent bariatric surgery? Curr Obes Rep. 2017;6:278–85.

35. Hofmann B. Bariatric surgery for obese children and adolescents: a review of the moral challenges. BMC Med Ethics. 2013;14:18.

36. van Geelen SM, Bolt IL, van der Baan-Slootweg OH, van Summeren MJ. The controversy over pediatric bariatric surgery: an explorative study on attitudes and normative beliefs of specialists, parents, and adolescents with obesity. J Bioeth Inq. 2013;10:227–37.

37. Snyder J, Crooks VA. Medical tourism and bariatric surgery: more moral challenges. Am J Bioeth. 2010;10:28–30.

38. Snyder J, Crooks VA. New ethical perspectives on medical tourism in the developing world. Dev World Bioeth. 2012;12:iii–vi.

39. Gilmartin J, Bath-Hextall F, Maclean J, Stanton W, Soldin M. Quality of life among adults following bariatric and body contouring surgery: a systematic review. JBI Database System Rev Implement Rep. 2016;14:240–70.

40. Altieri MS, Yang J, Park J, et al. Utilization of body contouring procedures following weight loss surgery: a study of 37,806 patients. Obes Surg. 2017;27:2981–7.

41. Athavale R. Morbidly obese and super-obese women should have surgery refused for benign conditions: FOR: a holistic approach is required to tackle the obesity pandemic and its effects. BJOG. 2016;123:224.

42. Mahoney C, Vincent K, Crosbie EJ. Morbidly obese and super-obese women should have surgery refused for benign conditions: AGAINST: refusing surgery is both legally and morally wrong. BJOG. 2016;123:224.

How to Solve Ethical Conflicts in Everyday Surgical Practice: A Toolbox

Darren S. Bryan and Peter Angelos

Introduction

The study of ethics has historically been the domain of philosophers and theologians, but this has dramatically changed in medicine over the last half century. Initially, there was strong resistance from clinicians to the field of bioethics in part because the field was theoretical and in part because physicians considered nonclinical bioethicists to be imposing their views on practicing clinicians. For centuries, clinicians have made decisions guided by experience, training, and scientific data. Largely born out of the patients' rights movement of the 1960s, the shift in medical education and provision of care from a physician-centric to a patient-centric model accompanied a growing recognition of a new way for clinicians to approach ethically challenging situations.

The field of clinical medical ethics was named and started in the 1970s [1–3]. Clinical medical ethics is not a subset of philosophy, theology, or law but rather is born from medicine and centers on the doctor-patient relationship. It is meant to be more practically applicable to address problems commonly encountered in practice. Clinical

D. S. Bryan · P. Angelos (✉)
Department of Surgery, The University of Chicago, Chicago, IL, USA

The MacLean Center for Medical Ethics, The University of Chicago, Chicago, IL, USA
e-mail: pangelos@surgery.bsd.uchicago.edu

medical ethics guidelines for issues such as truth-telling, informed consent, and confidentiality have now become the legal and professional "standard of care" in the United States.

There are many ways to approach an ethical question, situation, or conflict in the care of the surgical patient. This chapter is intended to help formulate and analyze surgical ethics questions in a systematic fashion, aiming to arrive at a conclusion that provides practical guidance to clinicians. First, we present a fictional patient scenario with several ethical issues commonly encountered by surgical providers. We then introduce three well-known published methods for the evaluation of the ethical problem – the Beauchamp and Childress model, the "Four-Topic (or Four-Box)" approach, and the Pellegrino approach – and apply each method to the theoretical case.

Case Presentation

Ms. A is a 78-year-old female presenting with confusion and recent melena to the emergency department of Mercy Hospital, a medium-sized, community-based facility outside of a major metropolitan area. Ms. A resides in a nursing facility near Mercy Hospital. Her caretakers, present at the bedside, report that she is normally interactive and participatory in self-care; however on evaluation she is only oriented to self and appears pale. Initial vital signs reveal tachycardia and hypotension, while laboratory values are significant for a hemoglobin concentration of 8 g/dL. She receives a blood transfusion with minimal improvement in hemoglobin concentration and is admitted to the intensive care unit. Her family arrives shortly thereafter, and her adult son consents for esophagogastroduodenoscopy (EGD), which reveals several large, slowly bleeding gastric ulcers. She undergoes localized control of the bleeding with epinephrine injection and remains intubated post-procedurally. Over the next 2 days, she continues to require large amounts of blood products and support of blood pressure with vasoactive medications. Repeat EGD is performed; however, visualization is hindered by large amounts of clotted blood within the stomach, and no intervention to control hemorrhage is possible. Surgical consultation is obtained.

The surgical team evaluates Ms. A, who is now critically ill. They advise that surgical intervention (a laparotomy and gastric resection) is extremely high risk and however agree that without an operation, she is unlikely to survive. As an alternative, or possible adjunct to an operation, an early consultation with palliative care is obtained, and her family is presented with options for comfort care. Her daughter, who has acted as primary decision-maker and has signed consent forms earlier in the hospital course, relates that her mother had previously expressed that she would not want "heroic measures" should she become ill. The daughter, therefore, suggests comfort care would be best for her mother. The patient's son however disagrees. He reports that his mother has an excellent quality of life and, if the source of bleeding were to be controlled, would have a chance to return to her previous level of function. He also recounts conversations in which she has spoken against "heroics" but disagrees that the proposed surgery should be classified as such. Neither child has been appointed a durable power of attorney nor does a formal advance directive exist. An ethics consultation is obtained by the clinical team for assistance in determining the best course of action.

Preparing to Address an Ethical Dilemma

The Gathering of Information

The basis of the physician-patient relationship and nature of the practice of medicine dictate the fundamental way in which challenging questions are addressed. How physicians, patients, and society view the interactions that occur between a patient and doctor has spanned a spectrum from a paternalistic, Hippocratic notion in which the physician is all-knowing and prescribes treatment as seen fit, to a Lockean model, in which the physician and patient are independent contractors bartering for service [4].

Ethical dilemmas, once recognized as such, should be approached in a systematic fashion. By doing so, and beginning each discussion or deliberation with a similar methodology, the practitioner or Ethics consultant minimizes bias and ensures that each decision or recommendation is given equal opportunity for consideration [5].

When eliciting the reason for the ethical challenge, it is often helpful to determine what seems to be the overarching ethical question. This is best posed in simple terms that avoid detail or complication. The question at hand in the case of Ms. A could be: "Should the patient undergo an operation?" While such phrasing does not address the *ethical* issues at hand and potential complexities of family dynamics associated with decision-making, it does provide a starting point as well as point for reference later in conversation when such complex discussions are taking place. In the provided example, as with most cases, there are multiple stakeholders involved in the care of the patient, and early identification of their relative roles is important. Stakeholders, in addition to the patient, typically include the care team (both the primary team and other consulting providers) and involved family members and however in certain cases may also contain close friends, clergy, or other individuals or entities important to the patient or involved with the case. In the example provided, in addition to Ms. A, her son, daughter, primary care team, palliative care providers, and surgical team can all be considered stakeholders with an interest in the care of the patient.

Next, the medical facts of the case should be gathered and reviewed. Medical care in the twenty-first century has evolved to a degree of complexity and nuance that it is rare for a single person to grasp all the issues at hand; therefore, the surgeon or an ethics consultant may need to enlist the help of colleagues or contacts to be able to understand the necessary details. In clinical medical and surgical ethics, one's understanding and ability to subsequently communicate medical facts of the case are often of absolute necessity to be able to reduce a multifaceted, unmanageable clinical picture to the correct ethical questions and discuss the issues at play in order to arrive at a decision or recommendation. In the case presented, the medical facts include the patient's baseline health including psychological and social factors, her illness from inception until present, the recommended treatments by the involved parties, and possible alternative treatments. As the case progresses and further conversations take place, the surgeon, when necessary, will gather and synthesize more detailed information.

If the question at hand is to be approached with the lens of clinical ethics, as opposed to a philosophical discussion, the surgeon must understand and know the patient with more depth than can be gathered through a medical chart or phone conversation. After a prior review and gathering the facts of the case, which can often be done remotely, the surgeon should meet with the patient and/or surrogate decision-makers. We advocate for the surgeon as well as other consultants to take an involved role, discussing the patient's quality of life, interests, short- and long-term goals, and values. Such information is invaluable and frequently becomes relevant, even when unpredicted. In the case of Ms. A, it would be important to discuss her baseline quality of life, her interests including how she fills her day, as well as her relationship to her family members.

Development and Framing of the Ethical Question

While the goal of the surgeon or clinical ethics consultant is to answer a question, usually regarding patient care, the central question itself rarely comes packaged and ready for discussion when there are conflicting views about what is best for a patient. More frequently, the *issue at hand* is readily apparent, but the ethical issues underlying the controversy may be less apparent. With a systematic approach, it is often possible for the surgeon or ethics consultant to facilitate discussion and deconstruct complex medical cases with multiple involved parties into more manageable questions. In the case presented above, the *issue at hand* is whether or not Ms. A should undergo an operation. The team presents two main options: comfort care or an operation with a high level of risk, neither of which is medically superior to the other. As she is unable to participate in decision-making, the care team has rightly involved her surrogate decision-makers, in this case her adult children, each of whom, under the Illinois Healthcare Surrogacy Act, has equal legal standing as the surrogate decision-maker.[1] As her children disagree with regard to the correct course of clinical action, the ethical issues at hand revolve around mediation and the ability of the physician to lead the decision-makers to a conclusion. In doing so, patient, surrogate decision-maker, and care team values must be balanced and discussed.

Models for Analysis

Below, we review three common methods for approaching an ethical scenario. Each has been used extensively in practice and is well established with both critics and proponents. While

---

[1] It is important to identify which healthcare surrogacy laws apply in the specific setting and location. In this patient case, the Illinois Healthcare Surrogacy Act is applied.

each method is applied to the provided case of Ms. A, we discuss these multiple methods of approach in recognition that an ethical physician must have a quiver of tools for analysis of a case at his or her disposal.

Beauchamp and Childress "Four Principles"

In 1979, Beauchamp and Childress published the first edition of what would become a landmark work in biomedical ethics, *Principles of Biomedical Ethics* [6]. Within its pages, they championed principlism as an approach to moral and ethical decision-making. Based on the notion that several ethical principles can be applied to a variety of situations, principlism is recognized by many to stand as a practical way to approach decision-making in the face of the great diversity of experiences in today's society. Beauchamp and Childress outlined based on the previous work of Donald Ross (The Good and the Right, 1930) four principles to which decision-making in biomedical ethics could be ascribed: respect for autonomy, beneficence, non-maleficence, and justice. Also in 1979, the Belmont report was released, detailing principle-based guidelines for the ethical treatment of human subjects in research [7].

While each must be considered and upheld, the four principles discussed in *Principles of Biomedical Ethics* are said to be nonhierarchical in nature. Depending on the scenario, multiple principles may conflict with one another, requiring specific facts of the case and a risk-benefit analysis to be applied in order to arrive at an acceptable, moral conclusion.

Autonomy

The principle of *respect for autonomy* is based upon the concept of individual autonomy, that is, that an individual has a right to be self-determining in action and decision, free from controlling interference by others and from limitations that prevent meaningful choice [6]. Two components of autonomy that frequently arise in discussions surrounding clinical ethics are agency and authenticity [8]. Agency can be

defined as the distinctively human ability for an individual to have the capacity for choice [9]. Of note, and importantly, this includes recognizing the ability for individuals to make a "bad choice." If an adult, capable of decision-making, is fully aware of health hazards associated with a diet heavy in wine and cheese yet continues to overindulge, society overwhelmingly respects that person's agency, assuming that they have weighed options, even though realizing that the decision is a "bad one." Patient authenticity, an individual's decision-making pattern based on their experiences, thoughts, and personally held beliefs, is something built over time [8]. Occasionally, acquaintances of a patient may observe that a choice or decision that a patient makes is "out of character" or incongruent with the choices they would normally expect the patient to make, based on prior knowledge and relationships. When this is the case and a patient's authenticity is in question, clinicians may search for causes of the deviation from normal; however only rarely will steps be taken to curtail a patient's agency. In clinical medicine, although the four principles are nonhierarchical, the principle of respect for autonomy is generally held to the highest of degrees.

Beneficence

The *principle of beneficence* is the moral obligation to act for the benefit of others. It is to be distinguished from *beneficence*, the action of doing good or affecting positive change, as well as from *benevolence*, the virtue or characteristic of being disposed to act for the benefit of others [6]. The concept of acting with a purpose to affect welfare is at the root of morality and utilitarian theory; however, the principle of utility should be distinguished from the *principle of beneficence*, as applied by Beauchamp and Childress. While strict utilitarianism may be criticized for placing the value to society over value to the individual, the *principle of beneficence* avoids this through interpretation simultaneously along with the other principles.

Non-maleficence

The *principle of non-maleficence* prescribes that one does not bring harm to an individual [6].

Historically and clinically, non-maleficence is balanced with beneficence, and the two are frequently in conflict. For this reason, many philosophers and ethicists consider the two together, arguing that beneficence cannot exist without first paying homage to non-maleficence. As with the other *principles*, there are those who argue that the concept of *non-maleficence* ought to take moral priority over others [10]. However, numerous counterexamples exist, challenging this thought. In the field of surgery, for example, interventions meant to bring about positive outcomes are often laced with possible complications or, at a minimum, expected pain. The only way for patients to have benefit is through the risk of harm. Beauchamp and Childress argue that when analyzing *beneficence* and *non*-maleficence as separate entities, they must, as with the other principles, be considered prima facie duties. Most interpret the *principle of non-maleficence* to be at least somewhat dependent upon the act being carried out, and the intention of the agent. The act must be good, and the agent must intend for the outcome to be positive [6].

Justice

The *principle of justice* is based on the concept of justice, revolving around fairness, entitlement, what is deserved, and equality. In clinical medical ethics, issues surrounding justice frequently are related to distribution of scare resources, claims on rights to healthcare, and subsequent potential allocation that must take place. When compared to the other *principles*, the principle of justice less frequently conflicts in a moral hierarchy yet should not be ignored.

Case Application

In the highlighted case of Ms. A, the principles of respect for autonomy, beneficence, and non-maleficence must be considered carefully. Given the nature of her illness, the patient lacks the ability to make a choice for herself; therefore, her agency and hence autonomy are limited. In this situation, assuming good faith, her son and daughter are both attempting to guide medical treatment as surrogate decision-makers, acting on the principle of substituted judgment, that is,

decisions as their mother would make if she were able. From the viewpoint of the providers, the principles of beneficence and non-maleficence must be balanced. The choice to act (surgical intervention in the form of gastrectomy) offers an opportunity for hemorrhage control and potentially eventual recovery, however not without serious risk of complications. While principlism helps to guide the conversations that must take place to evaluate individual stakeholders' point on the risk-benefit spectrum, it has been criticized for failing to provide a sufficiently actionable blueprint to guide courses of action.

Four-Box Model

In 1982, Jonson, Siegler, and Winslade first published *Clinical Ethics: A Practical Approach to Ethical Decisions in Clinical Medicine*, proposing a systematic method for addressing ethically challenging issues in the clinical care of patients. They argue that, similar to the approach of clinical medicine, the ethical care of a patient should be addressed systematically, resulting in a useful and practical pathway to resolution [11].

Within clinical ethics, the authors introduce four topics that constitute the ethical framework of most clinical encounters. Often referred to as the "four boxes," they include *Medical Indications*, *Patient Preferences*, *Quality of Life*, and *Contextual Features*. Each is relatable to the four principles of biomedical ethics as described by Beauchamp and Childress. However, with a different method of organization and combination, the four-box model attempts to help identify important principles with relation to each case, allowing principles to become more or less "weighty." Furthermore, the four-box model provides structure for conversation as it goes forward, giving credence to the often-difficult interpersonal issues that surround ethical deliberation. A checklist is provided for clinicians to review when disagreements between stakeholders arise, helping to provide a path forward.

Medical Indications

The first topic—*Medical Indications*—focuses primarily on the fact-gathering phase of a new encounter. Before making ethical recommendations or decisions, according to the four-box model, providers must first understand medical information relevant to a case. Questions such as "what is the patient's primary medical problem?" and "what are the goals of treatment?" are addressed, helping to guide further conversation as it occurs. Here, the ethical principles of beneficence and non-maleficence are considered as clinicians balance decisions on a spectrum of risk and benefit.

Preferences of Patients

If *Medical Indications* is approached from the provider's point of view, the topic *Preferences of Patients* interprets care from the perspective of the patient. While biomedical indications may be clear and leave no obvious space for interpretation or decision-making, individual patient goals may not align or even provide for the most complete definition of health. In this topic, the provider or consultant is chiefly concerned with the ethical principle of *respect for autonomy* as they elicit patient values, beliefs, and wishes on an individual level.

Quality of Life

With the topic *Quality of Life*, the authors address the ability of decisions to impact patient satisfaction, both in an overall and more focused point of view. The way in which individuals live day to day is taken into account and questions such as: "What are the prospects, with or without treatment, for a return to normal life and what physical, mental, and social deficits might the patient experience even if treatment succeeds?" As in the topic *Preferences of Patients*, the concept of health must encompass psychological and social needs, moving beyond biological needs alone. Once again, individual patient values are highly variable, placing emphasis on the importance of developing rapport and a strong provider-patient relationship. The ethical principles of *beneficence*, *non-maleficence*, and *respect for autonomy* must all be considered.

Contextual Features

The principles of *justice* and *fairness*[2] are taken into account when considering contextual features. All decisions, clinical and otherwise, are made in the context of the surrounding environment. Frequently, such settings are not carefully considered and are simply wrapped into the decision itself. It is however, important to be aware of the effect of the environment, particularly when considering complex ethical decisions. Legal, religious, racial, and socioeconomic factors can have a significant impact on clinical care and are addressed by this topic.

Pellegrino Approach

Edmund Pellegrino was the founding editor of *The Journal of Medicine and Philosophy* and an early thoughtful leader in the field of medical ethics. While bioethics had existed previously, Pellegrino pioneered the concept that medical ethics was, in and of itself, a field that was separate and required dedicated, philosophical energy. Through the course of his life and career, he espoused the idea that ethics of medicine was based upon the philosophy of medicine and the practice of doctoring, centering around the "good of the patient," and thus placed the *principle of beneficence* above the others. Therefore, when approaching ethical quandaries, defining and identifying "the good" on an individual patient level became of utmost importance in outlining recommendations and a path forward.

In defining "the good," Pellegrino identified four components of successively greater importance which the physician can only place in hierarchical order by cultivating a deeper relationship with the patient. The least important was "biomedical good," and the most important was whatever the patient self-identified the "highest good" to be. For example, in treating a patient who chose to forgo treatment on the basis of deeply

---

[2]Fairness is not one of the four *principles of biomedical ethics* as described by Beauchamp and Childress, however, is important for consideration and closely associated with the principle of *justice*.

held spiritual beliefs, the physician should concur with the patient and not provide treatment. In an essay for the Kennedy Institute of Ethics outlining the work of Pellegrino, Daniel Sulmasy writes, "Pellegrino would not interpret respect [by the physician] for refusal [of treatment] as the triumph of autonomy over beneficence, but the normatively correct decision that flows from a fuller understanding of what it means to be beneficent—to promote the good of the patient in this complete and richer sense" [12].

In the presented case, Pellegrino would likely argue that the providers should, as best as possible, get to know Ms. A through her family, identifying goals and relying on the principle of substituted judgment to arrive at a decision. Whatever decision was reached, while taking into consideration biomedical practicalities, should be first and foremost based on the goals of the individual.

Conclusion

With a systematic approach to patient care and ethical deliberation, complex situations become manageable. After fact gathering and framing an ethical question of the issue at hand, several approaches have been popularized and provide a framework for discussion.

With their landmark work, *Principles of Biomedical Ethics*, Beauchamp and Childress address the nonhierarchical ethical principles of *respect for autonomy*, *beneficence*, *nonmaleficence*, and *justice*. These principles should be considered, addressed, and weighed in each case to help come to a conclusion. The "four-topic" or "four-box" model introduced by Jonsen, Siegler, and Winslade focuses more heavily on the physician-patient relationship and provides a practical approach to commonly encountered scenarios, taking into account the frequent communication difficulties that occur in such situa-

tions. Finally, the ethos of Edmund Pellegrino and his approach to medical ethics, heavily centered on the physician-patient relationship, introduce the concept that medical ethical decisions can be normative and thus right or wrong.

In practice, a variety of approaches must be utilized when approaching ethically challenging cases. With a constant methodology and broad armamentarium, challenging topics can be given equal consideration and appropriate weight, ensuring thoughtful decisions, and case-to-case consistency for the ethical physician.

References

1. Siegler M. A legacy of Osler. Teaching clinical ethics at the bedside. JAMA. 1978;239(10):951–6.
2. Siegler M. Clinical ethics and clinical medicine. Arch Intern Med. 1979;139(8):914–5.
3. Siegler M, Pellegrino ED, Singer PA. Clinical medical ethics. J Clin Ethics. 1990;1(1):5.
4. Pellegrino ED. Toward a reconstruction of medical morality. Am J Bioeth. 2006;6(2):65–71.
5. Lo B. Resolving ethical dilemmas: a guide for clinicians. 5th ed. Philadelphia: Wolters Kluwer Health/Lippincott Williams & Wilkins. x; 2013, 369 p.
6. Beauchamp TL. Principles of biomedical ethics. Array ed. In: Childress JF, editors. New York: Oxford University Press; 2009.
7. Health, N.I.o., *The Belmont Report*. Belmont Rep. Ethical Princ. Guidel. Prot. Hum. Subj. Res. 1979. p. 4–6.
8. Brudney D, Lantos J. Agency and authenticity: which value grounds patient choice? Theor Med Bioeth. 2011;32(4):217–27.
9. Kant I. Groundwork for the metaphysics of morals. 1785.
10. Gillon R. "Primum non nocere" and the principle of non-maleficence. Br Med J (Clin Res Ed). 1985;291(6488):130–1.
11. Jonsen AR. Clinical ethics a practical approach to ethical decisions in clinical medicine. Array ed. In: Siegler M, Winslade WJ, editors. McGraw-Hill's AccessMedicine. New York: McGraw Hill Medical; 2015.
12. Sulmasy DP. Edmund Pellegrino's philosophy and ethics of medicine: an overview. Kennedy Inst Ethics J. 2014;24(2):105–12.

Afterword

"The voice of our patients"

The word patient means in its original form "the one who suffers". Its root can be traced to the latin word *patiens*, present participle of the verb *patior*, whose meaning is "I am suffering" and linked to the greek term *pashkein* ("to suffer) and also to *pathos*, which is considered a quality that evokes pity, sadness, or sympathy.

Nonetheless, the term patient seems to have been replaced by others such as health consumer or client, situation that implies a business relationship. Surgery should be far from being a trade or a commerce, highlighting the role of John Gregory (1724–1773) and his contributions. He was the first one to develop an ethical system of physician's behavior and conduct and a decisive influence in Medicine as a fiduciary profession.

What do us as patients request from our surgeons?

Firstly, deep concern for our situation and empathy. Clear, faithful and loyal communication should also be a must between a surgeon and his or her patient.

Communication needs to be "tête-à-tête"; we as patients want and need to look at our surgeons in the eyes, discussing our situation and our prognosis. When we pose the question "Doctor, what would you do?", we do not want to hear shortcuts or evasive responses. We just want an honest, upfront and altruistic advice from an expert, taking into consideration our set of values and surrounding circumstances and weaknesses. Patients, most times, want to know if you would choose the treatment options you are recommending for you or your relatives.

We also need that our surgeons acknowledge their errors, making the dictum "Primum non nocere" a mandate in everyday surgical care. Most times the harm is not just limited to the outcome, but to us and our families as a whole.

We also need that our surgeons engage themselves affectively with us, because they tend to become one of the most important people in the world for cancer survivors. Trust is a 2-way path, and we need to feel empowered and motivated by our treating surgeons. There has always been a traditional sense of belonging to a physician, but this sense of property has been lately lost due to multidisciplinary teams and the lack of an "orchestra director" and no one seems to "own" the patient, exception made of the payers.

At the end of the day, we request time from our surgeons that the surgical profession understands that we suffer with our pain, our disease, our weaknesses and our burdens.

In summary, we, as patients, want our surgeons to provide the best care and expertise but also to communicate effectively, be sympathetic and do not fear to show your feelings, understand your patient, put yourself in the patient's shoes and give the time we and our relatives deserve.

A. R. Ferreres (ed.), *Surgical Ethics*, https://doi.org/10.1007/978-3-030-05964-4

Index

© Springer Nature Switzerland AG 2019
A. R. Ferreres (ed.), *Surgical Ethics*, https://doi.org/10.1007/978-3-030-05964-4

Printed by Printforce, the Netherlands